Forest Nutrition Management

Forest Nutrition Management

Editor: Emelia Mitchell

MURPHY & MOORE
www.murphy-moorepublishing.com

www.murphy-moorepublishing.com

MURPHY & MOORE

Cataloging-in-Publication Data

Forest nutrition management / edited by Emelia Mitchell.
p. cm.
Includes bibliographical references and index.
ISBN 978-1-63987-756-0
1. Forest soils--Fertilization. 2. Forest management. 3. Forest ecology. 4. Trees--Nutrition.
5. Forest management. I. Mitchell, Emelia.
SD408 .F67 2023
634.95--dc23

Murphy & Moore Publishing
1 Rockefeller Plaza,
New York City,
NY 10020, USA

ISBN 978-1-63987-756-0

Contents

Preface

It is often said that books are a boon to mankind. They document every progress and pass on the knowledge from one generation to the other. They play a crucial role in our lives. Thus I was both excited and nervous while editing this book. I was pleased by the thought of being able to make a mark but I was also nervous to do it right because the future of students depends upon it. Hence, I took a few months to research further into the discipline, revise my knowledge and also explore some more aspects. Post this process, I begun with the editing of this book.

Forests are areas of land that have a dense tree cover. They are the most common terrestrial ecosystem on Earth and are found all over the world. Trees and forests are vital to food production and nutrition. Management of soil fertility and forest nutrition is essential for ensuring the productivity of planted or natural forests. Forest management involves evaluating the land and the nutrient cycling process in order to ensure that the forest soil has sufficient nutrients and organic matter. This evaluation helps in revealing whether or not there are any deficiencies in the soil. Thereafter, fertilizers might be used for ensuring nitrogen fixation and enhancing nutrients in the soil. This book unfolds the different aspects of forest nutrition management. It aims to equip students and experts with the advanced topics and upcoming concepts in this area of forest management.

I thank my publisher with all my heart for considering me worthy of this unparalleled opportunity and for showing unwavering faith in my skills. I would also like to thank the editorial team who worked closely with me at every step and contributed immensely towards the successful completion of this book. Last but not the least, I wish to thank my friends and colleagues for their support.

Editor

1

Estimating Values of Carbon Sequestration and Nutrient Recycling in Forests: An Application to the Stockholm-Mälar Region in Sweden

Ing-Marie Gren

Department of Economics, Swedish University of Agricultural Sciences, Box 7013, 75007 Uppsala, Sweden; E-Mail: ing-marie.gren@slu.se

---

Abstract: We calculate values of forest carbon sequestration and nutrient recycling applying the replacement cost method. The value is then determined as the savings in costs by the replacement of more expensive abatement measures with these ecosystem services in cost-effective climate and nutrient programs. To this end, a dynamic optimization model is constructed, which accounts for uncertainty in sequestration. It is applied to the Stockholm-Mälar region in southeast Sweden where the EU 2050 climate policy for carbon emissions and the Baltic Sea action plan for nutrient discharges are applied. The results show that the value of carbon and nutrient sequestration can correspond to approximately 0.5% of the region's gross domestic product, or 40% of the value of productive forest. The largest part of this value is attributed to carbon sequestration because of the relative stringency in targets and expensive alternative abatement measures. However, sequestration is uncertain because of stochastic weather conditions, and when society has a large risk aversion for not attaining climate and nutrient targets, the values of the forest carbon and nutrient sequestration can approach zero.

Keywords: forest carbon and nutrient sequestration; values; replacement cost method; uncertainty; Stockholm-Mälar region

---

1. Introduction

Forests provide a multitude of marketed and non-marketed ecosystem services, where timber outputs provide a marketed service, and biodiversity, pollutant sequestration, and recreational opportunities are examples of non-marketed services. Forest recreational opportunities and biodiversity have been subjected to valuation in monetary terms in several studies, but there are only a few estimates of other ecosystem services, such as carbon sequestration and nutrient recycling (see survey in [1]). The benefits of carbon sequestration and nutrient recycling emerge from their ability to regulate the carbon content in the atmosphere and nutrient leaching at a lower cost than other measures, such as reductions in fossil fuels. Valuation of carbon sequestration for combatting climate change has been carried out at the regional EU and global scale [2–4]. However, afforestation (plantation of forest on arable land), which is the least expensive measure for carbon sequestration [5], also implies a reduction in nutrient leaching. Unbalanced nutrient loads can lead to eutrophication of water, as well as increased frequency of harmful algal blooms, sea bottom areas without biological life, toxic cyanobacteria, and decreases in water transparency and populations of commercial fish species [6,7]. There is a complementarity between carbon and nutrient transformations in boreal forests [8], which implies that increases in carbon sequestration reduce nutrient leaching from soil. The purpose of this study is to calculate values of forest carbon sequestration and nutrient recycling in the Stockholm-Mälar region in southeast Sweden. The choice of these services and region is based on the existence of actual targets for carbon and nutrient emissions, and of numerical models allowing for the calculations [9,10].

To the best of our knowledge, there are no published studies that calculate the value of a forest's ecosystem services in terms of carbon and nutrient sequestration. On the other hand, there is a relatively large body of literature calculating the value of especially carbon sequestration [2–4]. There are also a relatively large number of studies on the value of nutrient sequestration by wetlands and coastal ecosystems [11–14], but not on forest ecosystem nutrient sequestration. Similar to these studies, we apply the so-called replacement cost method for calculating values. The basis for this method is the existence of environmental targets, and the value of a new technology is then measured as the decrease in total costs associated with achieving these targets when the technology is included as a measure. It has then been shown that inclusion of forest carbon sequestration can reduce the total cost of achieving the global target of a maximum of 550 ppmv (parts per million by volume) by 40% [15,16] and the EU 2020 climate target of reducing emissions to 20% from the 1990 emission level in 2020 by 30% [3].

However, pollutant sequestration may have a cost disadvantage when considering uncertainty. Uncertainty arises from stochastic weather conditions, which affect biomass growth, and thereby carbon sequestration and the release of carbon and nutrients from soil. This, in turn, implies uncertainty in reaching stipulated targets, which constitutes a cost associated with this risk. Such costs have been calculated, where cost per unit pollutant sequestration is increased by a risk discount [4,17–19]. The calculation of this is, in turn, based on the so-called safety-first principle, which has a long tradition in economics [20]. It is then assumed that decision makers are risk averse against non-attainment of stipulated environmental targets, and the risk discount is determined by this risk aversion and the variability in sequestration. As shown by [4] and [21], the value of forest carbon sequestration in the EU 2020 climate policy is reduced and can approach zero for high levels of the risk aversion.

The current study calculates the value of forest pollutant sequestration with and without uncertainty when we apply the safety-first principle. To this end, we construct a dynamic optimization model with uncertainty, which, in addition to pollutant sequestration, includes abatement measures reducing emissions from fossil fuels and nutrient leaching. A dynamic model is needed because of the focus on future climate and nutrient targets, as well as consideration of depreciation in investment in renewable technologies and nutrient cleaning facilities, and development over time of carbon sequestration in afforested land. In our view, the main contribution of this study is the calculation of values of forest sequestration of several pollutants. Another is the consideration of uncertainty, and a third is the calculation of values of forest nutrient recycling, which have been made mainly for wetlands in other studies.

The study is organized as follows. The conceptual approach and the dynamic optimization model are presented in Section 2. Next, data retrieval is shown in Section 3, and the results are presented in Section 4. The results are discussed in Section 5, and the study ends with concluding comments.

2. Conceptual Framework and Cost-Minimizing Model

The replacement cost method assigns a value to a cleaning technology only if this technology is less costly than other abatement measures for reaching certain emission targets in a cost effective solution. The cost effective solutions are, in turn, determined by a numerical model for minimizing total costs of achieving carbon and nutrient emission targets in a certain time and space. The construction of such a model is thus crucial for the determination of the value of forest ecosystem services in terms of carbon and nutrient sequestration. In the following, we therefore first provide a more intuitive presentation of the replacement cost method, which is followed by a description of the structure of the cost effectiveness model that is used to derive conditions for a positive value of forest carbon and nutrient sequestration and the magnitude of this value.

2.1. Conceptual Framework of the Replacement Cost Method

The value of the services of land use in terms of carbon and nutrient sequestration is determined by their cost in relation to the costs of other abatement measures in a cost effective achievement of certain emission targets. The higher the cost of other abatement measures, the larger is the value of the ecosystem service. This simple principle for determining value is illustrated in Figure 1 for carbon sequestration.

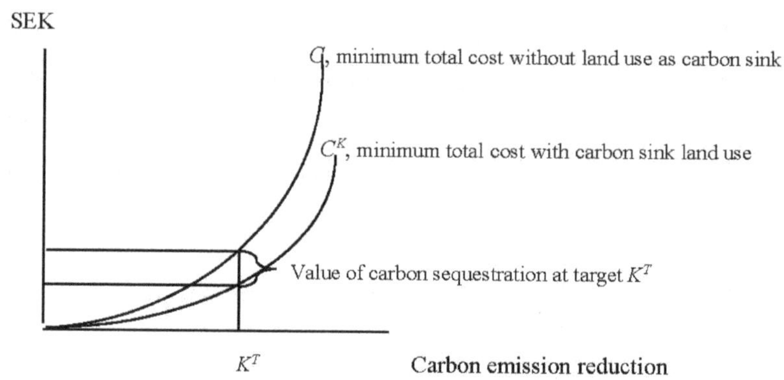

Figure 1. Illustration of the calculation of the value of land use as a carbon sink in a cost effectiveness framework. SEK: Swedish crown; K^T: carbon emission reduction target.

The horizontal axis illustrates carbon emission reductions, and K^T is the target to be achieved. The vertical axis shows the cleaning cost for different emission reduction levels. The curve C illustrates the minimum costs for achieving different emission reduction targets when the carbon sink provided by forests is not included as an abatement option in the cleaning program, and C^K illustrates the minimum costs when the carbon sink is included. Each point on C and C^K respectively reflects the allocation of all abatement measures that reach the target at minimum cost.

The value of carbon sequestration at the target K^T is now determined by the difference in total minimum costs with and without carbon sequestration, which corresponds to the distance C-C^K in Figure 1. The value of the carbon sink is then determined by its construction cost and the abatement costs of other measures. The larger the difference between the abatement costs of other measures and that of the carbon sink, the higher the value of the carbon sink. This is, in turn, determined by the stringency in the carbon target since the costs of all abatement measures increase at higher emission reduction levels.

The value of land use for nutrient sequestration is calculated in the same way as for carbon sequestration. However, the value of both these sequestration options depends on whether carbon and nutrient targets are managed separately or simultaneously. When managed separately without consideration of the other pollutant target, the full cost of land use will be borne by the reduction in the pollutant in question. Under simultaneous management, land use measures will have a cost advantage compared with measures affecting only one pollutant and will therefore be implemented to a larger extent than under separate target management.

2.2. Description of the Dynamic Cost Minimizing Model

The model used for calculating the value illustrated in Figure 1 builds on the dynamic model developed by [10], and adds water emission targets. Cost effectiveness is then defined as the allocation of abatement measures over time and at different locations, which reaches predetermined targets in a specific time at minimum costs. The location of carbon emission reductions and sequestration do not affect the impact on the atmosphere. This is in contrast to nutrient abatement and sequestration, where the location matters for the impact on the coastal waters because of the retention of nutrients from the source to the waters. We therefore divide the region into $m = 1,..,n$ drainage basins.

In each time period t, the net business as usual (BAU) emission of pollutant p, where p is carbon dioxide equivalents (CO_2e), nitrogen (N), and phosphorus (P), in drainage basin m is determined by the emission of sources minus existing sequestration, X_t^{pmBAU} . These emissions can be reduced by decreases in the pollution at the emission sources, $a^{pmf} A_t^{pmf}$ where $f = 1,..,l$ are the different types of sources such as energy production, agriculture, and transport, and a^{pm} is the impact on the target. This is unity for CO_2e reductions since location of source does not matter for the impact, but $0 < a^{pm} \leq 1$ for nutrient since some sources are located in upstream drainage basins where the effect is below unity. For emission sources located at coastal waters, such as sewage treatment plants, the discharge reduction at the source is the same as the load reduction to the sea, which implies that $a^{pm} = 1$. Annual reductions in CO_2e are also obtained by existing renewable technologies, A_t^{pmr}, where $r = 1,..,q$ technologies. These are obtained from investments, I_t^{mr}, which have a maximum technological life time, and are then regarded as capital investments subject to depreciation according to

$$A_t^{pmr} = \sum_\tau (1-\delta^r)^{t-\tau} a^{pmr} I_\tau^{mr} \tag{1}$$

where a^{pmr} is the reduction in pollutant p of one unit abatement by technology r, and $0 < \delta^r < 1$ is the annual depreciation rate of technology r. Abatement by means of a renewable technology r in time t is thus determined by the accumulated number of investments prior to t, and investment in period t.

The third abatement option is the creation of the carbon and nutrient sink, L_t^{ms}, which can be formed by afforestation. However, it may take some time before the full potential of the sink is achieved. This growth depends on biomass growth and soil processes, which, in turn, are determined by type of land use change. Since there are few data on these processes, we assign a simple function where the rate of growth, δ^s, is constant over time and carbon and nutrient sequestration achieve a maximum. Recall that the pollutant sink in each time period is subject to stochastic weather conditions. We therefore assign the carbon and nutrient sink in a given period t, A_t^{pms}, as dependent of land use changes in prior periods, L_τ^{ms}, and an additive stochastic parameter, ε_t^{pms}, which is written as

$$A_t^{pms} = \sum_\tau (1-(1-\delta^s)^{t-\tau}) a^{pm} q^{pmr} L_\tau^{ms} + \varepsilon_t^{pms} \tag{2}$$

where q^{pms} is the maximum sequestration per unit of land area of pollutant p. The larger the $t-\tau$ value, the higher the A_t^{pms} for a given L_τ^{ms}. Similar to renewable technologies, total pollutant sink in a municipality is then determined by the accumulated carbon sink additions. The difference is that sink capacity is increasing and capacities of renewable technologies are decreasing over time.

Total net emissions of each pollutant, X_t^p are then the sum of emissions in BAU minus emission reduction obtained by the three classes of measures in each drainage basin:

$$X_t^p = \sum_m (X^{pmBAU} - \sum_f A_t^{pmf} - \sum_r A_t^{pmr} - \sum_s A_t^{pms}) \tag{3}$$

The exercise of each abatement measure is subject to a cost, which is described by the cost functions $C^{pmf}(A_t^{pmf})$, $C^{pmr}(I_t^{pmr})$ and ,$C^{pms}(L_t^{pms})$ which are assumed to be increasing and convex in their arguments, *i.e.*, costs are increasing at a higher rate in the use of the measure.

Additional assumptions are that abatement in each period is subject to restrictions where only part, b^f, of the BAU emissions can be reduced, part of total emissions from fossil fuel or nutrient use can be replaced by a specific technology energy, b^r, in each period, and that part of agriculture land under BAU, b^s, can be converted into forests in each period. The reasons for these restrictions are that drastic reductions in emissions and land use are difficult to implement in a very short time period. The capacity restrictions are then written as

$$A_t^{pmf} \leq b^f X^{pmfBAU} \tag{4}$$

$$I_t^{ymr} \leq b^r \sum\nolimits_f X^{pmfBAU} \tag{5}$$

$$L_t^{yms} \leq b^s L^{ymsBAU} \tag{6}$$

The decision maker at the Stockholm-Mälar region is now assumed to apply a safety-first approach in reaching targets on total emission where they formulate a minimum probability or reliability level, α^p, of achieving the maximum emission target, $\overline{T}_t^p$, for each pollutant, which is written as

$$prob(\sum\nolimits_m X_t^{pm} \leq \overline{T}_t^p) \geq \alpha^p \tag{7}$$

This can be expressed in terms of mean emissions or leaching, $\mu_{T^p}^p$, risk aversion, ϕ^{α^p}, and variance in loads $\sigma_{T^p}^p$ as [22]

$$\mu_t^p + \varphi^{\alpha^p} (\sigma_t^p)^{1/2} \leq \overline{T}_t^p \tag{8}$$

where $\mu_t^p = E\left[X_t^p\right]$ with E as expectation operator; ϕ^{α^p} shows the choice of α^p as the acceptable deviation of the load from the mean. The level of ϕ^{α^p} depends on the shape of the probability distribution, which is discussed in Section 3. The left hand side of Equation (8) thus shows that reliability in achieving the target is obtained at a cost, which increases with reliability concern or the probability of achieving the target, *i.e.*, in ϕ^{α^p}, and in σ_t^p.

Recall from Section 2.1 that the value of L_τ^{ms} is calculated as the difference in cost for achieving the targets in Equation (7) with and without the inclusion of L_τ^{ms} as an abatement option. Given Equations (1)–(8), the decision maker is then assumed to choose among available options, A_t^{pmf} and I_t^{mr}, or A_t^{pmf}, I_t^{mr}, and L_τ^{ms}, in order to minimize total cost in present terms for achieving the carbon emission and nutrient leaching targets. When all options are available, this is written as

$$Min \quad C = \sum\nolimits_t \sum\nolimits_m (\sum\nolimits_f C^{mf}(A_t^{mf}) + \sum\nolimits_r C_t^{mr}(A_t^{mr}) + \sum\nolimits_s C_t^{ms}(L_t^{ms}))\rho_t \tag{9}$$

s.t. Equations (1)–(9)

where $\rho^t = \dfrac{1}{(1+i)^t}$ with i as the discount rate.

2.3. Determinants of the Value of Carbon and Nutrient Sink in Forests

The determinants of the magnitude of the value of forests as pollutant sinks can be found by solving for the decision problem in Equation (9). We solve for the cost effective solution by formulating the Lagrange expression, L, which is written as

$$L = C + \sum_p (\lambda_{T^p}(\overline{T}^p - \mu_{T^p}^p - \varphi^{\alpha^p}(\sigma_T^p)^{1/2}) + \sum_m (\sum_f \gamma_t^{pmf}(b^f X^{pmfBAU} - a^{pmf} A_t^{mf})) + \sum_r \gamma^{pmr}(b^r X^{pmfBAU} - a^{pmr} I_t^{mr}) + \sum_s \gamma^{pms}(b^s L^{msBAU} - L_t^{ms}))) \tag{10}$$

where $\lambda_{T^p} \leq 0$ is the Lagrange multiplier for the targets, and $\gamma^{pmf} \leq 0$, $\gamma^{pmr} \leq 0$, and $\gamma^{pms} \leq 0$ are those for the different capacity constraints on abatement measures. The Lagrange multipliers have an interesting interpretation; they provide information on the change in total discounted minimum cost for a relaxation of the constraint. For example, a relaxation of the carbon emission target in 2050 by one unit will decrease the cost corresponding to $\lambda_T^p \leq 0$ where *p* is CO_2e.

We derive the conditions for a positive value from the first-order condition of a cost-effective solution, *i.e.*, the marginal abatement cost for obtaining a unit reduction in the targets shall be equal for the dynamic and spatial allocation for all measures and equal to $\lambda_{T^p}^p$. For ease of exposition but without loss of generality, we assume interior solutions where the capacity constraints are not binding. The first-order conditions for the cost effective solution are then written as

$$\rho_t \frac{\partial C_t^{mf}}{\partial A_t^{mf}} = \sum_p \lambda_{T^p}^p a^{pmf} \tag{11}$$

$$\rho_t \frac{\partial C_t^{mr}}{\partial I_t^{mr}} = \sum_p \lambda_{T^p}^p \sum_t a^{pmr}(1-\delta^{pr})^{T^p - t} \tag{12}$$

$$\rho_t \frac{\partial C_t^{pms}}{\partial L_t^{ms}} = \sum_p \lambda_{T^p}^p \sum_t (a^{pms} q^{pms} - \frac{\varphi^{\alpha p}}{2} \frac{\partial \sigma_t^{pms}}{\partial L_t^{ms}})(1-(1-\delta^{ps})^{T^p - t}) \tag{13}$$

All three conditions state that the marginal abatement cost in the present value, *i.e.*, the discount factor times the marginal cost, on the left hand side shall equal the weighted sum of marginal impacts on the different targets on the right hand side. The Lagrange multipliers, $\lambda_{T^p}^p$, constitute the weights that reflect the marginal cost of changing the respective target by one unit. A measure has a cost advantage for a relatively low marginal abatement cost and high weighted impact.

Starting by comparing the cost of abatement of sources and investments in renewable energy and nutrient cleaning, *i.e.*, Equations (11) and (12), we can see that investment has a marginal cost advantage for a low depreciation rate, δ^{pr}, and long time period, $T^p - t$, when the marginal decrease in investment, a^{pmr}, acts. Next we compare the marginal cost of reaching the target by investment in cleaning technologies with increased forest area, *i.e.*, Equations (12) and (13). Disregarding uncertainty for the moment, we can then see that the larger the pollutant sink per area of land, $a^{pms}q^{pms}$, the higher the impact on the respective target. This is also the case for the accumulated sink over time, $(1-(1-\delta^{pms})^{T^p - t})$ which is increasing because of the growth of the planted forest trees. A comparative advantage with pollutant sink compared with investment in renewable energy and nutrient cleaning is then the growth over time in the sink capacity, instead of a decline, which needs replacement of new technology when the old is worn out. This cost advantage is counteracted when we consider the uncertainty term, $\frac{\varphi^{\alpha p}}{2} \frac{\partial \sigma_t^{pms}}{\partial L_t^{ms}}$, which instead reduces the marginal impact when uncertainty in the pollutant sink is increasing in the area of afforestation and, hence, increases the marginal cost for reaching the target.

We can also note that the marginal impact increases for all three types of measures when more than one target is affected. A marginal increase in the forest area, which affects all three targets, then has a cost advantage when the other measures affect only one or two targets. The relative advantage of growth in sink per unit of land is then enhanced. On the other hand, the disadvantage of the uncertainty discount is also increased. When the advantages exceed the disadvantages, the value of the services is also determined by simultaneous or separate management of the targets. In practice, each pollutant target is treated separately, at least in Sweden. This means that the advantages of multifunctional measures are not fully utilized since their cost advantages of impacts of several targets are not accounted for.

Based on this simple dynamic model with uncertainty in carbon and nutrient sinks, we can thus conclude that investment in forest ecosystem services has an advantage over other abatement and investment measures when

- the cost of forest plantations is relatively low
- the weighted marginal impact of forest on carbon and nutrient sequestration is high
- the annual growth in sink capacity and depreciation of investment in other technologies are large
- the uncertainty discount of the pollutant sink is low
- the management of all targets occurs simultaneously instead of separately

As will be shown in the following, these factors can have a considerable effect on the calculated value of afforestation in the Stockholm-Mälar region.

3. Data Retrieval

The theoretical model presented in Section 2 shows that we need data on pollutant emissions from all sources under BAU, abatement capacities and impacts on targets of different measures, cost functions for all measures, depreciation rates of investment in cleaning technologies, growth rates of pollutant sinks, variance in pollutant sink, risk aversion, and discount rate. All data are obtained from two existing cost minimizing models of the Stockholm-Mälar region. One is a dynamic model for achieving 80% reduction in carbon emission from the 1990 emission level by 2050 [10] and the other is a static model for achieving nutrient targets in the coastal area of the region [9]. Detailed documentations of the data retrieval are found in [9,10]. In the following, we therefore give a brief summary of these data items.

3.1. Carbon Emission and Nutrient Leaching

The Stockholm-Stockholm-Mälar region covers an area of approximately 34,000 km^2. Two thirds of the area, or 22,000 km^2, is covered by forests and 6400 km^2 by arable land. The region contains the capital of Sweden, Stockholm, and 50% of the total Swedish population of 9.5 million people. Sweden and hence the Stockholm-Mälar region is guided by two international commitments; the EU climate policy [23] and the Baltic Sea Section Plan (BSAP) [24]. According to the EU 2050 climate policy, the CO_2e emissions should be reduced by 80% from the emission level in 1990, to be reached by 2050, and the BSAP envisages different reductions for the marine basins of the Baltic Sea. The Stockholm-Mälar region is located at the coastal zone of the marine basins with the highest target stringency, *i.e.*, the Baltic Proper, where the requirements are 23% nitrogen and 61% phosphorus reductions.

When calculating leaching of nitrogen and phosphorus to the Baltic Proper, retention of nutrients during the transport from the emission sources located upstream in the drainage basins are deducted since they do not reach the sea. All sources located at the coastal waters, mainly sewage treatment plants and industry, have direct discharges into the Baltic Proper. The numerical model includes 36 drainage basins in the Stockholm-Mälar region with retention rates, *i.e.*, the share of nutrient emission that does not reach the sea, ranging between 0 and 0.9 [9]. A minor part of the nitrogen deposition originates from air-borne emissions from fossil fuel combustion in the Stockholm-Mälar region (Table 1).

Table 1. Carbon dioxide equivalent (CO_2e) emissions, and nitrogen (N) and phosphorus (P) leaching from different sources in the Stockholm-Mälar region into the Baltic Proper.

	N, ton [a]	**P, ton [a]**	**CO_2e Emission , kton [b]**	**CO_2e Sequestration kton [b]**
Forest land	773	19		3531
Arable land	1825	156	840	
Sewage treatment plants	5613	94		
Transport	82		5451	
Industry	23		1527	
Energy production	70		4688	
Other CO_2e sources			1677	
Total	8386	269	14183	3531
Targets [c,d]	6457	105	3438	

[a] [9]; [b] [10]; [c] Targets for N and P are 77% of N and 39% of P loads in 2008 but without target year [24];
[d] Target for CO_2e is 20% of 1990 emission to be reached by 2050 [23].

Arable land and sewage treatment plants are the largest sources of nutrients, and transport and energy production are the major sources of CO_2e emission. Existing forest carbon sequestration corresponds to 25% of total emission.

3.2. Abatement Measures and Costs

In addition to afforestation, we include reductions in the use of fossil fuel, replacement of fossil fuel driven cars with electrics cars, and investment in solar and wind power for reducing carbon emissions. These measures also affect the emission of nitrogen oxides and associated deposition in the Baltic Proper. Measures reducing nutrient leaching include investment in sewage treatment plants, cultivation of catch crops, and reductions in the use of fertilizers. All of these measures reduce both nitrogen and phosphorus. Afforestation affects all three pollutants. The basis for choosing these measures is that they turned out to be relatively inexpensive for combatting CO_2e emissions and nutrient loads into the Baltic Sea in the Mälar-region [9,10]. Our estimates of the value of the forest as a pollutant sink are then likely to be conservative.

The effects of afforestation on pollutants are calculated as the difference in the carbon and nutrient sink per unit area of forest and arable land. Given the relatively short period until 2050, afforestation requires fast-growing tree varieties to provide carbon sequestration. According to [25], the sink coefficient for forest land is 1.81 ton CO_2e/ha and −0.73 ton CO_2e/ha for arable land, which gives a net effect of 2.54 ton CO_2e/ha afforestation. This effect is assumed to be the same for all drainage basins, obtained after 20 years, and with a constant growth rate during the years.

The cost of afforestation consists of the opportunity cost of land, *i.e.*, differences in annual profit between agriculture and forestry. This information is obtained from a supply curve of arable land, which shows the profits foregone for transferring arable land into other uses [10]. In a similar vein, costs of fossil fuel reductions are calculated as associated decreases in consumer surplus, which are obtained from demand functions of different fossil fuel products (heating oil, coal, gasoline, and diesel).

Data on investment cost functions for wind and solar power are found in [26], who estimated quadratic cost functions for these energy sources for the whole of Sweden. The cost of electric cars is assumed to be constant per 10 km and correspond to the difference in driving costs between gasoline and diesel cars. This cost depends on the type of car and the distance driven per year; the longer the distance, the lower the difference in driving cost. We evaluate the cost for the average distance in Sweden [10]. It is also assumed that the rates are calculated from assumptions of technical life lengths of 40 years for wind and solar power, and 25 years for electric cars. With respect to capacity constraints, assumptions are imposed on the replacement of fossil fuel with renewable energy, and of the existing car fleet with electric cars.

Concerning measures reducing nutrient deposition into the Baltic Proper, considerable investments in sewage treatment plants have already been made, and we therefore assume that further cleaning is carried out by implementing tertiary cleaning. The average investment costs of tertiary cleaning at sewage treatment plants are obtained from [9] and converted into 2011 prices. It is more difficult to obtain information on the technical life length of the investment, but a 50-year perspective is usually assumed. Costs on reductions in the use of fertilisers are calculated as associated decreases in farmers' profits, or consumer surplus, data on which are found in [9]. Costs of cultivation of catch crops, which are sown at the same time as the ordinary crop is harvested in autumn and thereby prevent nutrient leaching from the soil during autumn and winter seasons, are also found in [9].

3.3. Uncertainty, Targets, and Discount Rate

Recall from Section 2 that data are needed on variance in nutrient and risk aversion in order to calculate the risk discount. There are no data on any of these parameters. We therefore use the standard deviation in carbon sink for Sweden as calculated by [27], and assume that this is the same for nitrogen and phosphorus loads into the Baltic Proper. The standard deviations in [28] include measurement errors in data on carbon sinks for different land uses and uncertainty related to weather variability, which affects biomass growth and thereby carbon sequestration.

The effect on the risk discount of a certain choice of probability of achieving the targets depends on the underlying probability distribution of the pollutant sink. However, there is no information on the probability distributions. In principle, $\phi^{\alpha p}$ can be quantified in two ways: by assigning a specific or parameter free distribution. The normal distribution is most commonly applied, and $\phi^{\alpha p}$ is then a standard number such that $\int_{-\infty}^{\varphi^{\alpha p}} f(\varphi)d\varphi = \alpha^p$, where ϕ is the standardized distribution of A_i^{pms}, *i.e.*, pollutant sequestration, and $f(\phi)$ is the probability density function for ϕ [22]. This approach is frequently applied in the literature on policy instruments for stochastic water pollution [9,17,29,30]. Another way of quantifying $\phi^{\alpha p}$ is more flexible, which is based on Chebyshev's inequality where no assumptions are made with respect to the probability distribution [28]. The value of $\phi^{\alpha p}$ is then determined where $\varphi^{\alpha p} = (1-\alpha)^{-1/2}$. There are considerable differences in the calculated $\phi^{\alpha p}$ depending on the assumption of distribution. For example, at a probability level $\alpha = 0.95$, $\phi^{\alpha p}$ is 1.67 for the normal and 4.47 for Chebyshev's distribution. We have no prior expectations for the shape of the probability distribution and we therefore make calculations with both distributions.

The EU 2050 climate policy is quite clear; 80% reduction in the 1990 emissions should be obtained in 2050. This is not the case with the nutrient targets set by BSAP, where reductions in nitrogen and phosphorus are specified for different marine basins but not the timing of their achievement. It is only stated that preparedness for implementing the targets should by obtained in 2020. Since it takes some time between implementation of nutrient reductions and final achievement, we therefore simply assume that the targets of 23% reduction in nitrogen and 61% reduction in phosphorus loads into the Baltic Proper should be achieved in 2035.

Regarding the choice of discount rate, there is a large body of literature on the appropriate level of the social discount rate, which is determined by pure time preferences, growth in consumption opportunities, and utility of consumption [31]. It is generally also suggested to use a hyperbolic discount rate, *i.e.*, a time-declining rate, for long-term projects exceeding 50–100 years. A simplification is made in this paper by assigning a uniform discount rate for all municipalities, counties and time periods. Following recommendations made for discounting future streams of net benefits, calculations of cost-effective solutions are made for a relatively low level of 0.015 [32].

With respect to choice of software, all calculations are made with the GAMS optimization program, Solver Conopt2 [33].

4. Results: Value of the Forest for Provision of Carbon and Nutrient Sequestration

The existing carbon sequestration corresponds to 24% of the total emissions in 2011. The inclusion of this sink into the Stockholm-Mälar climate and water policy would generate cost savings since part of the emission target is obtained by sequestration free of charge. However, there is currently a debate on whether carbon sequestration should be included in the EU climate policy, which focuses solely on increases in the carbon sink from a certain reference value, which is denoted additional sinks. One argument against including the existing carbon sink is that this is already considered when setting targets, and its inclusion would tighten the emission target. In the following, we will therefore focus on the value of the additional sink, but also present results on the values of the existing sink. In our model, the only additional sink option is forest plantations on agricultural land. As will be shown, the value of the existing sink available free of charge can be zero depending on risk attitudes and assumptions that stem from uncertainty in carbon sequestration.

4.1. Value of Pollutant Sink at Different Emission Targets

The main focus of this study is the value of pollutant sinks in achieving the EU 2050 climate policy and BSAP nutrient targets. As shown in Section 2, the value is determined by, among others, the stringency in emission targets, and we therefore present results for different levels of carbon and nutrient emission targets. Starting with carbon emission, the value of forest sequestration ranges from zero to 767 billion SEK (1 Euro = 9.36 SEK, 13 July 2015), depending on the emission target and the inclusion of existing or additional sequestration (Figure 2).

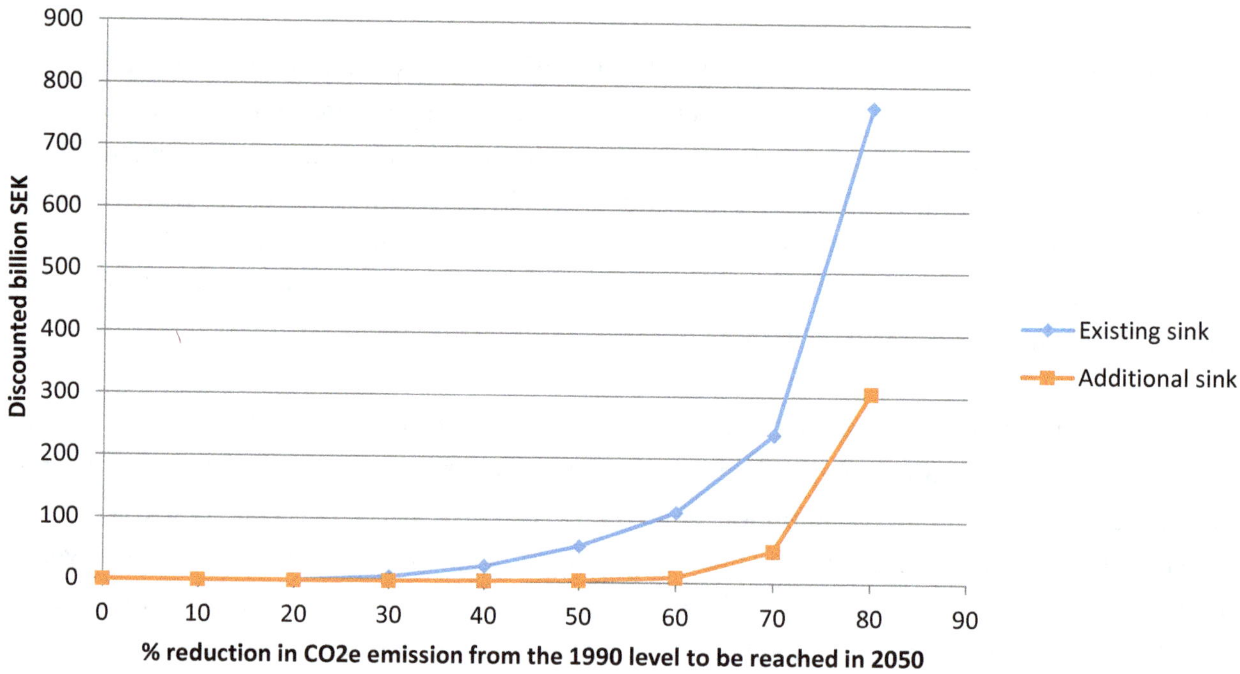

Figure 2. Value of forests for carbon sequestration when only existing or additional carbon sinks are included at different reduction levels from the 1990 emission, to be achieved by 2050.

The zero value arises from the reduction in emission of 15% in the Stockholm-Mälar region from 1990 to 2011 [10]. When existing sinks are included, the zero value increases to approximately 40% of the emissions in 1990. The total abatement cost for reaching the emission target in 2050 without the inclusion of any sink option amounts to 900 billion SEK. The average annual cost of 24 billion SEK corresponds to approximately 1.7% of the gross regional product (GRP) of the Stockholm-Mälar region. This is reduced to 0.3% when only the existing sink is included and to 1.2% when only the additional sink is considered in the climate program.

As expected from the illustration in Figure 1, the carbon sink value increases at higher emission reduction levels because of the higher abatement costs for emission reductions. The carbon sink values are relatively low at reduction levels up to 50%, and then increase rapidly. At the EU target of 80% emission reduction, the value when only existing sinks are included corresponds to approximately 767 billion SEK, and when only the additional sink is included, to 307 billion SEK. The values thus decrease the total cost without any sink option by 84% or 35%.

The corresponding values of forests for reaching different reductions in both nutrients to the Baltic Proper are presented in Figure 3.

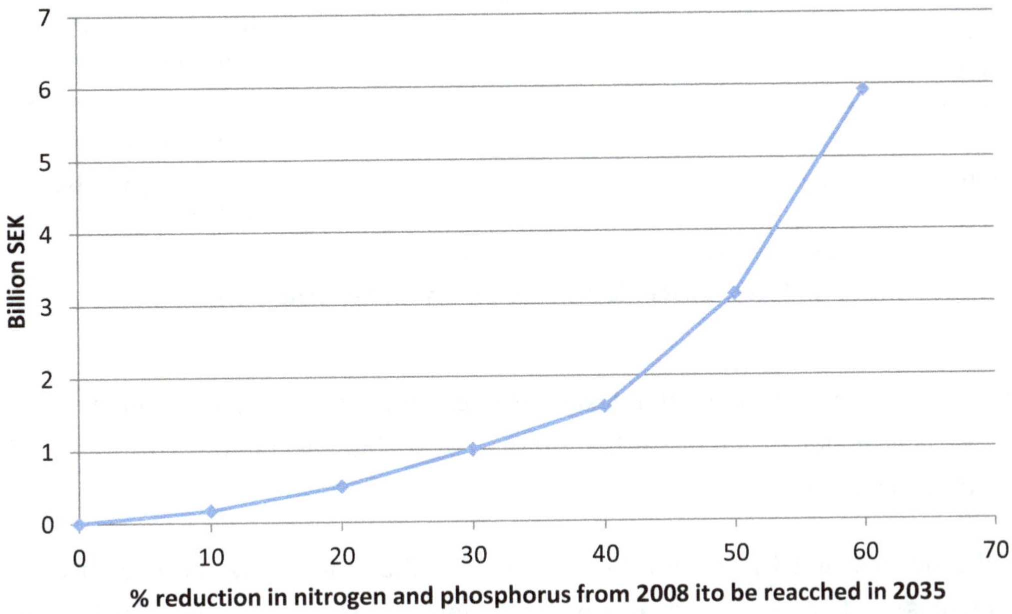

Figure 3. Value of the forest for nutrient sequestration in reducing nitrogen and phosphorus loads from the Stockholm-Stockholm-Mälar region to the Baltic Proper in 2035.

The value of the forest for nutrient sequestration is much lower than that for carbon sequestration. One reason is that total costs without nutrient sequestration are lower, and amount to 17 billion SEK at 60% reduction in both nutrients. This corresponds to an average annual cost of 0.5 billion SEK, also obtained by [9]. The cost is reduced by 5.2 billion SEK, or approximately 35%, when forest nutrient sinks are included. This relative reduction is of the same order of magnitude as the additional carbon sinks.

4.2. Uncertainty in Forest Pollutant Sink

One of the main arguments for not introducing carbon sinks in any climate policy is the uncertainty associated with stochastic weather conditions affecting biomass growth and permanence in carbon sinks over time. Therefore, 1 ton CO_2e sequestration does not correspond to 1 ton CO_2e reduction in emissions. As shown in Section 2, this is treated in terms of an uncertainty discount, which is determined by the risk preference, *i.e.*, the chosen probability of reaching the targets, the variability in the sink, and the chosen probability distribution. We evaluate the value of the forest as a pollutant sink under simultaneous achievement of the EU 2050 climate policy [23] and the Baltic Sea Action Plan [24] at different degrees of risk aversion and for the normal and Chebyshev probability distributions (Figure 4).

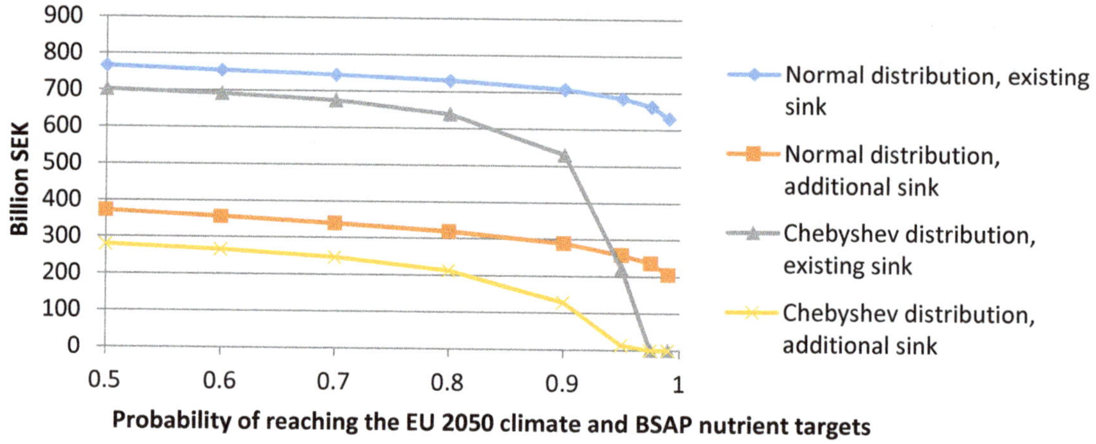

Figure 4. Value of forest pollutant sink for reaching the EU 2050 climate and BSAP nutrient targets under different probabilities of reaching the targets and assumptions of probability distributions.

For both the normal and Chebyshev probability distributions, the value decreases as the chosen probability increases. However, the decline is more drastic with the Chebyshev probability distribution, and the value is zero at probability levels exceeding 0.95. The maximum decline in the forest sink value differs for the different probability distributions: it is 166 billion SEK, or 45%, for the normal distribution and 279 billion SEK, or 100%, for the Chebyshev distribution, both of which occur at probability levels exceeding 0.95.

The maximum value of the forest carbon sink amounts to 767 billion SEK, which occurs when there is no risk aversion. The marginal costs for achieving the targets, the Lagrange multipliers in Section 2, then amount to SEK 2570/ton CO_2e reduction, and to SEK 10,300/kg phosphorus and SEK 231/kg nitrogen load reduction.

5. Discussion

The option of accepting an existing carbon sink within a climate program is not considered in other studies, in comparison to only additional carbon sinks. The calculated total cost of reaching the EU 2050 climate targets as measured in per cent of gross domestic product of the Stockholm-Mälar region, 1.12%, is within the range of that obtained for reaching the EU 2050 climate targets [34]. The calculated marginal cost of CO_2e reduction is also within the range of other studies. On the other hand, the marginal costs of the nutrient load reductions are slightly above the estimates obtained for the entire Baltic Sea and all riparian countries by [35]. One reason that these marginal costs differs from other studies could be the relatively large level of nutrient abatement already carried out in the Stockholm-Mälar region, which raises the marginal cost of further abatement.

The calculated maximum value of the existing sink of 767 billion SEK corresponds to an average forest value of SEK 22500/ha. This value can be compared with the average assessed value of productive forests, which varies between SEK 41,200/ha and SEK 58,800/ha in the different counties in the Stockholm-Mälar region [36]. The value of the existing carbon and nutrient sink can then correspond to 50% of the value for timber products. It is reduced to SEK 8700/ha, or maximum 21% of the value of productive forests, when only additional sinks are included.

As shown in Section 2, in addition to target stringency and uncertainty, there are other determinants of the value of forest pollutant sinks. These include the cost of afforestation, impact on the pollutant sink, and growth of forest sinks. It was also recognized that cost advantages of multifunctional measures materialize under simultaneous management of all targets, but not under separate management.

It is also of interest to investigate how the value is determined by the time path for decarbonization as envisaged by [23]. In order to reach the target in 2050, it is suggested that 20% of the 1990 emission should be reduced by 2020, 40% by 2030, 60% by 2040, and 80% by 2050. This path may not coincide with the cost effective path, the conditions of which were presented in Section 2, and the total cost will then be higher. The calculated total cost for achieving the EU and BSAP targets under optimal conditions amount to 900 billion, and under the EU budget, to 992 billion. This, in turn, affects the value of forest carbon and nutrient sequestration of both the existing and additional sink (Figure 5).

The value of the existing carbon sink is higher under the EU time path since the total abatement cost is larger. On the other hand, the value of the additional sink is slightly lower under the EU time path than under optimal conditions. This seems counter intuitive, but is explained by the gains made by the full use of the forest sinks' ability to grow over time for optimal solutions, which is limited by the EU suggested time path.

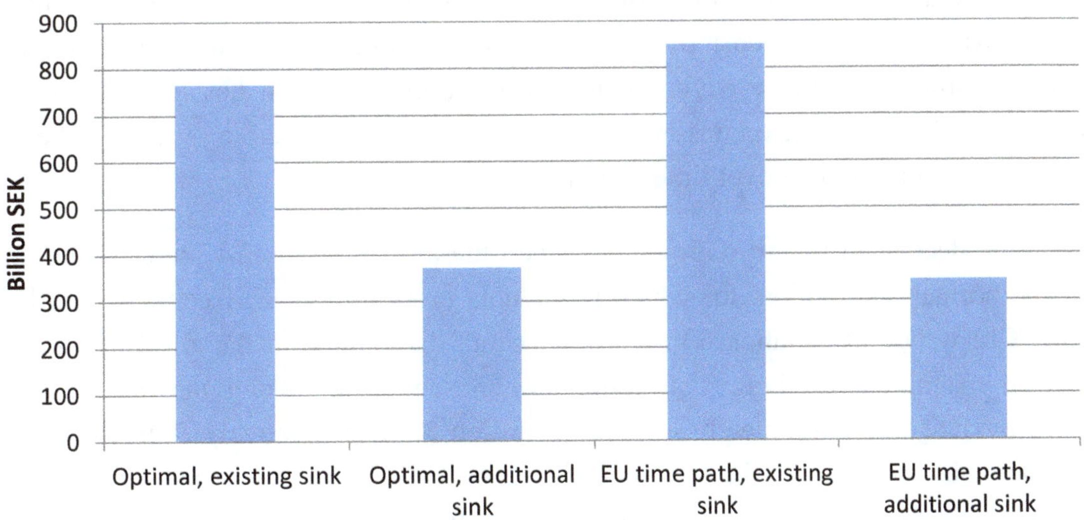

Figure 5. Value of forests as pollutant sinks for reaching the EU 2050 climate target and the BSAP nutrient targets in the Stockholm-Mälar region optimally or under the EU time path without uncertainty.

When considering uncertainty and the other factors influencing the calculated value of the forest as a pollutant sink, it is noted that the relative values, as measured in deviation from the reference values, are more sensitive to changes in parameter values under the Chebyshev than the normal probability distribution (Figure 6).

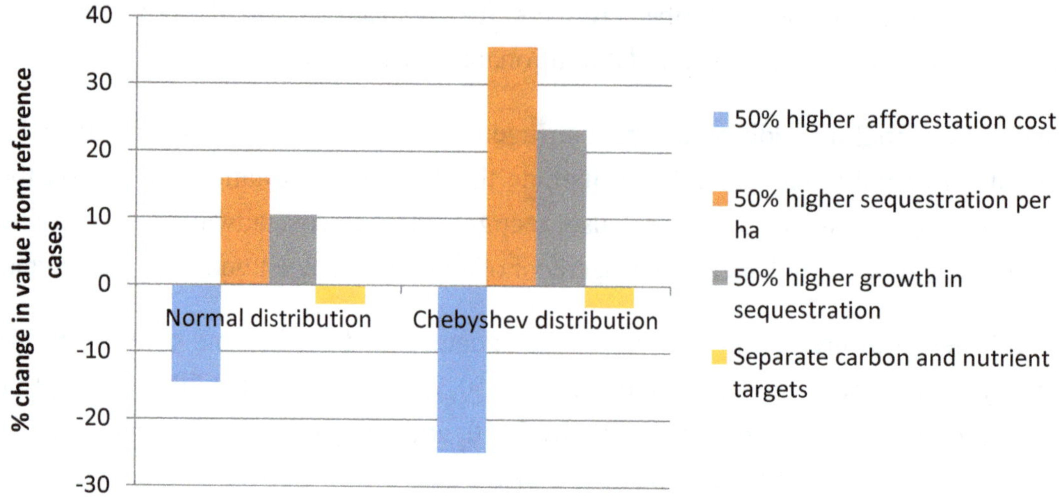

Figure 6. Impacts on the reference value of the forest as an additional pollutant sink under uncertainty with 0.9 probability (289 billion for the normal and 129 billion for Chebyshev the distribution) of different changes in afforestation cost, pollutant sequestration and growth, and separate management of carbon and nutrient targets.

As expected, the value decreases when the afforestation cost increases, which can be a result of higher future competition of land with food production. A 50% increase in the afforestation cost results in 15% and 25% decline in the reference value under the normal and Chebyshev distribution, respectively. The reason for the larger decline under the Chebyshev distribution is the relatively higher risk discount, which raises the cost of the forest as a pollutant sink compared with the normal distribution. A certain percentage deviation in afforestation costs then has a higher impact on the reference value. This also explains the higher increase in relative values when pollutant sequestration/ha and growth rate over time increases.

However, the decline in the absolute values when the afforestation cost increases by 50% is still higher for the normal distribution, for which it amounts to 43 billion SEK compared with 32 billion SEK for the Chebyshev distribution. On the other hand, the decline in the absolute values is of the same order of magnitude for both distributions when sequestration and the rate of sequestration increase and amount to approximately 46 and 30 billion SEK, respectively. The choice of separate *versus* simultaneous management has a minor impact on both distributions and raises the cost at the most by 8 billion SEK, which occurs for the normal distribution.

6. Conclusions

The calculation of values of forest carbon and nutrient sequestration in the Stockholm-Stockholm-Mälar region showed that these can be significant. They can create cost savings corresponding to 35% of the cost of achieving the EU 2050 climate and the Helcom BSAP nutrient targets in 2035. The cost saving in percentage for achieving the EU climate target is of the same magnitude as that obtained in other studies [34]. In absolute terms, the total value amounts to 307 billion SEK, which corresponds to approximately 0.45% of the region's gross domestic product.

However, this value declines when uncertainty in pollutant sink is considered, which is the main argument raised against its inclusion into, in particular, climate policy. Depending on risk aversion and the choice of the probability distribution, normal or Chebyshev distribution, the value can be reduced by 50% or 100%, respectively, at very high risk aversion, where the targets are to be achieved at probability levels exceeding 0.95.

The calculated values are affected not only by the level of uncertainty, but also by factors not considered in this study, which can result in higher or lower estimates (Table 2).

Table 2. Factors increasing or decreasing the calculated values of forest carbon sequestration and nutrient recycling, where × denotes the direction of impact.

Factor	Increased Calculated Value of Carbon and Nutrient Sinks	Decreased Calculated Value of Carbon and Nutrients
Uncertainty in all abatement measures and not only carbon sink	×	
Inclusion of more forest management option	×	
Inclusion of measures reducing uncertainty in the carbon sink	×	
Policy implementation of carbon and nutrient sink		×
Consideration of more non-marketed ecosystem services	×	×

Uncertainty in other abatement measures such as reduction in fossil fuels or investment in solar energy raises the cost of these measures, and, hence, the value of the forest carbon sink. This would also be the case if more forest management options were included, which would improve pollutant sink per area of forest and/or increase the rate of sequestration growth over time. Another factor contributing to increased values is the inclusion of management options for reducing uncertainty by, for example, increased monitoring of forest biomass growth and carbon sink content and/or improved management for reducing carbon releases from dead wood and soil.

The calculated values can be lower when considering the difficulties associated with implementing a multi-target strategy, which requires allowance for pollutant reduction in several policy systems. For example, afforestation must be deducted from the emission of nitrogen, phosphorus, and CO_2e. This can be obtained by an offset system where actors obtain credits for pollutant sequestration, which can be deducted from their pollutant emissions allowances. Another possibility is to include the abatement measures in the policy system where, for example, forest carbon sequestration can be traded on the EU ETS (Emission Trading System), or in a national tax system. Both options face problems when including land use measures because of the difficulties in monitoring and verifying sequestration, and to secure additionality and permanence in the sequestration. Associated transaction costs can be relatively high and the potential cost savings from pollutant sequestration are then smaller than pointed out in this study.

The results are also affected by the number of included non-marketed ecosystem services. This study considered only carbon sequestration and nutrient recycling. However, increases in these two services may result in decreases in other non-marketed services, such as biodiversity, which would reduce the calculated values in this study. On the other hand, if there is complementarity in the provisioning of these services, the calculated value can be raised since increases in the carbon and nutrient sink then provides a simultaneous improvement in biodiversity. Calculations of such trade-offs or complementarities in the provision of forest ecosystem services require quantified production functions for all of these services, which are not available for forests in Sweden.

The conclusion with respect to the value of forest carbon sequestration and nutrient recycling as abatement measures in climate and nutrient programs is thus ambiguous. Although our study shows considerable cost savings in the deterministic and relatively simple case, the consideration of uncertainty, policy implementation, and other ecosystem services can eliminate these savings. This inconclusiveness points out the need for further analysis and investigations of these factors for the potential value of the forest carbon sink and nutrient recycling as abatement options in climate and nutrient programs.

Acknowledgements

We are much indebted to two anonymous reviewers for valuable comments and to funding support from the EU 7th Framework Programme for the COMPLEX 'Knowledge based climate mitigation systems for a low carbon economy' project, grant agreement number 308601.

References

1. Pearce, D. The economic value of forest ecosystems. *Ecosyst. Health* **2001**, *7*, 284–296.
2. Lubowski, R.; Plantinga, A.; Stavins, R. Land-use change and carbon sinks: Econometric estimation of the carbon sequestration supply function. *J. Environ. Econ. Manag.* **2006**, *51*, 135–152.
3. Michetti, M.; Rosa, R.N. Afforestation and Timber Management Compliance Strategies in Climate Policy. A Computable General Equilibrium Analysis. *FEEM Working Paper No. 4.* **2011**. Available online: http://www.feem.it/userfiles/attach/2011118158514NDL2011-004.pdf (accessed on 14 July 2015).
4. Gren, I.-M.; Carlsson, M. Economic value of carbon sequestration in forests under multiple sources of uncertainty. *J. For. Econ.* **2013**, *19*, 174–189.
5. Aklilu, A.; Gren, I.-M. *Economic Incentives for Carbon Sequestration: A Review of the Literature. Working Paper 2014: 08*; Department of Economics, Swedish University of Agricultural Sciences: Uppsala, Sweden, 2014. Available online: https://ideas.repec.org/p/hhs/slueko/2014_008.html (accessed on 14 July 2015).
6. Gilbert, P.M. Eutrophication and harmful algal blooms: A complex global issue, examples from the Arabian Seas and including Kuwait Bay and an introduction to the global ecology and oceanography of harmful algal blooms (GEOHAB) Programme. *Int. J. Oceans Oceanogr.* **2007**, *2*, 157–169.
7. Heisler, J.; Glibert, P.M.; Burkholder, J.M.; Anderson, D.M.; Cochlan, W.; Dennison, W.C.;

Dortch, Q.; Goble, C.J.; Heil, C.A.; Humphries, E.; *et al.* Eutrophication and harmful algal blooms: A scientific consensus. *Harmful Algae* **2008**, *8*, 3–13.

8. Boberg, J.; Finley, R.; Stenlid, J.; Ekblad, A.; Lindahl, B. Nitrogen and carbon reallocation in fungal mycelia during decomposition of boreal forest litter. *PLoS ONE* **2014**, *9*, e92897.
9. Gren, I.-M. Climate change and the Water Framework Directive: Cost effectiveness and policy design for water management in the Swedish Stockholm-Mälar region. *Clim. Chang.* **2010**, *100*, 463–484.
10. Gren, I.-M.; Marbuah, G.; Tafesse, W. *Cost-Effective Land Use Dynamics towards a Low Carbon Economy in the Stockholm-Mälar Region*; Report for COMPLEX: Newcastle, UK, September 2015. Available online: http://owsgip.itc.utwente.nl/projects/complex/complex_files/D4.2%20land%20use%20simulation%20with%20report.pdf (accessed on 14 July 2015).
11. Gosselink, J.G.; Odum, E.P.; Pope, R.M. *The Value of the Tidal Marsh. Center for Wetland Resources*; Louisiana State University: Baton Rouge, LA, USA, 1972.
12. Breux, A.; Farber, S.; Day, J. Using natural coastal wetland systems for waste water treatment: An economic benefit analysis. *J. Environ. Manag.* **1995**, *44*, 285–291.
13. Byström, O. The replacement value of wetlands in Sweden. *Environ. Resour. Econ.* **2000**, *16*, 347–362.
14. Gren, I.-M. The economic value of coastal waters as nutrient filters for the Baltic Sea. *Reg. Environ. Chang.* **2013**, *13*, 695–703.
15. Tavoni, M.B.; Sohngen, B. Forestry and the carbon market response to stabilise climate. *Energy Policy* **2007**, *35*, 5346–5353.
16. Anger, N.; Sathaye, J. Reducing Deforestation and Trading Emissions: Economic Implications for the Post-Kyoto Carbon Market. ZEW-Centre for European Economic Research Discussion Paper (08–016), 2008. Available online: http://ftp.zew.de/pub/zew-docs/dp/dp08016.pdf (accessed on 14 July 2015).
17. Byström, O.; Andersson, H.; Gren, I.-M. Economic criteria for restoration of wetlands under uncertainty. *Ecol. Econ.* **2000**, *35*, 35–45.
18. Kurkalova, L. Carbon sequestration in agricultural soils: Discounting for uncertainty. *Can. J. Agric. Econ.* **2005**, *53*, 375–384.
19. Kim, M.; McCarl, B. Uncertainty discounting for land based carbon sequestration. *J. Agric. Appl. Econ.* **2009**, *41*, 1–11.
20. Tesler, L.G. Safety-first and hedging. *Rev. Econ. Stud.* **1955**, *23*, 1–16.
21. Gren, I.-M.; Carlsson, M.; Munnich, M.; Elofsson, K. The role of stochastic carbon sink for the EU emission trading system. *Energy Econ.* **2012**, *34*, 1523–1531.
22. Taha, H.A. *Operations Research, an Introduction*, 2nd ed.; Macmillan Publishing, Inc.: New York, NY, USA, 1976.
23. European Commission. Climate Action. Roadmap for Moving to a Low-Carbon Economy in 2050. Available online: http://ec.europa.eu/clima/policies/strategies/2050/faq_en.htm (accessed on 1 October 2015).
24. Helcom HELCOM Baltic Sea Action Plan. Nutrient Reduction Scheme. Available online: http://www.helcom.fi/baltic-sea-action-plan/nutrient-reduction-scheme/ (accessed on 7 January 2014).

25. SEPA (Swedish Environmental Protection Agency) 2014. Utsläpp av Växthusgaser Från Markanvändning. Available online: http://www.naturvardsverket.se/sa-mar-miljon/statistik. A-O/vaxthusgaser-utslapp-och-upptag-fran-markanvändning/ (accessed on 24 February 2014).
26. Munnich Vass, M. *Can Renewable Energies with Learning-by-Doing Compete with Forest Sequestration to Cost-Effectively Meet the EU Carbon Target for 2050?* Working Paper 2015:04, Department of Economics, SLU: Uppsala, Sweden, 2015. Available online: http://pub.epsilon. slu.se/11905/ (accessed on 14 July 2015).
27. Janssens, L.; Freibauer, A.; Schlamadinger, B.; Ceulemans, R.; Ciais, P.; Dolman, A.; Heimann, M.; Nabuurs, G.-J.; Smith, P.; Valentini, R.; *et al.* The carbon budget of terrestrial ecosystems at country-scale a European case study. *Biogeosciences* **2005**, *2*, 15–26.
28. McCarl, B.A.; Spreen, T.H. *Applied Mathematical Programming Using Algebraic Systems.* Texas A&M University: Austin, TX, USA, **2010**, pp. 14.1–14.18. Available online: ftp://178.213.241.34/pub/Library/BOOKS/heap/SCI%20BOOKS/books2/_djvu/M_Mathematics/ MOc_Optimization%20and%20control/McCarl%20B.A.,%20Spreen%20T.H.%20Applied%20ma thematical%20programming%20using%20algebraic%20systems%28567s%29.pdf (accessed on 3 January 2015).
29. McSweeny, W.T.; Shortle, J.S. Probabilistic cost effectiveness in agricultural nonpoint pollution control. *Southern J. Agric. Econ.* **1990**, *22*, 95–104.
30. Shortle, J. The allocative efficiency implications of water pollution abatement cost comparisons. *Water Resour. Res.* **1990**, *26*, 793–797.
31. Weitzman, M. Gamma Discounting. *Am. Econ. Rev.* **2001**, *91*, 260–271.
32. Newell, R.; Pizer, W. Discounting the distant future. How much do uncertainty increase valuations? *J. Environ. Econ. Manag.* **2003**, *46*, 52–71.
33. Rosenthal, R. *GAMS—A User's Guide*; GAMS Development Corporation: Washington, DC, USA, 2008.
34. Capros, P.; Paroussos, L.; Fragkos, P.; Tsani, S.; Boitier, B.; Wagner, F.; Busch, S.; Resch, G.; Blesl, M.; Bollen, J. European decarbonisation pathways under alternative technological and policy choices: A multi-model analysis. *Energy Strategy Rev.* **2013**, *2*, 231–245.
35. Gren, I.-M.; Savchuck, O.; Jansson, T. Cost effective spatial and dynamic management of a eutrophied Baltic Sea. *Mar. Resour. Econ.* **2013**, *28*, 263–284.
36. Average Assessed Value of Standing Forest and Forest Land by Ownership Class and County, Year 2011. Available online: http://www.skogsstyrelsen.se/Myndigheten/Statistik/Amnesomraden /Ekonomi/Tabeller--figurer/ (accessed on 13 July 2105).

2

Carbon Budgets for Caribbean Mangrove Forests of Varying Structure and with Phosphorus Enrichment

Catherine E. Lovelock [1,*], Lorae T. Simpson [2,3], Lisa J. Duckett [2] and Ilka C. Feller [2]

[1] School of Biological Science, The University of Queensland, St Lucia, Queensland 5072, Australia

[2] Smithsonian Environmental Research Center, 647 Contees Wharf Rd, Edgewater, MD 21037, USA; E-Mails: simpsonl@si.edu (L.T.S.); duckettl@si.edu (L.J.D.); felleri@si.edu (I.C.F.)

[3] Smithsonian Marine Station at Fort Pierce, 701 Seaway Dr, Fort Pierce, FL 34949, USA

* Author to whom correspondence should be addressed; E-Mail: c.lovelock@uq.edu.au

---

Abstract: There are few detailed carbon (C) budgets of mangrove forests, yet these are important for understanding C sequestration in mangrove forests, how they support the productivity of the coast and their vulnerability to environmental change. Here, we develop C budgets for mangroves on the islands of Twin Cays, Belize. We consider seaward fringing forests and interior scrub forests that have been fertilized with phosphorus (P), which severely limits growth of trees in the scrub forests. We found that respiration of the aboveground biomass accounted for 60%–80% of the fixed C and that respiration of the canopy and aboveground roots were important components of respiration. Soil respiration accounted for only 7%–11% of total gross primary production (GPP) while burial of C in soils was ~4% of GPP. Respiration by roots can account for the majority of soil respiration in fringing forests, while microbial processes may account 80% of respiration in scrub forests. Fertilization of scrub forests with P enhanced GPP but the proportion of C buried declined to ~2% of GPP. Net ecosystem production was 17%–27% of GPP similar to that reported for other mangrove forests. Carbon isotope signatures of adjacent seagrass suggest that dissolved C from mangroves is exported into the adjacent ecosystems. Our data indicate that C budgets can vary among mangrove forest types and with nutrient enrichment and that low productivity mangroves provide a disproportionate share of exported C.

Keywords: *Rhizophora mangle*; respiration; fertilization; carbon burial; Belize

1. Introduction

Mangrove forests are important sinks of carbon (C) throughout the tropics and subtropics [1–3] due to their high C burial rates within soils [3–5]. These observations have led to a heightened level of interest in conserving and restoring mangroves for their C values [6,7]. Additionally, mangrove forest primary production is responsible for approximately 10% of the global export of both particulate and dissolved organic C to the oceans, which supports production in near shore waters [8,9]. However, there are very few detailed C budgets available for mangrove forests [3,10,11]. Here, we describe a carbon budget for mangrove forests in Belize. Our aims are to increase knowledge of the global variation in C budgets of mangroves and the understanding of the factors giving rise to variation in C stocks, rates of C burial and the role of mangrove forests in supporting productivity of the coasts on which they occur.

Gross primary productivity (GPP) and the size of the detrital pool in ecosystems are strongly influenced by variation in vegetation structure and rates of turnover [12]. Although little is known of variation in mangrove tree turnover rates [13], mangrove forests vary widely in structure and productivity across large (e.g., latitude [13]) and small spatial scales (e.g., ecotones [14]). Some of the major features of the few existing C budgets of mangrove forests indicate that despite being rooted in saline soils, they are highly productive forests, with similar productivity to terrestrial forests, and that a large proportion of fixed C is respired by the canopy [3,13]. Carbon burial rates of detritus have been estimated to be approximately 2%–4% of GPP, but this could vary with age of the forest and productivity [3,11]. At our study site, C burial is particularly important because of the dependence on the burial of autochthonous organic matter for maintaining the soil elevation relative to sea level [15,16]. Finally, approximately 40% of the fixed C has not yet been accounted for in the existing budgets and is proposed to be exported from the ecosystem [10,13,17], although how the size of the putatively exported fraction varies among different types of mangrove forests is also not known.

Mangrove forests are threatened by agriculture, industry and urbanization [18] which can lead to losses in ecosystem C stocks and alterations to C budgets. Not only are mangrove forests cleared, which results in losses of aboveground and belowground C [2,19,20], but anthropogenic nutrient enrichment alters a range of processes that could affect C budgets [21]. In terrestrial ecosystems, increases in the availability of nutrients have been shown to enhance primary productivity, reduce allocation to roots and increase rates of decomposition of soil organic matter [22,23]. Mangroves also show strong sensitivity to nutrient additions for a range of processes, including above and belowground growth and decomposition [14,24,25]. These studies have demonstrated that scrub mangrove growth can be severely limited by phosphorus (P) availability. However, little is known of how nutrient enrichment influences the efficiency of C burial within soils through reductions in allocation to roots or enhanced rates of decomposition of soil organic matter.

We used a mass balance approach of C inputs (autotrophic components) and outputs (the fate of autotrophic production) similar to those presented in Alongi [11] to identify the major C sources and sinks of mangrove forests of Twin Cays, Belize. Twin Cays is a small archipelago comprised of 92 ha of mangrove forest that lies within the lagoon of the Meso-American Barrier Reef [26]. In the Caribbean region, mangrove forests often grow on peat soils that overlie limestone [27–29]. Unlike mangrove forests in deltaic systems, where forests often grow on mineral soils, mangrove forest soils in oceanic settings in the Caribbean are often comprised of mangrove roots that have accumulated during the rise in sea level over the Holocene [15]. Accumulation of these peat soils is facilitated by inundation and anoxia which limits decomposition [30]. The mangrove forests of Twin Cays consist of a range of vegetation zones with varying levels of productivity and which are differentially limited by nutrient availability [14]. Tall seaward fringing forests are adjacent to stunted dwarf or scrub forests in the interior of the islands [31]. In addition to stunted trees, microbial mats are well developed in the interior of the islands and have been proposed to be important sites of biogeochemical cycling, including proposed sources of C and nutrients [32]. Dwarf or scrub trees within the islands interiors show pronounced growth enhancements when fertilized with P, indicating P limitations to primary productivity in these areas, in contrast to nitrogen limitation to growth observed in the taller seaward fringing forests [14]. Using data from a range of studies, including those focused on long-term fertilization experiments on Twin Cays [24], we assess how forest C budgets vary among different types of mangrove stands and how they are influenced by nutrient enrichment. The intensive research activity at Twin Cays over decades provides an unparalleled, comprehensive insight into elements of the C budget of these mangrove forests that have rarely been assessed simultaneously in other forests.

2. Experimental Section

2.1. Study Site

Twin Cays, Belize (16°50′ N, 88°06′ W) is a small archipelago comprised of 92 ha of mangrove forest that lies within the lagoon of the Meso-American Barrier Reef [26]. The mangroves forests of Twin Cays can be divided into two main vegetation zones: (1) Forests fringing the ocean dominated by *Rhizophora mangle*, which are approximately 3 to 6 m tall and occupy approximately 100,000 m^2 of the islands; and (2) Dwarf or scrub stands (less than 1.5 m) of *R. mangle* and well developed microbial mats that are associated with ponded areas of the interior of islands which cover ~190,000 m^2 of the islands area [31,32] (Figure 1). *Avicennia germinans* and *Laguncularia racemosa* also occur on the islands in areas of higher elevation, but are excluded from this study because sufficient data are not available to create detailed C budgets for these vegetation zones. Dwarf or scrub trees within the islands interiors show pronounced growth enhancements when fertilized with P (see insert in Figure 1), indicating P limitations to primary productivity in these areas, in contrast to nitrogen limitation to growth observed in the taller seaward fringing forests [14].

Data for the construction of C budgets for Twin Cays, Belize have been acquired over the last 25 years due to the presence of the Smithsonian's field station at nearby Carrie Bow Cay [26] and a range of research initiatives, including the Caribbean Coastal Marine Productivity program (CARICOMP [33]), that have been supported by the Smithsonian Institution and other funding agencies. In order to construct a C budget for different forest zones (seaward fringing forests and landward scrub forests) and for nutrient enriched (fertilized) and control (non-fertilized) plots, we have collated and integrated results from a range of studies that have used a wide array of approaches (Table 1). Much of the data are from long term fertilization experiments that were commenced on Twin Cays in scrub forest in 1988 [34] and extended to cover both fringe and scrub forests in 1995 [14]. Briefly, the fertilization experiments were conducted at three locations in both fringing forests and scrub forests (Figure 1), At each location and in each forest zone nine plots were fertilized with nitrogen (N, applied as urea) and nine fertilized with slow release phosphorus fertilizer (P) with nine remaining as unfertilized controls [14]. In each plot a focal tree was assessed to determine rates of tree growth. As there were no differences among control and P fertilized trees in the fringing forests for many measured parameters and relatively small differences among N fertilized and control trees in the fringing forests [14,15,35], and because in the scrub forests there were small differences between N fertilized and control trees, but very large differences between P fertilized trees and control trees [14,15,35,36], we chose to focus on contrasts among fringe and scrub forests and within the scrub forest between P fertilized plots and control plots (Table 1).

Figure 1. Aerial image of the study site, Twin Cays, Belize in 2006. Study locations are indicated by yellow arrows. The inset provides a detailed view of one of the study locations. Scrub and fringe forests are indicated as well as the phosphorus (P) fertilized scrub trees.

Table 1. Parameters used in estimating components of the carbon budget of mangrove forests of Twin Cays, Belize. Density of trees is from Koltes *et al.* [33]; leaf photosynthetic carbon gain is from Lovelock *et al.* [35] and Cheeseman and Lovelock [36]. Values are means ± standard errors.

Forest Attributes	Forest Type		
	Fringe	**Scrub**	**P Fertilized Scrub**
Density of trees No. m^{-2} ($N = 5$)	0.51 ± 0.23	0.23 ± 0.09	0.23 ± 0.09
Tree Height ($N = 9$)	3.3 ± 0.2	1.19 ± 0.10	2.24 ± 0.22
LAI $m^2 \cdot m^{-2}$ ($N = 9$)	1.82 ± 0.07	0.39 ± 0.06	1.70 ± 0.77
Direct measure of canopy leaf area ($m^2 \cdot m^{-2}$) ($N = 9$)	2.62 ± 0.30	1.17 ± 0.33	4.61 ± 0.94
Surface area stems (m^2 tree)	1.3 ± 0.3	0.06 ± 0.01	0.46 ± 0.09
Aboveground root length (m length per tree) ($N = 9$)	62 ± 20	7.4 ± 4.0	39 ± 16
Surface area aboveground roots (m^2 root area m^{-2})	5.4 ± 1.7	0.59 ± 0.29	2.8 ± 1.1
Biomass of macroalgae $g \cdot m^{-2}$ ($N = 9$)	40 ± 20	8 ± 8	24 ±13
Surface area of habitat for macroalgae (m^2 root area m^{-2}) ($N = 9$)	0.68 ± 0.14	0.22 ± 0.02	0.53 ± 0.08
Leaf photosynthetic carbon gain $\mu mol \cdot m^{-2} \cdot s^{-1}$ ($N = 9$)	10.3 ± 0.4	7.4 ± 0.1	12.1 ± 0.7
Leaf respiration $\mu mol \cdot m^{-2} \cdot s^{-1}$ ($N = 9$)	1.41 ± 0.06	1.61 ± 0.04	1.98 ± 0.12

P: phosphorus; LAI: leaf area index.

2.2. Primary Productivity and Respiration of the Canopy, Mats and Macroalgae

Tree gross primary production (GPP) was estimated by addition of canopy net primary productivity (NPP) to canopy respiration [13]. Leaf area index (LAI) of all forest plots was determined using hemispherical photography analysed with the Hemiview program (Delta T Devices, Cambridge, UK). Canopy NPP was estimated by multiplying the LAI by the mean photosynthetic rate measured in the field (obtained from [35]), which was then scaled up to an annual NPP per plot assuming 10 h per day of photosynthetic activity. Respiration rates of three canopy leaves from each plot were measured at night (in darkness) with the LiCor 6400 standard leaf chamber (Li-Cor, Lincoln, NE, USA). Because LAI incorporates variations in leaf angle (assesses the projected leaf area), and mangroves are known to have steep leaf angles to avoid excess solar radiation and limit water loss [37,38] we also assessed the total leaf area per m^2 by harvesting all leaves within a 1 m^2 quadrat through the canopy in the fertilization experimental plots (9 per treatment for each forest zone within each of the three study locations). We used this direct measure of leaf area to scale up leaf respiration rates to an annual rate canopy respiration rate by multiplying leaf respiration per leaf area (Table 1) by total leaf area per m^2 × 24 h × 365 days.

GPP of the microbial mats was obtained from [39,40]. We assumed 100% cover of the soil surface by these microbial communities [3], but differing productivity over forest zones due to variation in mangrove canopy cover and thus light reaching the benthos [39]. No values of GPP for the mats were available for the P fertilized trees. We therefore used values for the fringe forest zone anticipating that the similar canopy cover (LAI) limited productivity of the mats in the treatment. GPP of the macroalgaal community that is associated with mangrove roots (comprised mainly of *Bostrychia tenella*, *B. montagnei* and *Catenella repens*) was estimated from measures of maximum rates of electron transport from chlorophyll fluorescence using a PAM 2100 (H. Walz, Effeltrich, Germany) and the relationship between carbon gain and electron transport described by Genty *et al.* [41]. There were no significant differences among photosynthetic rates of the macroalgae over fertilization treatments thus, we used a mean photosynthetic rate of 4.75 μmol C $m^{-2}·s^{-1}$ to estimate GPP. Mean photosynthetic rate was scaled up to mol C $m^{-2}·year^{-1}$ in a similar way to canopy photosynthesis. Respiration of algae was measured in air using the Licor 6400 Portable Photosynthesis System (Li-Cor, Lincoln, NE, USA) with the standard leaf chamber in darkness. Values were scaled to GPP on an area basis by multiplication by the area of root macroalgal habitat (aboveground root surface area) measured in 0.25 m^2 plots within each experimental plot (Table 1).

2.3. Other Outputs: Litter Export, Herbivory, Respiration of Woody Tissues and Soils

Litter fall, litter turnover and forest structure data were obtained for un-manipulated forests from the CARICOMP data sets, which were part of a Caribbean-wide program to assess variation in marine productivity [33]. The CARICOMP procedure includes measurement of all trees within replicate 10 × 10 m plots, the deployment of litterfall traps sampled monthly and periodic sampling of leaf litter on the soil surface used to calculate leaf litter turnover. The CARICOMP data set was augmented with another year of data collected on Twin Cays in 2003 using the same procedures Litter fall data was not available from the trees within the fertilization experimental plots and was therefore estimated from leaf initiation rates published in Feller *et al.* [14]. We assumed a relatively constant number of leaves in the canopy and scaled up to litterfall per m^2 area by multiplying leaf production per tree by the density of trees in each vegetation zone. Litterfall estimated in this way was comparable to litterfall measured in the CARICOMP plots. For the C budget leaf litter was presumed to be exported from the system consistent with the high rates of litter turnover observed at this site [34] and the minor component that leaves contribute to the peat [15]. Losses of leaf material due to herbivory were sourced from Feller and Chamberlain [42].

Wood production was calculated from the mean increment of woody tissues from nine dendrometer bands deployed on stems within the fringing forest for 1 year. We detected no differences among fertilization treatments in wood production in the fringing forest and thus assumed a mean rate of basal area increase of 2.9 $cm^2·year^{-1}$. Tree stems in the fringe had a mean diameter of 11.2 cm. Wood production in the scrub forests was assumed to be low and assigned a zero value. We used a simple approach to convert basal area increment to woody biomass increment of stems by assuming stems are cylindrical with a height of 3 m. To estimate C accumulation from woody biomass increments we used literature values of wood biomass density for *R. mangle* (0.86 $g·cm^{-3}$) and C content of wood (46% carbon) [43]. Respiration rates of aboveground woody tissues (roots and stems) were calculated from non-destructive measurements of surface area of woody tissues and measurements of respiration from detached tissues harvested from within the experimental plots. Respiration was measured on detached tissues that were approximately 1–2 cm diameter and 7 cm in length using the 1 L conifer chamber attached to the LiCor 6400 Portable Photosynthesis System. Respiration was then expressed on a surface area basis. Although previous studies in terrestrial trees have indicated that sap wood volume is the most appropriate way to express respiration of woody stems [44] and others have suggested mass is more appropriate [45], we used surface area as a proxy for the amount of respiring tissues (cambium) rather than volume or mass because a study of whole tree respiration in mangroves from Japan indicated that while respiration was explained adequately by mass in rapidly growing trees, surface area may be a better predictor in slower growing trees. That is, respiration is better explained by mass to the exponent 2/3 or 3/4 rather than 1 [46,47]. Respiration of woody tissues was scaled to a rate $m^{-2}\cdot year^{-1}$ by multiplication by the surface area of aboveground roots and stems. Total surface area of woody tissues per tree was estimated by field measurements of tree architecture. Diameter, length of branches, and number of branches arising from each branch was recorded for representative stems and aboveground roots at each branching order on each of the focal trees within the fertilization experiments. The number of main stems (trunks) was assessed then for each branching order it was assumed, based on our observations, that only one branch (of possible three) continued to support the next branching order. To obtain the total area of stem surface area per tree the area of stem from all stems arising from the main stems were multiplied by the number of main stems. To estimate root surface area we followed a similar process assessing the number of main roots arising from the base of the stem and then assessing the diameter and length of all branching orders for the focal trees within the fertilization experiments.

Soil respiration was measured when soils were exposed in air using the LiCor 6400 with soil flux chamber attached. The chamber was placed at the sediment surface, penetrating only 5 mm of the sediment, to avoid severing surface roots [48,49]. The production of fine and coarse roots was obtained from McKee [50] who measured root production using in-growth bags within the fertilization experimental plots. Carbon burial within soils was calculated from mean increases in soil surface elevation over time, which was measured over the peat depth of 9–12 m (depth to the benchmark) [15] multiplied by the mean C density of soils obtained from [50]. Carbon density in this study was assessed from small cores (5 cm in diameter and 20 cm depth) from each plot which were dried and finely ground before analysis using an elemental analyser (EA 1110, CE Instruments, Milan, Italy) [50]. Root respiration from each plot was measured on detached fine roots in moist air using a custom built root respiration chamber attached to the LiCor 6400 Portable Photosynthesis System [51].

2.4. Data Analysis

Statistical tests were not appropriate for the descriptive C budgets presented as values were collated over a many different studies using a range of different approaches that did not always adhere to a similar experimental design. Regression analysis was used to assess the relationship between GPP and C burial in soils.

3. Results and Discussion

3.1. Gross Primary Productivity

Gross primary production was higher in fringing forests than scrub forests and higher still in the P fertilized scrub forests (Table 2). A larger portion of GPP was attributed to the microbial mat production in the scrub (~27%) compared to the fringing forest (~2.5%) and the P fertilized trees (~2%). The contribution to GPP of the epiphytic macroalgal community that adheres to the roots (*i.e.*, *Bostrychia* and *Catenella*) was low in the interior scrub forest and highest in the seaward fringing forest where tidal inundation is frequent. The mean biomass of these communities was 8 ± 8 $g{\cdot}m^{-2}$ (mean ± SE) for the scrub forests and 40 ± 20 $g{\cdot}m^{-2}$ in the fringing forest (Table 1). The macroalgal community contributes 3.7% of total community GPP in P fertilized scrub, 5.6% in the scrub and 8.9% in the fringing forest.

Table 2. Summary carbon budget (mol C $m^{-2}{\cdot}year^{-1}$) of fringing and scrub *Rhizophora mangle* forests and phosphorus fertilized scrub forests for Twin Cays, Belize.

Budget Components	Forest Type		
	Fringe	Scrub	Fertilized P Scrub
INPUTS			
Mangrove trees	378.0	89.0	555.0
Epiphytic algal community	38.0	7.5	22.0
Microbial mat	11.0	36.5	11.0
TOTAL INPUTS	**427.0**	**133.0**	**588.0**
OUTPUTS			
Canopy respiration	132.0	51.0	286.0
Soil respiration	47.0	8.6	81.5
Aboveground root respiration	108.0	29.0	168.0
Stem respiration	18.6	3.2	36.5
Algal respiration	8.2	1.6	4.8
Wood production	11.6	0.1	15.0
Herbivory	1.2	0.4	1.6
Burial in soil	17.6	2.8	12.9
Export (litter)	13.1	0.8	13.1
TOTAL OUTPUTS	**357.3**	**97.5**	**619.4**
Net Ecosystem Production (NEP) (% GPP)	69.6 (16.3)	35.5 (26.7)	−31.4 (−5.3%)
% C of GPP buried	4.1	2.1	2.2
Production:Respiration ratio (P:R)	1.19	1.36	0.95

Within the same island, different habitats (seaward fringing forests and scrub or dwarf habitats) can have vastly different levels of productivity and have differences in the sources and fates of fixed C. The mangrove canopy is responsible for most GPP, but soil surface microbial communities accounted for 27% of GPP in the scrub compared to 2.5% in the fringing forest. These values are consistent with the limited development of the canopy in scrub forests, which at this site is due to P limitations to tree growth [14,34,52].

3.2. Autotrophic Respiration

There were differences between aboveground root tissue and aboveground stem tissues in respiration rates (Table 1). We used a respiration rate of 3.3 μmol CO_2 $m^{-2}·s^{-1}$ for aboveground roots and 1.27 μmol $m^{-2}·s^{-1}$ for woody stem tissues in order to scale-up aboveground woody tissue respiration to the whole tree and stand level. Respiration of aboveground components (leaves, stems, aboveground roots) accounted for a large proportion of GPP in seaward fringing forests (68%), scrub forests (94%) and P fertilized scrub (88%). These values reflect the large aboveground root biomass that occurs in these forests (Table 1). We estimated 7.1 ± 3.9 m of aboveground root length per tree in scrub trees (0.6 m^2 surface area), 61 ± 20 m (5.4 m^2 surface area) in seaward fringing trees and 39 ± 16 m (2.8 m^2 surface area) for P fertilized scrub trees.

Our C budget indicates, similar to other forests that respiration of the biomass accounts for the largest fraction of GPP [13]. We found that aboveground roots and stems are a significant respiratory sink with values similar to canopy respiration in the fringing forest, and values proportionally lower in the scrub and P fertilized scrub. *Rhizophora mangle* under flooded (and nutrient limited) conditions produces a large biomass of aboveground roots [53]. This component of mangrove tree metabolism has rarely been explicitly included in C budgets but could be a large factor in some forests, particularly where primary production is low but aboveground root biomass is high [54].

The magnitude of our estimate of respiration of woody tissues could be overestimated because: (1) Aboveground roots and twigs of mangroves are photosynthetically active [55,56], which at high levels of solar radiation (as occurs in the scrub forest) may reclaim respired CO_2. Measurements in other mangrove species indicate the epidermis of roots and twigs remove 90% of incident photosynthetic radiation [57], thus, over estimation of respiration from woody tissues would most likely occur in the scrub habitats where woody tissues are exposed to high levels of solar radiation compared to fringing habitats where aboveground roots are shaded by the canopy; (2) Rates of respiration measured in detached tissues may be higher than that for intact trees. Rates of stem respiration measured in pine species [41] were lower than the values we observed. Ryan *et al.* [43] observed values of 0.2–0.6 μmol CO_2 $m^{-2}·s^{-1}$ while our aboveground tissues had mean values of 1.44 μmol CO_2 $m^{-2}·s^{-1}$ and aboveground roots even higher at 3.8 μmol CO_2 $m^{-2}·s^{-1}$. In contrast, fine root respiration rates of *R. mangle* were lower than most other species [50]. But, high levels of aerenchyma tissues that occur in aboveground roots of mangroves may impose a high metabolic cost as well as high respiration rates in aboveground roots and stems [58]. Given the sensitivity of respiration to temperature and rising levels of atmospheric CO_2 [59], further work is required to fully understand this component of mangrove forest C budgets. Over-estimation of the respiration of woody tissue would increase the unaccounted for fraction of C in the budget (the net ecosystem production, NEP), potentially leading to an underestimation of C available for export (described below).

Other output components of the C budget are comparatively small and within the range observed in other studies. Litter production was similar to that predicted by the relationship between precipitation and export of litter [60]. Wood production was also similar to that measured in other low productivity mangrove forests [61], accounting for 3% of GPP in fringing forest and P fertilized scrub but lower (<1%) in the scrub forests. Losses to canopy herbivory are <1% of GPP. Wood feeding insects contribute to woody tissue turnover [62] and may therefore contribute to the wood production term in the budget (7% of fringing forest GPP) and to C burial due to incorporation of dead wood into the soil profile [49,63].

3.3. Soil Respiration

Soil respiration accounted for 11% in fringing forests, 6.5% in fringing forests and 14% of GPP in P fertilized scrub forests (Table 2). Using values of root respiration and annual root production rates from [15], we estimated the proportion of soil respiration that may be due to root metabolism (Table 3). Approximately 100% of soil respiration could be due to root respiration in the seaward fringing forest and P fertilized scrub forests, while only 20% of soil respiration can be accounted for by root respiration in the scrub habitat, the balance (80%) is then attributed to respiration of the microbial community.

Table 3. Rates of soil respiration in mol C $m^{-2}\cdot year^{-1}$ from mangrove forests of Twin Cays, Belize and estimates of the components of soil respiration due to root respiration and respiration of the microbial community.

Budget Components	Forest Type		
	Fringe	**Scrub**	**P fertilized Scrub**
Soil Respiration	61.2	34.7	92.7
Root respiration (stock × respiration rate)	64.2	6.9	102.4
Microbial respiration (by difference) (% of soil respiration)	−3.0 (−5%)	27.8 (80%)	−9.7 (−10%)

That respiration of roots could completely account for soil respiration in both the seaward fringing forests and P fertilized scrub forest suggests low activity of the heterotrophic microbial community, consistent with high level of accumulation of organic matter in mangrove ecosystems [1–5]. Our estimates of the amount of live roots are probably underestimated because under control conditions (not fertilized) root mortality was not observed during the experiments of McKee *et al.* [15]. In Florida, fine to small roots had longevities of 1.7 to 4.4 years [64], indicating that the abundance of live roots may be much larger than annual production and thus root respiration could be ~2 to 4 times higher than our conservatively estimated value. Studies in terrestrial forests have also found that root respiration can be a major component of soil respiration [65]. However, respiration of the microbial community contributed 80% of the CO_2 flux in the scrub habitats. The C sources and nutrients supporting microbial respiration in the scrub forests are derived from litter and the microbial mats [37,38] but may also be from the mangrove peat. High rates of CO_2 emissions were observed following clearing of mangroves at Twin Cays [18] suggesting that the peat may be decomposed under some conditions [66] and also that living tree roots may suppress decomposition of soil organic matter [67]. Although anoxia due to tidal flooding contributes to slow rates of decomposition of soil C in mangroves, other factors, potentially associated with productivity of trees, e.g., competition for limiting P, may also play important roles in suppressing microbial metabolism and thus decomposition of soil organic matter [67].

3.4. Carbon Burial

Carbon burial in soils, calculated from annual rates of soil surface accretion and C density of soils was higher in the seaward fringing forests (17.6 mol C $m^{-2} \cdot year^{-1}$) than in the scrub forests (2.8 mol C $m^{-2} \cdot year^{-1}$) and the P fertilized scrub (12.9 mol C $m^{-2} \cdot year^{-1}$). These were 4.1% of GPP in the seaward fringe and 2.1% of GPP in both scrub and P fertilized scrub (Table 2).

Carbon burial in soils in the fringing forest and P fertilized scrub trees were similar to the global mean of 10.8 mol C $m^{-2} \cdot year^{-1}$ [4]. Extrapolation of these rates using the area of habitat on Twin Cays indicates that the *R. mangle* dominated fringe and scrub forests sequesters ~27.5 Mg C $year^{-1}$ (or about 100 Mg CO_2 equivalents per year) which is ~0.02% of the total CO_2 emissions of Belize in 2008 (116 Gg of C [68]). Using areas of scrub and tall/fringing mangrove habitat for the whole country of Belize [69] suggests that nation-wide C burial by these mangrove habitats is 61,600 Mg C $year^{-1}$, which is ~50% Belize's nation-wide reported CO_2 emissions. The sheer magnitude of the C burial by mangrove habitats in Belize warrants their conservation and restoration where they have been lost.

By combining our values of C burial with those in the literature [11] and unpublished values, we found that C burial tends to increase with increasing productivity of the forests (Figure 2). Moreover, fringing forests in Mexico (similar to Belize) were observed to have higher stocks of soil C in fringe (987 Mg C ha^{-1}) compared to scrub (381 Mg C ha^{-1}) [28]. These findings indicate that soil C stocks accumulated over time may be a function of stable, long term patterns in environmental conditions and productivity over mangrove landscapes in this region.

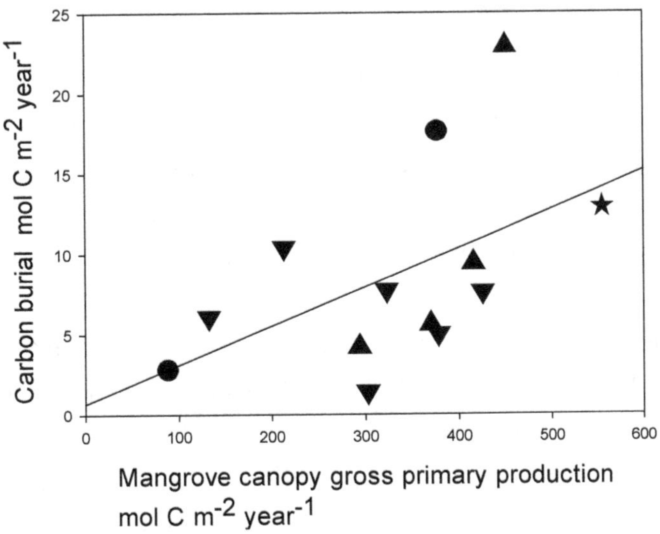

Figure 2. Variation in carbon burial with gross productivity of mangrove canopies. Data is from Alongi [11] (upward triangles), Lovelock *et al.* [70,71] Moreton Bay, downward triangles) and this study (scrub and fringe forests are circles, P fertilized scrub a star). The form of the curve is $y = 0.682 \times 0.0242x$, $R^2 = 0.27$.

The proportion of GPP that was buried in the fringing forests (4.1%) was higher than that of the scrub forest (2.1%), suggesting more favorable conditions for C burial in fringing forests. However, given the flooded nature of the scrub soils and the low redox potential in these areas [52], we expected the scrub forests to have higher proportional C burial as anoxia reduces decomposition rates of organic matter. Instead, the higher productivity fringing forest had higher proportional rates of C burial. Alongi [66] has suggested that some, as yet unidentified aspects of productive mangrove forests, may decrease decomposition of organic matter and increase C burial.

Fertilization with P, despite higher productivity of trees, resulted in a smaller proportional burial of carbon (2%) compared to the fringing forest (4%). This observation is consistent with lower proportional allocation of GPP to roots in fertilized plants that has been observed in other "wild" plants [22]. However, the ratio of root production to GPP is 4% in scrub forest and 6% in both fringing forests and P fertilized scrub trees. Therefore, rather than differences in proportional allocation to roots, alternative hypotheses to explain differences in proportional C burial between forests zones and with fertilization include: (1) Differences in rates of decomposition may be responsible for the reduced proportional C burial in P fertilized plants compared to fringing forests; or (2) Differences in root structure may contribute to reduced proportional C burial in P fertilized trees. Using a standardized test comparing the tensile strength of buried cotton strips, decomposition was found to be enhanced in P fertilized plots compared to control sites [14], supporting the hypothesis that rates of decomposition increases with P fertilization. Phosphorus (and iron) limitation of the microbial community has been suggested previously in other mangrove forests [72]. The addition of excess P may lead to increased decomposition of soil C and thereby reduce rates of C burial. Variation in root structure in mangroves has not been described over nutrient gradients and thus remains to be assessed as a factor influencing C burial in mangroves.

3.5. Variation in Net Ecosystem Production

In the seaward fringing forest, 16% of GPP and 27% of scrub forests GPP could not be accounted for and thus is assumed to be NEP in the mass balance budget (Table 2). This contrasted with the P fertilized scrub forest where the budget was balanced within 5% (inputs of fixed C was totally accounted for in the outputs measured or estimated). Correspondingly, production to respiration (P:R) ratios in the seaward fringing forest were 1.2, 1.4 and 0.95 for the fringing, scrub and P fertilized scrub, respectively.

The presence of well-developed surface microbial mats is an important feature of mangrove habitats contributing both to GPP [38,39] and to reducing diffusion of respired CO_2 across the soil surface thereby leading to under estimations of soil respiration [47,48,68]. Experiments where surface films were removed (scraped off) indicated that soil respiration increased by approximately 60% compared to when the surface was intact [47,68,73]. Possibly as a consequence of this reduced rate of diffusion, very high concentrations of dissolved inorganic C and low pH (6.5) are observed in the soil porewater (16 mmol in scrub forests compared to 1.5 mmol in the fringe soils [32,38]). These high levels of dissolved inorganic carbon (DIC) could be due to root respiration as well as decomposition of microbial (and algal) detritus and also decomposition of soil organic matter derived from mangrove tissues (Table 1). Bouillon *et al.* [74] estimated that porewater contributed a large component (18%–87%) of creek water. The interior ponds at Twin Cays are known to drain through creeks [75]. Photographic evidence indicates drainage of dissolved coloured organic material from the mangrove islands via creeks and subterranean channels [76]. Additionally, the C isotopic signature ($\delta^{13}C$) of seagrass on the edges of the mangrove islands is 7 per mil less (−14) compared to seagrass associated with the nearby coral reef [77]. The low value ($\delta^{13}C$ of −14) indicates that seagrass on the edge of the mangrove are using an additional DIC source that is more negative than that in the seawater and may be mangrove in origin [78]. Mangroves have a $\delta^{13}C$ of approximately −26 to −29, while algae and cyanobacteria are substantially higher [79]. Thus, mangrove root respiration or decomposition of mangrove derived soil organic matter is likely to contribute to DIC exported from the scrub forests and interior ponds.

Despite the potential errors using mass balance approaches (because of errors in large fluxes e.g., GPP and respiration [11]), our estimates of NEP were similar to that observed in other mangrove forest C budgets [3,10,11,13]. NEP is proposed to be exported from the ecosystem as DIC rather than dissolved organic carbon (DOC), which is known to be low [59]. Measurements of NEP of mangroves in Florida using eddy-covariance techniques gave a value of 97.5 mol $m^{-2}·year^{-1}$ of which they estimate 25%–70% of NEP is likely to be exported [80]. This is similar in magnitude to our estimates of NEP for fringing mangroves in our study (70 mol C $m^{-2}·year^{-1}$) but higher than that measured for scrub mangroves (35 mol C $m^{-2}·year^{-1}$). Our values of NEP for Twin Cays are slightly lower than values reported for Australian and Asian mangrove forests which range from 101–178 mol C $m^{-2}·year^{-1}$ [11,16] and globally, 118 mol C $m^{-2}·year^{-1}$ [10]. Maher *et al.* [81] bserved that DIC export from mangroves in subterranean groundwater flow was driven by tidal pumping. Despite the very low tidal range at Twin Cays (<50 cm), photographic evidence and observed $\delta^{13}C$ of adjacent seagrass suggest that export of DIC from porewater is a plausible fate for the unaccounted for carbon at Twin Cays. Given the area of fringing forests is ~100,000 m^2 ha and scrub forest and interior ponds comprise ~190,000 m^2 of the island vegetation [31], and if we assume 100% of NEP is exported, annual rates of DIC export from forests at Twin Cays could be 835 Gg·$year^{-1}$ for fringing habitats and 809 Gg·$year^{-1}$ for scrub forests, indicating that despite lower productivity, scrub forests may be an important source of exported C. Export of DIC may have influence on other habitats within the lagoon including the lagoon microbial community, seagrass beds, sponges and coral reefs.

4. Conclusions

There are few detailed C budgets for mangrove forests and the impact of nutrient enrichment, which is a serious threat in the coastal zone, has not yet been explicitly considered. From our C budget for Twin Cays, we contribute to the overall understanding of mangrove C cycling by finding that microbial mats can provide a significant proportion of GPP in scrub habitats, respiration of aboveground roots and stems may be a significant source of respired C, and that the processes contributing to soil respiration differ across habitats, with the microbial community making a greater contribution in the scrub forests than in more productive fringing forests. Additionally, we found that C burial is a small proportion of GPP (2%–4%) which is correlated with GPP of forests, and which may, with more data collected over time, lead to development of predictive relationships between GPP and rates of soil C burial in mangrove ecosystems. But the proportion of GPP that is buried in soils decreased with fertilization, in this case with P, suggesting that the efficiency of C burial to soils in these habitats may have declined where nutrient enrichment of the coastal zone has occurred. The impacts of nutrient enrichment may therefore include reduced soil surface elevation gains, which may increase vulnerability of mangrove ecosystems to sea level rise. Finally, we find NEP is 16%–27% of GPP, similar to that observed in other mangrove C budgets. Scrub forests may contribute a high proportion of NEP because of enhanced decomposition of organic matter compared to fringing forests. There is isotopic evidence to suggest NEP is exported to the adjacent waters around the islands where it may contribute to productivity of adjacent ecosystems.

Acknowledgments

We thank Anne Chamberlain, Rainer Feller, the participants of the Mangrove Biocomplexity project and the staff of the Carrie Bow Cay Research Station and Pelican Beach Resort. This research was funded by the National Science Foundation (DEB-9981535) and the Smithsonian Institution's Marine Science Network. We thank the Australian Research Council (awards DP0774491, DP0986179, DP1096749 and FS100100024) and the Marine and Coastal Carbon Biogeochemistry Cluster for their support.

Author Contributions

All authors contributed data. C.E.L. wrote the manuscript while L.T.S., L.J.D. and I.C.F. edited the manuscript.

References and Notes

1. Chmura, G.L.; Anisfeld, S.C.; Cahoon, D.R.; Lynch, J.C. Global carbon sequestration in tidal, saline wetland soils. *Glob. Biogeochem. Cycles* **2003**, *17*, 1111, doi:10.1029/2002GB001917.
2. Donato, D.C.; Kauffman, J.B.; Murdiyarso, D.; Kurnianto, S.; Stidham, M.; Kanninen, M. Mangroves among the most carbon-rich forests in the tropics. *Nat. Geosci.* **2011**, *4*, 293–297.
3. Alongi, D.M. Carbon cycling and storage in mangrove forests. *Annu. Rev. Mar. Sci.* **2014**, *6*, 195–219.
4. Duarte, C.M.; Middelburg, J.J.; Caraco, N. Major role of marine vegetation on the oceanic carbon cycle. *Biogeosciences* **2005**, *2*, 1–8.
5. Mcleod, E.; Chmura, G.L.; Bouillon, S.; Salm, R.; Bjork, M.; Duarte, C.M.; Lovelock, C.E.; Schlesinger, W.H.; Silliman, B. A Blueprint for Blue Carbon: Towards an improved understanding of the role of vegetated coastal habitats in sequestering CO_2. *Front. Ecol. Environ.* **2011**, *9*, 552–560.
6. Pendleton, L.; Donato, D.C.; Murray, B.C.; Crooks, S.; Jenkins, W.A.; Sifleet, S.; Craft, C.; Fourqurean, J.W.; Kauffman, J.B.; Marba, N.; *et al.* Estimating global "Blue Carbon" emissions from conversion and degradation of vegetated coastal ecosystems. *PLoS ONE* **2012**, *7*, e43542.
7. Siikamäki, J.; Sanchirico, J.N.; Jardine, S.L. Global economic potential for reducing carbon dioxide emissions from mangrove loss. *Proc. Natl. Acad. Sci. USA* **2012**, *109*, 14369–14374.
8. Jennerjahn, T.C.; Ittekkot, V. Relevance of mangroves for the production and deposition of organic matter along tropical continental margins. *Naturwissenschaften* **2002**, *89*, 23–30.

9. Dittmar, T.; Hertkorn, N.; Kattner, G.; Lara, R.J. Mangroves, a major source of dissolved organic carbon to the oceans. *Glob. Biogeochem. Cycles* **2006**, *20*, doi:10.1029/2005GB002570.
10. Bouillon, S.; Borges, A.V.; Castaneda-Moya, E.; Diele, K.; Dittmar, T.; Duke, N.C.; Kristensen, E.; Lee, S.Y.; Marchand, C.; Middelburg, J.J.; *et al.* Mangrove production and carbon sinks: A revision of global budget estimates. *Glob. Biogeochem. Cycles* **2008**, *22*, doi:10.1029/2007GB003052.
11. Alongi, D.M. Carbon sequestration in mangrove forests. *Carbon Manag.* **2012**, *3*, 313–322.
12. Cebrian, J.; Duarte, C.M. Plant growth-rate dependence of detrital carbon storage in ecosystems. *Science* **1995**, *5217*, 1606–1608.
13. Alongi, D.M. *Energetics of Mangrove Ecosystems*; Springer Science & Business Media: New York, NY, USA, 2009.
14. Feller, I.C.; McKee, K.L.; Whigham, D.F.; O'Neill, J.P. Nitrogen *vs.* phosphorus limitation across and ecotonal gradient in a mangrove forest. *Biogeochemistry* **2003**, *62*,145–175.
15. McKee, K.L.; Cahoon, D.R.; Feller, I.C. Caribbean mangroves adjust to rising sea level through biotic controls on change in soil elevation. *Glob. Ecol. Biogeogr.* **2007**, *16*, 545–556.
16. Wooller, M.J.; Behling, H.; Smallwood, B.J.; Fogel, M. Mangrove ecosystem dynamics and elemental cycling at Twin Cays, Belize, during the Holocene. *J. Quat. Sci.* **2004**, *19*, 703–711.
17. Maher, D.T.; Eyre, B.D. Carbon budgets for three autotrophic Australian estuaries: Implications for global estimates of the coastal air-water CO_2 flux. *Glob. Biogeochem. Cycles* **2012**, *26*, doi:10.1029/2011GB004075.
18. Alongi, D.M. Mangrove forests: Resilience, protection from tsunamis, and responses to global climate change. *Estuar. Coast. Shelf Sci.* **2002**, *76*, 1–13.
19. Lovelock, C.E.; Ruess, R.W.; Feller, I.C. CO_2 efflux from cleared mangrove peat. *PLoS ONE* **2011**, *6*, e21279.
20. Kauffman, J.B.; Heider, C.; Norfolk, J.; Payton, F. Carbon stocks of intact mangroves and carbon emissions arising from their conversion in the Dominican Republic. *Ecol. Appl.* **2014**, *24*, 518–527.
21. Reef, R.; Feller, I.C; Lovelock, C.E. Nutrition of mangroves. *Tree Physiol.* **2010**, *30*, 1148–1160.
22. Chapin, F.S. Mineral nutrition of wild plants. *Annu. Rev. Ecol. Syst.* **1980**, *11*, 233–260.
23. Tilman, D. Secondary succession and the pattern of plant dominance along experimental nitrogen gradients. *Ecol. Monogr.* **1987**, *57*, 189–214.
24. Feller, I.C.; Lovelock, C.E.; Piou, C. Growth and nutrient conservation in *Rhizophora mangle* in response to fertilization along latitudinal and tidal gradients. *Smithson. Contrib. Mar. Sci.* **2009**, *38*, 345–358.
25. Feller, I.C.; Lovelock, C.E.; Berger, U.; McKee, K.L.; Joye, S.B.; Ball, M.C. The biocomplexity of mangrove ecosystems. *Annu. Rev. Mar. Sci.* **2010**, *2*, 395–416.
26. Ruetzler, K.; Feller, I.C. Caribbean mangrove swamps. *Sci. Am.* **1996**, *274*, 94–99.
27. Macintyre, I.G.; Littler, M.M.; Littler, D.S. Holocene history of Tobacco Range, Belize, Central America. *Atoll Res. Bull.* **1995**, *43*, 1–18.
28. Woodroffe, C.D. Mangrove vegetation of Tobacco Range and nearby mangrove ranges, central Belize barrier reef. *Atoll Res. Bull.* **2005**, *427*, 1–35.

29. Adame, M.F.; Kauffman, J.B.; Medina, I.; Gamboa, J.N.; Torres, O.; Caamal, J.P.; Reza, M.; Herrera-Silveira, J.A. Carbon stocks of tropical coastal wetlands within the karstic landscape of the Mexican Caribbean. *PLoS ONE* **2013**, *8*, e56569.
30. Middleton, B.A.; McKee, K.L. Degradation of mangrove tissues and implications for peat formation in Belizean island forests. *J. Ecol.* **2001**, *89*, 818–828.
31. Rodriguez, W.; Feller, I.C. Mangrove landscape characterization and change in Twin Cays, Belize using aerial photography and IKONOS satellite data. *Atoll Res. Bull.* **2004**, *513*, 1–24.
32. Lee, R.Y.; Joye, S.B. Patterns and controls on nitrogen fixation and denitrification in intertidal soils of a tropical oceanic mangrove island. *Mar. Ecol. Prog. Ser.* **2006**, *307*, 127–41.
33. Koltes, K.; Tschirky, J.; Feller, I.C. Carrie Bow Cay, Belize. In *Caribbean Coastal Marine Productivity (CARICOMP): Caribbean Coral Reef, Seagrass and Mangrove Sites Characteristics*; Coastal Region and Small Island Papers 3; Kjerfve, B., Ed.; UNESCO: Paris, France, 1998; pp. 79–94.
34. Feller, I.C. Effects of nutrient enrichment on growth and herbivory of dwarf red mangrove *Rhizophora mangle*. *Ecol. Monogr.* **1995**, *65*, 477–505.
35. Lovelock, C.E.; Ball, M.C.; Choat, B.; Engelbrecht, B.M.J.; Holbrook, N.M.; Feller, I.C. Linking physiological processes with mangrove forest structure: Phosphorus deficiency limits canopy development; hydraulic conductivity and photosynthetic carbon gain in dwarf *Rhizophora mangle*. *Plant Cell Environ.* **2006**, *29*, 793–802.
36. Cheeseman, J.C.; Lovelock, C.E. Photosynthetic characteristics of dwarf and fringe *Rhizophora mangle* in a Belizean mangrove. *Plant Cell Environ.* **2004**, *27*, 768–780.
37. Ball, M.C.; Cowan, I.R.; Farquhar, G.D. Maintenance of leaf temperature and the optimisation of carbon gain in relation to water loss in a tropical mangrove forest. *Aust. J. Plant Physiol.* **1988**, *15*, 263–267.
38. Lovelock, C.E.; Clough, B.F. Influence of solar radiation and leaf angle on leaf xanthophyll concentrations in mangroves. *Oecologia* **1992**, *91*, 518–525.
39. Lee, R.Y.; Porubsky, W.P.; Feller, I.C.; McKee K.L.; Joye, S.B. Porewater biogeochemistry and soil metabolism in dwarf red mangrove habitats (Twin Cays; Belize). *Biogeochemistry* **2008**, *87*, 181–198.
40. Joye, S.B.; Lee, R.Y. Benthic microbial mats: Important sources of fixed nitrogen and carbon to the Twin Cays, Belize ecosystem. *Atoll Res. Bull.* **2004**, *528*, 1–24.
41. Genty, B.; Briantas, J.-M.; Baker, N.R. The relationship between quantum yield of photosynthetic electron transport and quenching of chlorophyll fluorescence. *Biochim. Biophys. Acta* **1989**, *990*, 87–92.
42. Feller, I.C.; Chamberlain, A.H. Herbivore responses to nutrient enrichment and landscape heterogeneity in a mangrove ecosystem. *Oecologia* **2007**, *153*, 607–616.
43. Kauffman, J.B.; Heider, C.; Cole, T.; Dwire, K.A.; Donato, D.C. Ecosystem carbon pools of Micronesian mangrove forests: Implications of land use and climate change. *Wetlands* **2011**, *31*, 343–352.
44. Ryan, M.G.; Gower, S.T.; Hubbard, R.M.; Waring, R.H.; Gholz, H.L.; Cropper Jr, W.P.; Running, S.W. Woody tissue maintenance respiration of four conifers in contrasting climates. *Oecologia* **1995**, *101*, 133–140.

45. Reich, P.B.; Tjoelker, M.G.; Machado, J.L.; Oleksyn, J. Universal scaling of respiratory metabolism, size and nitrogen in plants. *Nature* **2006**, *439*, 457–461.
46. Hoque, A.T.M.R.; Sharma, S.; Suwa, R.; Hagihara, A. Seasonal variation in the size-dependent respiration of mangroves *Kandelia obovata*. *Mar. Ecol. Prog. Ser.* **2010**, *404*, 31–37.
47. Mori, S.; Yamaji, K.; Ishida, A.; Prokushkinc, S.G.; Masyaginac, O.V.; Hagiharad, A.; Hoqued A.T.M.R.; Suwad, R.; Osawae, A.; Nishizono, T. *et al.* Mixed-power scaling of whole-plant respiration from seedlings to giant trees. *Proc. Natl. Acad. Sci.* **2010**, *107*, 1447–1451.
48. Lovelock, C.E. Soil respiration in tropical and subtropical mangrove forests. *Ecosystems* **2008**, *11*, 342–354.
49. Lovelock, C.E.; Feller, I.C.; Reef, R.; Ruess, R.W. Variable effects of nutrient enrichment on soil respiration in mangrove forests. *Plant Soil* **2015**, *379*, 135–148.
50. McKee, K.L. Biophysical controls on accretion and elevation change in Caribbean mangrove ecosystems. *Estuar. Coast. Shelf Sci.* **2011**, *91*, 475–483.
51. Lovelock, C.E.; Ruess, R.W.; Feller, I.C. Root respiration in *Rhizophora mangle* over variation in forest stature and nutrient availability. *Tree Physiol.* **2006**, *26*, 1601–1606.
52. McKee, K.L.; Feller, I.C.; Popp, M.; Wanek, W. Mangrove isotopic fractionation (δ^{15}N and δ^{13}C) across a nitrogen *versus* phosphorus limitation gradient. *Ecology* **2002**, *83*, 1065–75.
53. Ellison, A.M.; Farnsworth, E.J. Spatial and temporal variability in growth of *Rhizophora mangle* saplings on coral cays: Links with variation in insolation, herbivory, and local sedimentation rate. *J. Ecol.* **1996**, *84*, 717–731.
54. Lugo, A.E. Old-growth mangrove forests in the United States. *Conserv. Biol.* **1997**, *11*, 11–20.
55. Kitaya, Y.; Yabuki, K.; Kiyota, M.; Tani, A.; Hirano, T.; Aiga, I. Gas exchange and oxygen concentration in pneumatophores and prop roots of four mangrove species. *Tree Struct. Funct.* **2002**, *16*, 155–158.
56. Schmitz, N.; Egerton, J.J.G.; Lovelock, C.E.; Ball, M.C. Light-dependent maintenance of hydraulic function in mangrove branches: Do xylary chloroplasts play a role in embolism repair? *New Phytol.* **2012**, *195*, 40–46.
57. Material not intended for publication: C.E. Lovelock (The University of Queensland, St Lucia, QLD). Transmittance of light through epidermal tissues of aboveground roots of *Rhizophora* sp., 2013.
58. Drew, M.C.; Saglio, P.H.; Pradet, A. Larger adenylate energy charge and ATP/ADP ratios in aerenchymatous roots of *Zea mays* in anaerobic media as a consequence of improved internal oxygen transport. *Planta* **1985**, *165*, 51–58.
59. King, J.S.; Hanson, P.J.; Bernhardt, E.; DeAngelis, P.; Norby, R.J.; Pregitzer, K.S. A multiyear synthesis of soil respiration responses to elevated atmospheric CO_2 from four forest FACE experiments. *Glob. Chang. Biol.* **2004**, *10*, 1027–1042.
60. Adame, M.F.; Lovelock, C.E. Carbon and nutrient exchange of mangrove forests with the coastal ocean. *Hydrobiologia* **2011**, *663*, 23–50.
61. Day, J.W.; Coronado-Molina, C.; Vera-Herrera, F.R.; Twilley, R.; Rivera-Monroy, V.H.; Alvarez-Guillen, H.; Day, R.; Conner, W. A 7 year record of above-ground net primary production in a southeastern Mexican mangrove forest. *Aquat. Bot.* **1996**, *55*, 39–60.

62. Feller, I.C. The role of herbivory by wood-boring insects in mangrove ecosystems in Belize. *Oikos* **2002**, *97*, 167–76.
63. Krauss, K.W.; McKee, K.L.; Lovelock, C.E.; Cahoon, D.R.; Saintilan, N.; Reef, R.; Chen, L. How mangrove forests adjust to rising sea level. *New Phytol.* **2014**, *202*, 19–34.
64. Castaneda-Moya, E.; Twilley, R.R.; Rivera-Monroy, V.H.; Marx, B.D.; Coronado-Molina, C.; Ewe, S.M. Patterns of root dynamics in mangrove forests along environmental gradients in the Florida Coastal Everglades, USA. *Ecosystems* **2011**, *14*, 1178–1195.
65. Hogberg, P.; Nordgren, A.; Buchmann, N.; Taylor, A.F.; Ekblad, A.; Högberg, M.N.; Nyberg, G.; Ottosson-Löfvenius, M.; Read, D.J. Large-scale forest girdling shows that current photosynthesis drives soil respiration. *Nature* **2001**, *411*, 789–90.
66. Keuskamp, J.A.; Schmitt, H.; Laanbroek, H.J.; Verhoeven, J.T.; Hefting, M.M. Nutrient amendment does not increase mineralisation of sequestered carbon during incubation of a nitrogen limited mangrove soil. *Soil Biol. Biochem.* **2013**, *57*, 822–829.
67. Alongi, D.M.; de Carvalho, N.A.; Amaral, A.L. Uncoupled surface and below-ground soil respiration in mangroves: implications for estimates of dissolved inorganic carbon export. *Biogeochemistry* **2012**, *109*, 151–162.
68. Anon. Carbon Dioxide Information and Analysis Center. Available online: http://cdiac.ornl.gov/trends/emis/blz.html (accessed on 1 September 2015).
69. Murray, M.R.; Zismana, S.A.; Furleya, P.A.; Munro, D.M.; Gibson, J.; Ratter, J.; Bridgewater, S.; Minty, C.D.; Place, C.J. The mangroves of Belize Part 1. Distribution, composition and classification. *For. Ecol. Manag.* **2003**, *174*, 265–279.
70. Material not intended for publication: C.E. Lovelock (The University of Queensland, St Lucia, QLD). Estimates of Gross Primary Production for mangrove forests in Moreton Bay, derived from assessment of LAI and photosynthetic rates of leaves; and estimates of carbon burial derived from assessment of soil accretion and soil carbon density, 2013.
71. Lovelock, C.E.; Adame, M.F.; Bennion, V.; Hayes, M.; O'Mara, J.; Reef, R.; Santini, N. Contemporary rates of carbon sequestration through vertical accretion of sediments in mangrove forests and saltmarshes of South East Queensland, Australia. *Estuar. Coasts*, **2013**, doi:10.1007/s12237-013-9702-4.
72. Alongi, D.M. Dissolved iron supply limits early growth of estuarine mangroves. *Ecology* **2011**, *91*, 3229–3241.
73. Leopold, A.; Marchand, C.; Debordeb, J.; Chaduteau, C.; Allenbach, M. Influence of mangrove zonation on CO_2 fluxes at the sediment-air interface (New Caledonia). *Geoderma* **2012**, *202–203*, 62–70.
74. Bouillon, S.; Middleburg, J.J.; Dehairs, F.; Borges, A.V.; Abril, G.; Flindt, M.R.; Ulomi, S.; Kristensen, E. Importance of intertidal sediment processes and porewater exchange on the water column biogeochemistry in a pristine mangrove creek (Ras Dege, Tanzania). *Biogeosciences* **2007**, *4*, 311–322.
75. Urish, D.W.; Wright, R.M.; Feller, I.C.; Rodriguez, W. Dynamic hydrology of a mangrove island: Twin Cays, Belize. *Smithson. Contrib. Mar. Sci.* **2009**, *38*, 473–490.
76. Feller, I.C. (Smithsonian Environmental Research Center, Edgewater, MD, USA). Aerial photographs from light aircraft. 2011.

77. Feller, I.C.; Baker, R. (Smithsonian Environmental Research Center, Edgewater, MD, USA). Stable isotope composition of seagrass at Twin Cays, 2013.
78. Lin, G.; Banks, T.; Sternberg, L.S.L.O. Variation in $\delta^{13}C$ values for the seagrass *Thalassia testudinum* and its relations to mangrove carbon. *Aquat. Bot.* **1991**, *40*, 333–341.
79. Raven, J.A.; Johnston, A.M.; Kübler, J.E.; Korb, R.; McInroy, S.G.; Handley, L.L.; Scrimgeour, C.M.; Walker, D.I.; Beardall, J.; Vanderklift, M.; *et al.* Mechanistic interpretation of carbon isotope discrimination by marine macroalgae and seagrasses. *Funct. Plant Biol.* **2002**, *29*, 355–378.
80. Barr, J.G.; Engel, V.; Fuentes, J.D.; Zieman, J.C.; O'Halloran, T.L.; Smith, T.J.; Anderson, G.H. Controls on mangrove forest-atmosphere carbon dioxide exchanges in western Everglades National Park. *J. Geophys. Res.* **2010**, *115*, doi:10.1029/2009JG001186.
81. Maher, D.T.; Santos, I.R.; Golsby-Smith, J.; Gleeson, J.; Eyre, B.D. Groundwater-derived dissolved inorganic and organic carbon exports from a mangrove tidal creek: The missing mangrove carbon sink? *Limnol. Oceanogr.* **2013**, *58*, 475–488.

3

Available Nitrogen and Responses to Nitrogen Fertilizer in Brazilian Eucalypt Plantations on Soils of Contrasting Texture

Ana Paula Pulito [1], José Leonardo de Moraes Gonçalves [2,*], Philip J. Smethurst [3], José Carlos Arthur Junior [4], Clayton Alcarde Alvares [5], José Henrique Tertulino Rocha [2], Ayeska Hübner [2], Luiz Fabiano de Moraes [6], Aline Cristina Miranda [7], Marcos Yassuo Kamogawa [8], José Luiz Gava [6], Raul Chaves [9] and Claudio Roberto Silva [1]

[1] Fibria S.A., Aracruz CEP: 29197-900, Brazil;
E-Mails: ana.pulito@fibria.com.br (A.P.P.); claudio.silva@fibria.com.br (C.R.S.)

[2] Department of Forest Science, Luiz de Queiroz College of Agriculture (ESALQ), University of São Paulo (USP), Piracicaba CEP:13418-900, Brazil; E-Mails: jlmgonca@usp.br (J.L.M.G.); rocha.jht@gmail.com (J.H.T.R.); ayeskahubner@yahoo.com.br (A.H.)

[3] CSIRO, Private Bag 12, Hobart TAS 7001, Australia; E-Mail: philip.smethurst@csiro.au

[4] Forest Science and Research Institute (IPEF), Piracicaba, CEP: 13400-970, Brazil;
E-Mail: arthur@ipef.br

[5] Department of Forestry and Environmental Resources, North Carolina State University, Raleigh, NC 27695, USA; E-Mail: calcard@ncsu.edu

[6] Suzano, Itapetininga CEP: 18207-780, Brazil; E-Mails: luizmoraes@suzano.com.br (L.F.M.); jgava@suzano.com.br (J.L.G.)

[7] Department of Forest Science, "Julio de Mesquita Filho" Universidade Estadual Paulista, Botucatu CEP: 18603-970, Brazil; E-Mail: miranda.acf@gmail.com

[8] Mathematics, Statistics and Chemistry Department, Luiz de Queiroz College of Agriculture (ESALQ), University of São Paulo (USP), Piracicaba CEP:13418-900, Brazil;
E-Mail: kamogawa@usp.br

[9] Duratex, Agudos CEP: 17120-000, Brazil; E-Mail: raul.chaves@duratex.com.br

* Author to whom correspondence should be addressed; E-Mail: jlmgonca@usp.br

to supply tree needs. However, subsequent crop rotations with low N fertilization may result in declining concentrations of organic and potentially mineralizable N (N_0), and consequent loss of wood productivity. This study investigated (a) *in situ* N mineralization and N_0 in soils of eucalypt plantations in São Paulo state, Brazil; (b) tree growth responses to N fertilizer applied 6–18 months after planting; and (c) the relationships between N_0, other soil attributes and tree growth. We established eleven N fertilizer trials (maximum 240 kg ha^{-1} of N) in *E. grandis* and *E. grandis* x *urophylla* plantations. The soil types at most sites were Oxisols and Quartzipsamments, with a range of organic matter (18 to 55 g kg^{-1}) and clay contents (8% to 67%) in the 0–20 cm layer. Concentrations of N_0 were measured using anaerobic incubation on soil samples collected every three months (different seasons). The samples collected in spring and summer had N_0 140–400 kg ha^{-1} (10%–19% total soil N), which were best correlated with soil texture and organic matter content. Rates of *in situ* net N mineralization (0–20 cm) ranged from 100 to 200 kg ha^{-1} $year^{-1}$ and were not correlated with clay, total N, or N_0. These high N mineralization rates resulted in a low response to N fertilizer application during the early ages of stand growth, which were highest on sandy soils. At the end of the crop rotation, the response to N fertilizer was negligible and non-significant at all sites.

Keywords: forest; productivity; potentially mineralizable N; sustainability; fertilizer application; nutrition

---

1. Introduction

Establishment of a forest plantation using the *Eucalyptus* genus for pulp and paper production and other products is justified by their high productivity across different soils and climates. However, sustainable wood production might be compromised in the short- or long-term when plantations are established on soils of low fertility, such as Oxisols and Quartzipsamments [1,2]. Productive plantations with high capacity for nutrient extraction can greatly impact N pools in the soil. Adequate management practices, including fertilizer application, are required to sustain tree growth rates and ensure soil quality over successive rotations [3].

Although trees require large amounts of N, some researchers found that eucalypts did not respond to N fertilizer application under tropical and subtropical climate conditions [1,4,5]. Commonly, N fertilizer application enhances the early growth of a eucalypt stand, but the response is not continued through the entire rotation [3–5]. The lack of response possibly occurs due to significant mineralization rates of organic N [3,6] and to atmospheric N deposition [7–10]. Mineralization of organic N is the main natural N source for plantations and appears to be sufficient to meet tree demand [3,11], which in southeast Brazil ranges from 20 to 50 kg ha^{-1} $year^{-1}$ [3,6]. However, intensively managed eucalypt plantations are expected to respond to N fertilizer application following several crop rotations because of high N outputs via harvesting [3,6,12], low N fertilizer rates applied [1,13], and depletion of organic N pools in soil [3]. Some studies carried out at sites with low soil organic matter (SOM) and N status showed that eucalypt stands might respond to N application under this condition [14,15]. Hence, the need for N fertilization in Brazilian plantations should be examined each rotation across a range of soil types.

The complexity of determining N fertilizer recommendations is attributed to the difficulty in accurately predicting the supply of available N (N mineralization). Gonçalves *et al.* [2] described a recommendation index for N fertilizer application to eucalypt plantations in Brazil based on SOM content. Although SOM is the main source of N, such an indicator does not consider the effect of SOM quality, climate and forest management [16,17].

The definition of a useful index of N availability is one that is practical, and chemical analyses can be faster than biological assays [18,19]. Several laboratory methods have been proposed to estimate potentially mineralizable N (N_0) in soils, and the anaerobic incubation method appears promising [11,20,21].

To better understand the interactions of N supply and tree growth at this stage of eucalypt plantation development in southeast Brazil,, this study investigated (a) *in situ* N mineralization and N_0 in soils of eucalypt plantations in São Paulo state, Brazil, (b) tree growth responses to N fertilizer applied 6–18 months after planting, and (c) the relationships between N_0, other soil attributes and tree growth.

2. Material and Methods

2.1. Site Description

We selected eleven *Eucalyptus grandis* and *Eucalyptus grandis* × *urophylla* sites in São Paulo state, Brazil, ranging from 1 to 11.4 years old. The sites belong to industrial companies and to the Itatinga Experimental Station of Forest Sciences, University of São Paulo and are representative of the majority of soils, climates and plantation management in this state (Figure 1). The climate is Cwa, according to Köppen classification at Altinópolis site; Cfa at Agudos, Angatuba, Capão Bonito, São Miguel Arcanjo and Votorantim sites; and Cfb at Botucatu, Itatinga and Paraibuna sites [22]. Cwa is a humid subtropical climate with a dry winter and a hot summer. Cfa is a humid subtropical climate without a dry season and with a hot summer. Cfb is a humid subtropical climate without a dry season and with a temperate summer. C climate types have an average annual temperature below 18 °C. Cw climates have rainfall of the driest month (R_{DRY}) more than 40 mm, while the Cf climates have R_{DRY} less than 40 mm. Average annual rainfall at the studied sites ranges from 1170 to 1517 mm (Table 1). Soils in the municipalities of Itatinga, São Miguel Arcanjo and Paraibuna are Typic Hapludox (Red-Yellow Latosol); at Altinópolis, Angatuba and Botucatu, Typic Quartzipsamment (Quartzarenic Neosol); at Agudos and Capão Bonito 2, Typic Hapludox (Red Latosol); at Capão Bonito 1, Typic Hapludox (Yellow Latosol); at Votorantim, Typic Paleudult (Red-Yellow Argisol); and at Capão Bonito 3, Typic Dystropept (Dystrophic Cambisol) [23]. These are the main soils used in forest plantations in São Paulo State [2]. Further soil details of Itatinga and Capão Bonito can be found in Gonçalves *et al.* [24] and Alvares *et al.* [25], respectively. The contents of SOM ranged from 18 to 55 g kg^{-1} and clay contents from 8% to 68% in the 0–20 cm layer (Table 2). They are also quite acidic with pH 3.9–4.9.

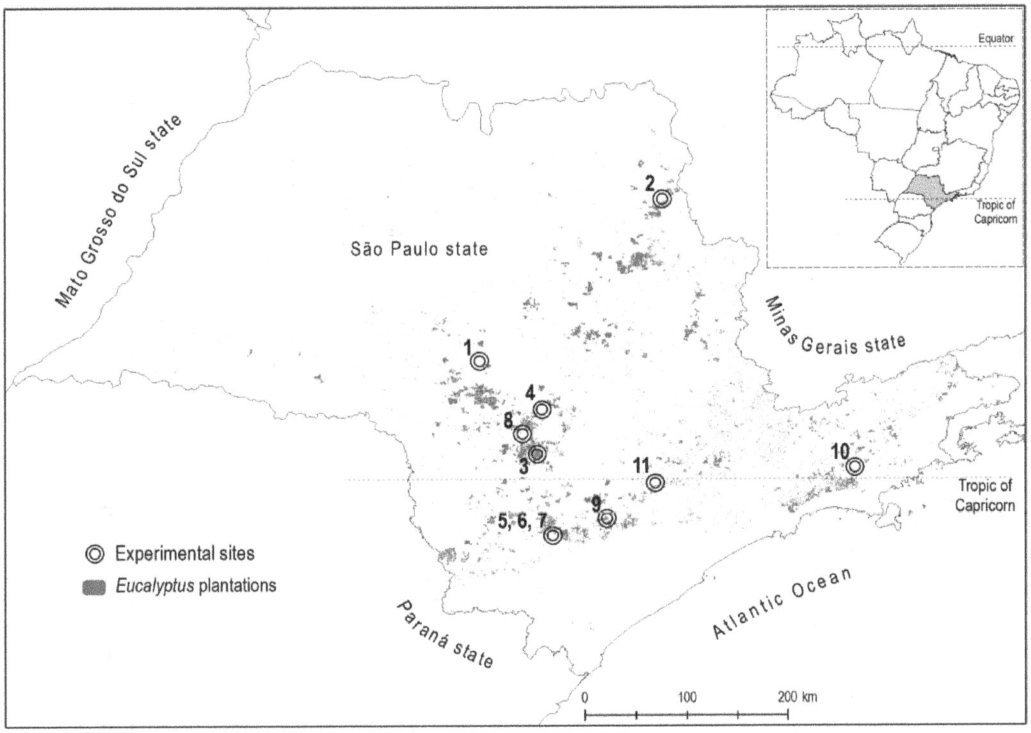

Figure 1. Map of São Paulo state showing locations of the experimental sites: 1—Agudos; 2—Altinópilis; 3—Angatuba; 4—Botucatu; 5, 6, 7—Capão Bonito; 8—Itatinga; 9—São Miguel Arcanjo; 10—Paraibuna; and 11—Votorantim.

At each site, we used a randomized block experimental design with three replicates. Measured plots were 10 m by 10 m, with 36 trees per plot, surrounded by a two-tree border. The treatments were: i) control (low N dose of fertilizer application, sufficient to provide good establishment of plants; *c.* 15 kg ha^{-1}); (ii) usual dose (N dose used by the companies; *c.* 100 kg ha^{-1}); and (iii) increased dose (100% higher than commercial doses; *c.* 200 kg ha^{-1}). The N fertilizer application schedule was 25% of total dose at planting, and the rest applied equitably at 6, 12 and 18 months after planting. The fertilizer application after planting was placed in a continuous line under canopy projection. Additional to N fertilization, all treatments received 30 kg ha^{-1} of P, 130 kg ha^{-1} of K, 230 kg ha^{-1} of Ca, 100 kg ha^{-1} of Mg, 3 kg ha^{-1} of B, 60 kg ha^{-1} of S, 1.5 kg ha^{-1} of Zn and 0.5 kg ha^{-1} of Cu. Calcium and Mg were supplied through lime application. Throughout the experiment, the stands were kept free of weed competition by herbicide application.

Table 1. Latitude (Lat), longitude (Long), altitude (Alt), mean annual temperature (T), mean annual pluviometric precipitation (PP), topography, planted genotype, tree spacing and planting date of each experiment.

Municipality	Site Code	Lat	Long	Alt	T [3]	PP	Geology [4]		Genotype	Tree Spacing	Planting
							Formation or Group	Lithotype			
		S	W	m	°C	mm				m	
Agudos	AGU	22°28′	48°59′	580	20.6	1170	Marília	Sandstone, sandy argillite and limestone	*E.grandis* [1]	3.0 × 2.0	Aug-2005
Altinópolis	ALT	21°01′	47°22′	889	19.4	1517	Botucatu	Quartz-sandstone	*E.grandis* × *urophylla* [2]	3.0 × 2.5	May-2002
Angatuba	ANG	23°17′	48°28′	649	19.7	1262	Pirambóia	Shale, thin sandstone and silty-clayey sandstone	*E.grandis* × *urophylla* [2]	3.0 × 2.0	Apr-2006
Botucatu	BOT	22°53′	48°26′	804	19.1	1302	Marília	Sandstone, sandy argillite and limestone	*E.grandis* [1]	3.0 × 2.0	Nov-2005
Capão Bonito 1	CB1	24°00′	48°20′	705	18.9	1210	Itararé	Sandstone, diamictite and shale	*E.grandis* × *urophylla* [2]	3.0 × 3.0	Jun-1999
Capão Bonito 2	CB2	24°00′	48°20′	705	18.9	1210			*E.grandis* × *urophylla* [2]	3.0 × 2.0	Feb-2007
Capão Bonito 3	CB3	24°00′	48°20′	705	18.9	1210			*E.grandis* × *urophylla* [2]	3.0 × 2.0	Dec-2006
Itatinga	ITA	23°06′	48°36′	845	18.8	1308	Marília	Sandstone, sandy argillite and limestone	*E.grandis* [1]	3.0 × 2.0	Apr-2002
São M. Arcanjo	SMA	23°51′	47°51′	715	18.9	1174	Itararé	Sandstone, diamictite and shale	*E.grandis* × *urophylla* [2]	3.0 × 2.0	Aug-2006
Paraibuna	PAR	23°23′	45°39′	634	19.2	1249	Natividade da Serra	Monzogranite, biotite and granite	*E.grandis* [1]	3.0 × 2.5	Mar-1997
Votorantim	VOT	23°32′	47°26′	570	19.8	1287	Granite Sorocaba	Granite, Granodiorite, monzogranite and syenogranite	*E.grandis* × *urophylla* [2]	3.0 × 2.0	Oct-2006

[1] Seedling plantations; [2] Clonal plantations; [3] Alvares *et al.* [26]; [4] IPT [27].

Table 2. Soil physical and chemical attributes (0–20 cm layer) at the eleven sites.

Site	Clay [1]	Silt [1]	Sand [1]		OM [2]	pH [3]	P-resin [4]	Cation Exchange [4]			
			Coarse	Fine				K	Ca	Mg	Al
	%				g kg^{-1}		mg kg^{-1}	mmol$_c$ kg^{-1}			
AGU	16.7	2.7	30.7	49.9	15	3.9	2.3	0.6	1.0	0.6	8.3
ALT	6.7	1.3	38.7	53.3	13	4.3	4.4	0.8	4.8	1.9	3.7
ANG	10.0	1.0	29.3	59.7	16	4.0	8.5	0.5	5.3	1.5	7.6
BOT	10.0	3.0	32.0	55.0	11	4.0	4.8	0.3	3.0	3.7	4.1
CB1	47.8	10.4	8.7	33.1	23	3.9	2.2	0.8	1.0	0.8	12.8
CB2	65.3	15.3	5.3	14.1	29	4.4	3.6	1.6	4.4	4.8	13.7
CB3	27.2	23.4	1.0	48.4	16	4.1	3.7	1.8	2.7	2.6	19.1
ITA	19.3	2.2	37.5	41.0	18	4,0	2.3	0.6	1.6	1.8	9.7
SMA	65.1	17.3	2.9	14.7	45	4.9	46.9	5.3	48.6	13.6	1.3
PAR	36.5	5.5	43.9	14.1	15	4.1	3.6	2.0	29.0	10.3	0.9
VOT	67.0	11.1	15.3	6.6	46	4.0	4.2	1.6	1.8	1.0	14.7

[1] Pipette method [28]; [2] Organic matter determined by potassium dichromate and sulfuric acid extraction; [3] $CaCl_2$ 0.01mol L^{-1} soil to solution ratio 1:2.5; [4] Ion Exchange resin [29].

2.2. Growth Assessment

In all plots, we measured diameter at breast height (DBH), total height and tree survival, and estimated the mean annual increment (MAI) of solid wood volume (SV) with bark. The SV was estimated by allometric equations using DBH and height developed by companies specifically for the genetic material used in each plantation. To compare eucalypt response to N fertilizer application in all sites, regardless of the growth increment, relative productivity (RP) increase was calculated as follows (Equation (1)).

$$\text{RP}\ (\%) = \frac{\text{SV}}{\text{SVmax}} \times 100 \tag{1}$$

where: “SV” is solid wood volume with bark of a given treatment and “SVmax” is the solid wood volume with bark of the highest dose treatment.

2.3. Soil Analysis

Soil chemical and physical properties were determined for the 0–20 cm layer at all sites. Ten soil samples were collected from each plot, in a diagonal transect across the inner part of the plot between planting lines. The samples were used to make one mixed sample per plot, which was air dried and homogenized. Next, the samples were sieved through a 2 mm mesh and mixed up again.

Particle size analysis (pipette method) of soil was performed according to Embrapa [28]. We determined the pH in $CaCl_2$ 0.01 mol L^{-1}, available phosphorus and exchangeable calcium, magnesium, potassium extracted by ion-exchange resins and aluminum extracted by KCl 1 M, according to Raij *et al.* [29]. We assessed soil total carbon (C_t) at all sites in the control treatment (0–20 cm layer) using the South Dakota method with modifications by Raij *et al.* [29]. This method consists of organic matter oxidation by dichromate ($K_2Cr_2O_7 + H_2SO_4$), and quantification by colorimetry. Total N (N_t) was determined by the micro-Kjeldahl method [30].

2.4. Assessment of N Mineralization

At all sites, soil samples from the 0–20 cm layer were collected to evaluate potential N mineralization under anaerobic conditions in the laboratory (N_0). The samples were collected in April (fall), July (winter), November (spring), 2007, and in January (summer), 2008. We performed anaerobic incubations (40 °C for 7 days) using chemical methods according to those proposed by Keeney and Bremner [20]. Prior to incubation, 30 mL of nutritional solution, consisting of: $MgSO_4$ (0.002 mol L^{-1}) and Ca $(H_2PO_4)_2$ (0.005 mol L^{-1}), was added. Jars were manually shaken until soil dispersion, and then covered with polyethylene film to avoid water loss by evaporation and algal growth [11]. N was extracted again 7 days after incubation by adding 4.47 g of KCl to each jar, which provide a concentration of 2 mol L^{-1} of KCl. Jars were then shaken manually for about 60 s, placed to rest 24 h and then filtered and analyzed for NH_4 as described above. To calculate potentially mineralizable N we subtracted the initial concentrations of mineral N.

To measure net N mineralization, the *in situ* method of Raison *et al.* [31] was used to in the 0–20 cm soil layer of control plots. Five pairs of steel tubes 30 cm long and 5 cm diameter were placed between tree rows in a diagonal transect across the inner part of the plot. The principle of the method is that inserting a sharpened corer severs roots, and adding capping during *in situ* incubation excludes rainfall. Net N mineralization thereby proceeds without leaching or uptake. By measuring mineral N initially (in additional representative soil samples) and again after incubation, net ammonification and nitrification can be calculated. For this study, one tube of each pair was immediately removed to assess initial mineral N concentrations (t_0) for all plots, and the remaining tubes were covered to avoid leaching of N, remaining in soil for 30 days on average. The amount of mineral N at t_0 was subtracted from the mineral N content after incubation to calculate *in situ* net N mineralization for each 30 day period. The same rate was assumed for other periods of the same season. Soil from each set of five tubes per plot was mixed to obtain a composite sample. The *in situ* method was used at all sites for January–December 2008.

For each collection, samples at all sites were collected within 8 days of each other to minimize time-related climatic variations between sites. The tubes were transported vertically to the lab, to minimize disturbance, and they were kept in thermal boxes (2–5 °C), individually wrapped in plastic bags. Refrigeration was used to minimize microbial activity, reducing mineralization that could occur prior to mineral N extraction [32]. Soil samples remained refrigerated until extraction, which was performed within two days of each collection.

For initial N (t_0) extraction, 10 g of soil were placed in 110 mL jars and 100 mL of KCl 2 mol L^{-1} added. Jars were manually shaken for *c.* 60 seconds for soil dispersion, and placed to rest for 24 h. Afterwards, the suspension was filtered through Whatman No. 42 filter paper, and the filtrate was analyzed to for NH_4^+ and NO_3^-, following the addition of 0.1 mL of microbial inhibitor (mercury acetate phenyl 0.5 mg L^{-1}).

For both N_0 and *in situ* N mineralization, NH_4^+ and NO_3^- concentrations were determined using the Analyze System of Injection in Automatic flow—ASIA (Ismatec, Glattbrugg, Switzerland) [33] and detected by colorimetry at 605 nm, which has a detection limit of 0.01 μg mL^{-1}.

2.5. Data Analysis

Data were checked for normality (Shapiro-Wilk) and homoscedasticity (Box-Cox). ANOVA was used to assess the significance of differences between means. Where there was a significant ($p < 0.05$) F test, a Tukey test (5% probability) was used for contrasting means. The relationship between dependent and independent variables was assessed by Pearson correlation and linear regression.

3. Results

Net N mineralization *in situ* rates ranged from 4.8 to 11.5 kg ha^{-1} month^{-1} (fall and winter) and from 10.6 to 15.0 kg ha^{-1} month^{-1} (spring and summer) for sandy soils, from 7.4 to 15.2 kg ha^{-1} month^{-1} (fall and winter) and from 11.0 to 17.6 kg ha^{-1} month^{-1} (spring and summer) for loamy soils, and from 3.6 to 19.1 kg ha^{-1} month^{-1} (fall and winter) and from 9.4 to 24.3 kg ha^{-1} month^{-1} (spring and summer) for clayey soils (Table 3). The average rates of N mineralization were highest in clayey soils, ranging from 110 to 207 kg ha^{-1} year^{-1}, and the lowest to sandy soils, ranging from 107 to 140 kg ha^{-1} year^{-1}. The average ratio of N-NH_4^+/N-NO_3 was 1.6 for sand soils, 1.7 for loamy soils and 1.9 for clayey soils.

Values of N_0 during summer ranged from 60 to 154 mg kg^{-1} of soil (190 and 398 kg ha^{-1}), with an average of 97 ± 11 mg kg^{-1} of soil (241 ± 18 kg ha^{-1}) (Table 4). Values of N_0 in sandy soils during summer were on average 168 kg ha^{-1} of N_0, 212 kg ha^{-1} in loamy soils, and 303 kg ha^{-1} in clayey soil. During winter, these values ranged from 20 to 91 mg kg^{-1} of soil (63 and 178 kg ha^{-1}). In sandy soils, N_0 corresponded to 19% of N_t, 14% in loamy soil, and 13% in clayey soils.

At all sites, tree survival was higher than 95%. On average, nitrogen fertilizer application resulted in an increase in MAI of 14% at early age, 6% at intermediate age and 0% at the end of the crop rotation (Table 5). Relative production (RP) at early age of control treatment ranged from 74% to 98% (average of 87% ± 2%), and from 90% to 98% (average of 95% ± 1%) at intermediate age. In the treatment that received commercial fertilizer application, RP ranged from 88% to 111% (average of 100% ± 2%) at early age, and from 98% to 107% (average of 100% ± 1%) at intermediate age. After five years of age (approximate harvesting age), RP ranged from 99% to 103% (average of 102% ± 1%).

The relative wood volume response to N application was greater in sandy soils, and, in absolute terms, at soils with higher clay content (Table 5 and Figure 2). During the first two years of age, RP of the control was about 16% lower than in the treatments that received commercial fertilizer application at sandy soils, and 9% to 10% at soils with higher clay content. Greater relative response to N application at early age occurred in soils with lower N_t, N_0 and clay contents (Figure 2). At the end of the rotation, regardless of soil texture, no significant responses to N application were found.

4. Discussion

4.1. N mineralization

Higher rates of N mineralization in clayey soils are attributed to higher stocks of organic N, due to the greater net primary productivity of the ecosystem, and to higher amounts of soil organic-mineral complexes (Tables 3 and 4). Under this condition, microbial activity increases due to higher amounts of substrate [34,35] and water availability. Eaton [36] reported that after two days of intensive rains, clayey soils in subtropical forests showed a significant increase in microbial C and N-NH_4^+ mineralization rates, compared to sandier soils. The author speculated that part of the active organic matter pool became detached from soil clay particles, and thereby became available to microorganisms.

Low rates of nitrification (Table 3) might be caused by high acidity and low fertility in these soils, and high NH_4^+ uptake under high growth rate stands could restrict substrate availability for nitrifying bacteria [3,37–39]. Under high acidity and low soil fertility, nitrifying bacteria (*Nitrobacter* and *Nitrosomonas*) have impaired active. Higher NH_4^+ than NO_3^- availability is usually not a limitation for eucalypt nutrition [40,41]. Gonçalves and Carlyle [39] studying N mineralization in *Pinus radiata* plantation soils under laboratory conditions reported that, despite the increased rate of nitrification during incubation, it was not proportional to the reduction of NH_4^+ concentration, showing the possibility of NH_4^+ immobilization and/or denitrification caused by soil moisture variations.

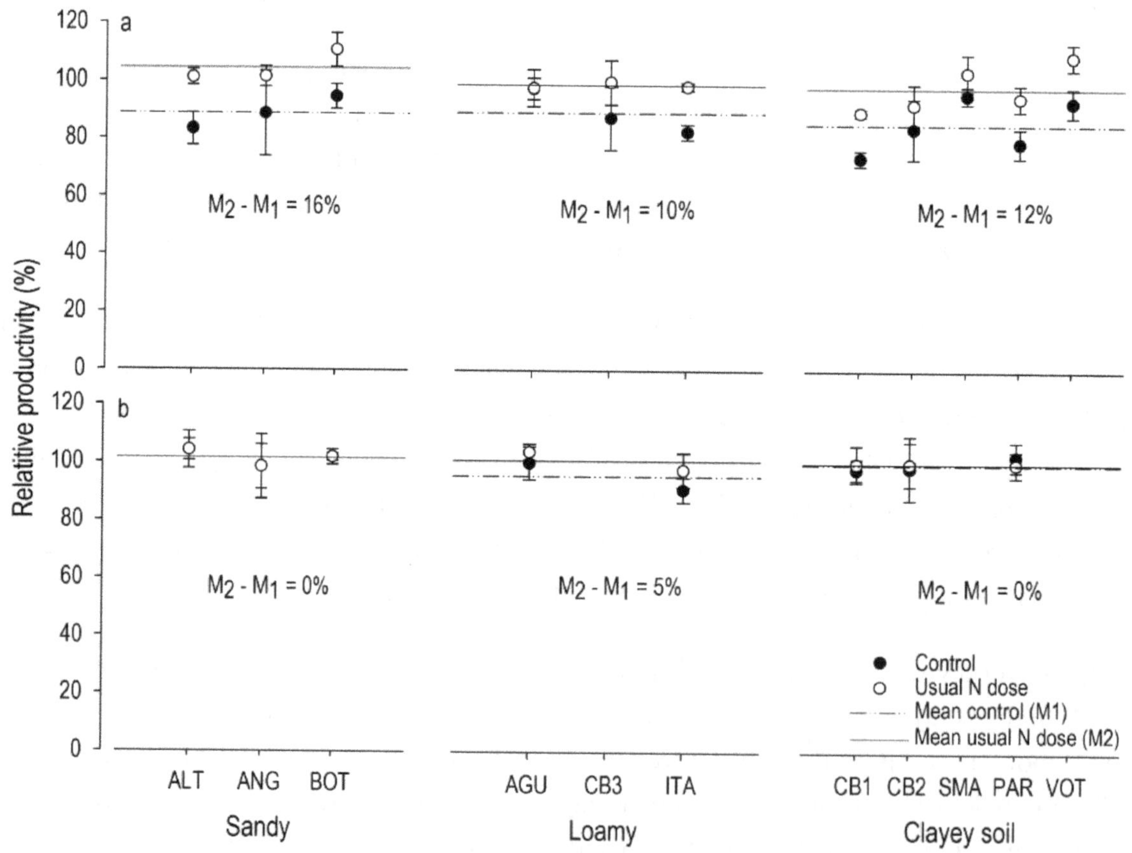

Figure 2. Relative productivity (RP) of the control treatments in relation to the usual and increased N rate treatments. (**a**) at young ages (less than two years) and (**b**) later ages (more than five years) at all sites, grouped according to soil texture. See Table 1 for *x*-axis site codes.

Table 3. Rates of net ammonification and net nitrification *in situ* and N-NH_4^+/N-NO_3^- ratio at each site (0–20 cm layer).

Site	Summer				Fall				Winter				Spring				Yearly Total			
	N-NH_4^+	N-NO_3^-	N-total	NH_4^+/NO_3^-	N-NH_4^+	N-NO_3^-	N-total	NH_4^+/NO_3^-	N-NH_4^+	N-NO_3^-	N-total	NH_4^+/NO_3^-	N-NH_4^+	N-NO_3^-	N-total	NH_4^+/NO_3^-	N-NH_4^+	N-NO_3^-	N-total	NH_4^+/NO_3^-
	kg ha^{-1} month^{-1}				kg ha^{-1} month^{-1}				kg ha^{-1} month^{-1}				kg ha^{-1} month^{-1}				kg ha^{-1} yr^{-1}			
									Sandy Soils											
ALT	5.7	8.0	13.7	0.7	4.6	4.1	8.7	1.1	6.6	2.6	9.2	2.6	10.1	4.8	15.0	2.1	81.2	58.6	139.8	1.4
ANG	8.6	3.8	12.4	2.3	3.3	1.6	4.8	2.1	4.3	1.3	5.5	3.4	8.0	5.1	13.1	1.6	72.4	35.2	107.6	2.1
BOT	5.6	7.6	13.2	0.7	6.9	4.5	11.5	1.5	4.1	3.7	7.8	1.1	9.1	1.5	10.6	6.1	77.0	52.0	129.0	1.5
Mean	6.6	6.5	13.1	1.2	4.9	3.4	8.3	1.6	5.0	2.5	7.5	2.4	9.1	3.8	12.9	3.3	76.9	48.6	125.5	1.6
s [1]	1.7	2.3	0.7	0.9	1.8	1.6	3.4	0.5	1.4	1.2	1.9	1.2	1.1	2.0	2.2	2.5	4.4	12.1	16.4	0.4
									Loamy Soils											
AGU	11.2	6.5	17.6	1.7	9.0	4.2	13.2	2.2	6.1	3.2	9.3	1.9	11.3	3.2	14.5	3.5	113.0	51.1	164.0	2.2
CB3	6.9	7.5	14.4	0.9	3.9	5.9	9.8	0.7	6.8	4.9	11.7	1.4	8.0	3.0	11.0	2.7	76.8	63.7	140.5	1.2
ITA	8.5	4.4	12.9	2.0	3.3	4.1	7.4	0.8	9.2	6.1	15.2	1.5	9.6	3.0	12.6	3.2	91.7	52.6	144.3	1.7
Mean	8.9	6.1	15.0	1.5	5.4	4.7	10.1	1.2	7.4	4.7	12.1	1.6	9.6	3.1	12.7	3.1	93.8	55.8	149.6	1.7
s	2.2	1.6	2.4	0.6	3.1	1.0	2.9	0.8	1.6	1.5	3.0	0.3	1.7	0.1	1.8	0.4	18.2	6.9	12.6	0.5
									Clayey Soils											
CB1	7.0	1.9	8.9	3.7	8.9	6.7	15.6	1.3	4.0	8.6	12.6	0.5	13.0	6.8	19.8	1.9	98.7	72.2	170.9	1.4
CB2	11.2	4.9	16.0	2.3	3.6	2.3	5.9	1.5	3.8	7.0	10.9	0.5	11.8	5.3	17.1	2.2	91.1	58.7	149.8	1.6
SMA	6.0	3.4	9.4	1.8	6.5	1.2	7.7	5.4	3.4	1.9	5.2	1.8	9.7	4.7	14.4	2.1	76.9	33.5	110.4	2.3
PAR	14.3	7.8	22.1	1.8	12.7	6.5	19.1	1.9	1.1	2.6	3.6	0.4	17.7	6.6	24.3	2.7	137.2	70.2	207.4	2.0
VOT	13.4	6.9	20.4	1.9	6.4	2.6	9.0	2.5	7.5	1.9	9.4	3.9	10.2	3.5	13.8	2.9	112.5	45.0	157.5	2.5
Mean	10.4	5.0	15.4	2.3	7.6	3.9	11.5	2.5	3.9	4.4	8.3	1.4	12.5	5.4	17.9	2.4	103.3	55.9	159.2	1.9
s	3.7	2.4	6.1	0.8	3.4	2.6	5.6	1.7	2.3	3.2	3.8	1.5	3.2	1.4	4.3	0.4	22.9	16.6	35.1	0.5

[1] Mean standard deviation.

Table 4. Seasonal values of potentially mineralizable N (N_0), total C (C_t), total N (N_t), C/N ratio and N_0/N_t ratio at each site grouped by texture class.

Site [1]	N_0								C_t [2]	N_t [3]	N_0 [4]	N_0/N_t	C/N
	Fall	Winter	Spring	Summer	Fall	Winter	Spring	Summer					
	mg kg^{-1}				kg ha^{-1}				mg kg^{-1}			%	
	Sandy Soils												
ALT	26	20	47	60	82	63	149	190	7723	387	60	15	20
ANG	28	46	39	61	63	103	88	137	7025	323	61	19	22
BOT	53	31	56	66	142	83	150	177	7090	301	66	22	24
Mean	36	32	47	62	96	83	129	168	7279	337	62	19	22
s [5]	15.0	13.1	8.5	3.2	41.2	20.0	35.5	27.6	385.6	44.7	3.2	3.5	2.0
	Loamy Soils												
AGU	30	42	73	60	84	118	205	169	9233	387	60	16	24
CB3	104	- [6]	95	115	279	-	255	309	9342	810	115	14	12
ITA	71	43	73	75	149	90	153	158	7850	589	75	13	13
Mean	68	43	80	83	171	104	205	212	8808	595	83	14	16
s	37.1	0.7	12.7	28.4	99.3	19.8	51.0	84.2	831.7	211.6	28.4	1.5	6.7
	Clayey Soils												
CB1	56	69	91	111	139	171	226	275	13,403	774	111	14	17
CB2	114	-	112	107	276	-	271	259	18,814	981	107	11	19
SMA	34	91	98	125	67	178	192	245	24,498	1267	125	10	19
PAR	125	83	108	138	361	240	312	398	10,837	893	138	15	12
VOT	158	-	134	154	344	-	292	336	29,724	1276	154	12	23
Mean	97	81	109	127	237	196	258	303	19,455	1038	127	13	18
s	51.1	11.1	16.4	19.4	129.3	38.0	49.0	63.6	7781.3	225.3	19.4	2.1	4.0

[1] Described in Table 1; [2] Determined by humid oxidation; [3] Bremmer [30] method; [4] adaptation of Keeney and Bremner [20] method; [5] Mean standard deviation; [6] Data not collected.

Gonçalves *et al.* [11] found similar values of N_0 in eucalypt stands ranging from 50 to 249 mg kg^{-1} of soil (average of 111 ± 23 mg kg^{-1} of soil). Variations in N mineralization rates under laboratory conditions due to season of sampling were also found by Adams and Attiwill [42], Khanna [43], Smethurst and Nambiar [38], and Theodorou and Bowen [44]. Those authors also found higher N mineralization rates during temperate summers coinciding with warm, moist soils. For Theodorou and Bowen [44,45], seasonal fluctuations in mineral N availability were related to microbial activities in the soil, which are mostly affected by temperature and soil moisture [39]. When incubating a soil sample collected during summer, a higher microbial population is also incubated, leading to a higher mineralization rate.

N_0 contents were highly correlated with N_t (r = 0.92; $p < 0.0001$), but less so with SOM ($r = 0.69$; $p = 0{,}0192$) and clay contents ($r = 0.83$; $p = 0{,}0015$) (Figure 3). This confirms that soil total N is a good indicator of potential N mineralization, as also found by Pottker and Tedesco [46] and Noble and Herbert [47]. However, N_0 was only weakly correlated with annual N mineralization *in situ* ($r = 0.35$), and the latter was not correlated with N_t, SOM, or clay content.

The N_0/N_t percentage varied mostly between 10 and 22% (Table 4), signifying the proportional amount of mineralizable organic N. Gonçalves *et al.* [11] found N_0/N_t percentages between 5% and 15%. The N_t in this study accounted for 3%–5% of SOM. That ratio decreased with increased clay content (Figure 4). Therefore, there may be greater proportional N availability in soils with lower clay content. High N mineralization in sandy soils is related to better soil aeration and less clay protection of SOM. In absolute terms though, soils with higher clay content have more potential availability of N, because N_t stocks are higher.

Table 5. Solid wood volume with bark (SV) and mean annual increment (MAI) in the different treatments and ages at each site.

Site	Treatment	SV ($m^3 ha^{-1}$)					MAI ($m^3 ha^{-1} year^{-1}$)				
		Sandy Soils									
	Age (year)	**1.8**	**4.0**	**5.7**			**1.8**	**4.0**	**5.7**		
ALT	Control	45 b[1]	192 a	285 a			25 b	48 a	50 a		
	Usual dose	55 a	198 a	285 a			31 a	50 a	50 a		
	Increased dose	55 a	198 a	275 a			30 a	49 a	48 a		
		2.0	**3.0**	**5.5**			**2.0**	**3.0**	**5.5**		
ANG	Control	96 a	175 a	352 a			48 a	58 a	64 a		
	Usual dose	109 a	181 a	351 a			55 a	60 a	64 a		
	Increased dose	108 a	183 a	356 a			54 a	61 a	65 a		
		2.0	**3.0**	**4.0**	**5.0**		**2.0**	**3.0**	**4.0**	**5.0**	
BOT	Control	40 a	172 a	209 a	267 a		20 a	57 a	52 a	53 a	
	Usual dose	47 a	187 a	211 a	266 a		23 a	62 a	53 a	53 a	
	Increased dose	42 a	176 a	197 a	260 a		21 a	59 a	49 a	52 a	
		Loamy soils									
	Age (year)	**2.0**	**3.0**	**4.0**	**5.0**	**6.0**	**2.0**	**3.0**	**4.0**	**5.0**	**6.0**
AGU	Control	52 a	132 a	186 a	257 a	313 a	26 a	44 a	47 a	51 a	52 a
	Usual dose	52 a	133 a	190 a	274 a	323 a	26 a	44 a	47 a	55 a	54 a
	Increased dose	53 a	135 a	183 a	249 a	312 a	27 a	45 a	46 a	50 a	52 a
		1.1	**2.0**				**1.1**	**2.0**			
CB3	Control	14 a	88 b				13 a	44 b			
	Usual dose	16 a	109 a				14 a	54 a			
	Increased dose	16 a	111 a				14 a	55 a			
		2.0	**4.0**				**2.0**	**4.0**			
ITA	Control	50 b	159 a				25 b	40 b			
	Usual dose	60 a	174 a				30 a	43 a			
	Increased dose	61 a	174 a				30 a	44 a			

Table 5. *Cont.*

Site	Treatment	SV m³ ha⁻¹						MAI m³ ha⁻¹year⁻¹					
		Clayey Soils											
	Age (year)	**2.0**		**4.0**		**9.0**		**2.0**		**4.0**		**9.0**	
CB1	Control	28	b	168	b	452	a	14	b	42	b	50	a
	Usual dose	33	ab	184	ab	460	a	17	ab	46	a	51	a
	Increased dose	38	a	187	a	455	a	19	a	47	a	51	a
		1.0		**2.0**		**4.7**		**1.0**		**2.0**		**4.7**	
CB2	Control	10	a	80	b	330	a	10	a	40	b	70	a
	Usual dose	11	a	92	a	332	a	11	a	46	a	71	a
	Increased dose	12	a	96	a	333	a	12	a	48	a	71	a
		1.5		**2.5**				**1.5**		**2.5**			
SMA	Control	68	a	187	a			45	a	75	a		
	Usual dose	73	a	185	a			49	a	74	a		
	Increased dose	71	a	200	a			47	a	80	a		
		2.2		**4.0**		**11.4**		**2.2**		**4.0**		**11.4**	
PAR	Control	33	a	146	a	439	a	15	a	36	a	39	a
	Usual dose	39	a	153	a	430	a	18	a	38	a	38	a
	Increased dose	42	a	153	a	430	a	19	a	38	a	38	a
		1.2		**3.1**				**1.2**		**3.1**			
VOT	Control	21	b	177	a			18	a	57	a		
	Usual dose	25	a	194	a			21	a	62	a		
	Increased dose	23	b	184	a			19	a	59	a		

[1] Values in the same column and site followed by the same letter do not differ statistically by Tukey mean test ($p = 0.05$).

The C/N ratio ranged from 12 to 24 (average of 19 ± 4) (Table 4). Maquere *et al.* [48], Montero [49] and Lima *et al.* [50] found higher ratios in eucalypt stands compared to other native vegetation. This explains in part the larger recalcitrance of SOM in eucalypt stands. N release slows as N mineralization proceeds, since reserves of labile N decrease along with microbial activity. This effect is unfavorable to plant nutrition in the short-term, but beneficial for N conservation in the long-term because it decreases N losses by leaching and volatilization [37,51,52]. The C/N ratio across sites was inversely correlated with N_0 ($r = 0.66$, $p = 0.036$), confirming that the more recalcitrant SOM has lower potential N availability. Although not quantified in the current study, temporal changes in C/N ratios at each site can be expected [53], whereby these ratios decrease during the early age of a rotation along with specific rates of mineralization (*i.e.*, rates of N mineralization per unit of carbon). In sandy surface soils (0–15 cm depth) supporting *Pinus radiata* stands in southeast Australia, the specific rates of N mineralization decreased by more than 50% (from 207 to 93 g N month^{-1} t^{-1} C) during the first four years after planting as C/N ratios also decreased in both the <2 mm and >2 mm soil fractions [53]. For unsieved soil, the C/N ratio decreased from 38 to 31 during this period. This slow down in specific rate of N mineralization might reflect a decrease in the labile pool of organic N and organic matter quality. Such changes can be expected in any stand prior to replenishment of this labile pool as organic matter and nutrient cycling is restored by above- and below-ground litter production later in the crop rotation. Hence, the relationship

C/N ratio and N mineralization depends on poorly understood temporal and spatial factors of SOM quality.

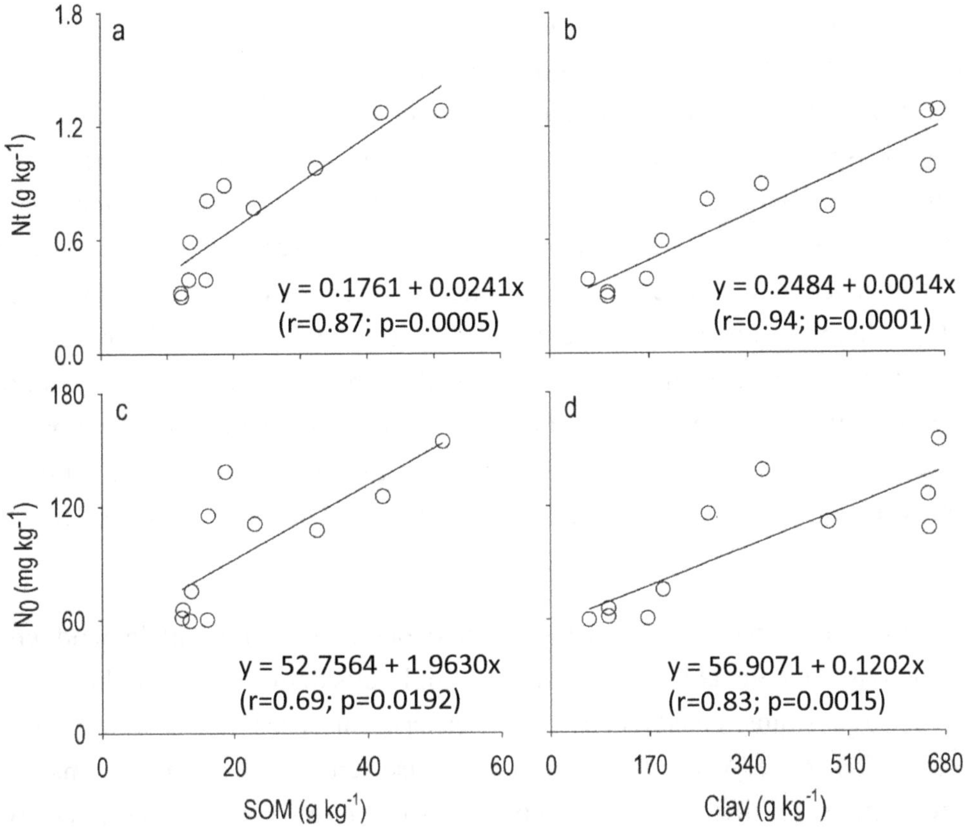

Figure 3. Correlations between soil organic matter (SOM), clay content, total N (N_t) and potentially mineralizable N (N_0).

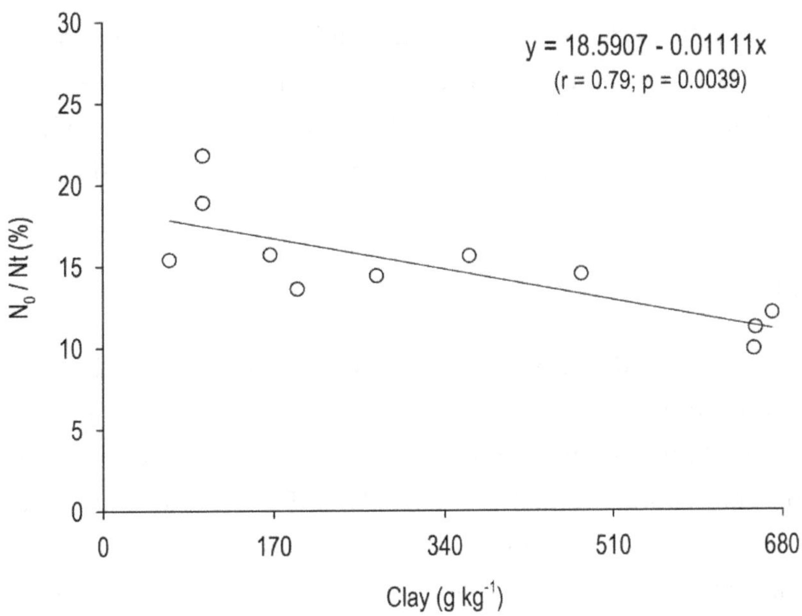

Figure 4. Correlation between clay content and N_0/N_t ratio across all sites.

4.2. Nitrogen Fertilizer Application Response

The response to N fertilizer application only occurred at an early stage of tree growth, when the canopy was in formation. For eucalypt stands in Brazil, around 70%–80% of N accumulates in aboveground components during the first two or three years of growth [1,54]. At this stage, N relative buildup is higher than biomass accumulation, due to formation of N-rich components (mainly, leaf and fine root), and N released from SOM by mineralization might not be enough to meet the N demand of trees [2]. After three years of age, competition among trees for water and light intensifies [55] and tree growth rates decline, leading to reduced N demand that is mostly supplied by N released through litter decomposition (biogeochemical cycling) and internal transfers (biochemical cycling) [1,54]. Gonçalves *et al.* [3] found a biochemical cycling of 54 kg N ha^{-1} $year^{-1}$ and biogeochemical cycling 42 kg N ha^{-1} $year^{-1}$ in *E. grandis* stands at seven years old, which are higher than the amounts required by trees (50 kg N ha^{-1} $year^{-1}$). Therefore, during early growth, N fertilizer application may accelerate tree growth rates by increasing N availability when N mineralization rates in soil and litter do not supply enough N to highly demanding trees. However, these responses disappear in subsequent years under the conditions examined here.

The mineral fertilizer requirements of any plantation depend on the nutrient demand required to reach an expected productivity, and the ability of the soil to supply this demand. When plant demand is greater than soil supply, fertilizers must be added. Thus, the criteria for N fertilizer application should involve only situations where response to N fertilizer exists, because fertilizer application aims to fill the deficit of N not released by soils. Fertilizer application practices must be linked with conservative methods of management to minimize N losses from the system, and thereby increasing sustainability [3].

Acknowledgments

The authors wish to thank Duratex S.A., Fibria Celulose S.A., International Paper of Brazil and Suzano Papel e Celulose S.A. for the financial support and making their areas available for the project. We also thank the Silviculture and Management Thematic Program (PTSM) of IPEF for financial support and assistance with for fieldwork. We also thank Antonio Bianchi for the revision of the English.

Author Contributions

Ana Paula Pulito, José Leonardo de Moraes Gonçalves, Luiz Fabiano de Moraes, José Luiz Gava, Raul Chaves and Claudio Roberto Silva designed the study and conducted of field trial. Ana Paula Pulito, José Carlos Arthur Junior, Aline Cristina Miranda and Marcos Yassuo Kamogawa were responsible for the samples collection and laboratory analysis. Ana Paula Pulito, José Carlos Arthur Junior, Clayton Alcarde Alvares and José Henrique Tertulino Rocha, were responsible for the statistical analyses, with contribution from José Leonardo de Moraes Gonçalves. Ana Paula Pulito, José Leonardo de Moraes Gonçalves, Philip J. Smethurst, Clayton Alcarde Alvares, José Henrique Tertulino Rocha and Ayeska Hübner wrote the paper.

References

1. Barros, N.F.; Novais, R.F.; Neves, J.C.L. Fertilização e correção do solo para o plantio de eucalipto. In *Relação Solo- Eucalipto*; Barros, N.F., Novais, R.F., Eds.; UFV: Viçosa, MG, Brazil, 1990; Volume 4, pp. 127–186.
2. Gonçalves, J.L.M.; Alvares, C.A.; Higa, A.R.; Silva, L.D.; Alfenas, A.C.; Stahl, J.; Ferraz, S.F.B.; Lima, W.P.; Brancalion, P.H.S.; Hubner, A.; *et al.* Integrating genetic and silvicultural strategies to minimize abiotic and biotic constraints in Brazilian eucalypt plantations. *For. Ecol. Manag.* **2013**, *301*, 6–27.
3. Gonçalves, J.L.M.; Stape, J.L.; Laclau, J.P.; Bouillet, J.P.; Ranger, J. Assessing the effects of early silvicultural management on long-term site productivity of fast growing Eucalypt plantations: The Brazilian experience. *South. For.* **2008**, *70*, 105–118.
4. Gonçalves, J.L.M.; Barros, N.F.; Nambiar, E.K.S.; Novais, R.F. Soil and stand management for short-rotation plantations. In *Management of Soil, Nutrients and Water in Tropical Plantation Forests*; Nambiar, S., Brown, A., Eds.; ACIAR Austrália/CSIRO Austrália/CIFOR Indonésia: Canberra, Australia, 1997; pp. 379–418.
5. Herbert, M.A.; Schönau, A.P.G. Fertilizing commercial forest species in southern Africa: Research progress and problems (part 1). *S. Afr. For. J.* **1989**, *151*, 58–70.
6. Reis, M.G.F.; Barros, N.F.; Kimmins, J.P. Acúmulo de nutrientes em uma sequência de idade de *Eucalyptus grandis* (ex-Maiden) plantado no cerrado, em duas áreas com diferentes produtividades, em Minas Gerais. *Rev. Árvore* **1987**, *11*, 1–15.
7. Laclau, J.-P.; Ranger, J.; Goncalves, J.L.M.; Maquere, V.; Krusche, A.V.; M'Bou, A.T.; Nouvellon, Y.; Saint-Andre, L.; Bouillet, J.-P.; Piccolo, M.D.C.; *et al.* Biogeochemical cycles of nutrients in tropical Eucalyptus plantations Main features shown by intensive monitoring in Congo and Brazil. *For. Ecol. Manag.* **2010**, *259*, 1771–1785.
8. Dovey, S.B.; du Toit, B.; Clercq, W. Nutrient fluxes in Rainfall, Throughfall and Stemflow in Eucalyptus Stands on the Zululand Coastal Plain, South Africa. *South. For.* **2011**, *73*, 193–206.
9. Sato, A.M.; Souza, A.A.; Coelho Netto, A.L. Spatial variability and temporal stability of throughfall in a eucalyptus plantation in the hilly lowlands of southeastern Brazil. *Hydrol. Process.* **2011**, *25*, 1910–1923.
10. Mayer, R.; Liess, S.; Lopes, M.I.M.S.; Kreutzer, K. Atmospheric pollution in a tropical rain forest: effects of deposition upon biosphere and hydrosphere II. Fluxes of chemicals and element budgets. *Water Air Soil Pollut.* **2000**, *121*, 79–92.
11. Gonçalves, J.L.M.; Mendes, K.C.F.S.; Sasaki, C.M. Mineralização de nitrogênio em ecossistemas florestais naturais e implantados do estado de São Paulo. *Rev. Bras. Cienc Solo* **2001**, *25*, 601–616.
12. Poggiani, F. Ciclagem de nutrientes em ecossistemas de plantações florestais de Eucalyptus e Pinus: Implicações silviculturais. *Tese (Livre Docência)—Escola Superior de Agricultura "Luiz de Queiroz"*; Universidade de São Paulo: Piracicaba, Brazil, 1985.
13. Godinho, V.P.C.; Barros, N.F.; Pereira, P.R.G.; Sediyama, C.S.; Fluxo parcial de nitrogênio e potássio em solo arenoso de cerrado, sob povoamentos de *Eucalyptus camaldulensis*, influenciado

pelo modo de aplicação de adubo nitrogenado e potássico. In Proceedings of the IUFRO Conference on silviculture and improvement of eucalyptus, Salvador, Brazil, 12–15 November 1997; pp. 125–133.

14. Smethurst, P.; Baillie, C.; Cherry, M.; Holz, G. Fertilizer effects on LAI and growth of four *Eucalyptus nitens* plantations. *For. Ecol. Manag.* **2003**, *176*, 531–542.
15. Laclau, J.P.; Ranger, J.; Deleporte, P.; Nouvellon, Y.; Saint-André, L.; Marlet, S.; Bouillet, J.P. Nutrient cycling in a clonal stand of Eucalyptus and an adjacent savanna ecosystem in Congo. 3. Input-output budgets and consequences for the sustainability of the plantations. *For. Ecol. Manag.* **2005**, *210*, 375–391.
16. Cantarella, H.; Van Raij, B. *Adubação nitrogenada no Estado de São Paulo*. In *Reunião brasileira de fertilidade do solo*; SBCS: Ilhéus, Brazil, 1986; pp. 47–79.
17. Larcher, W. *Ecofisiologia vegetal*; Rima: São Carlos, Brazil, 2004; p. 531.
18. Gianello, C.; Bremner, J.M. A simple chemical method of assessing potentially available organic nitrogen in soil. *Commun. Soil Sci. Plant Anal.* **1986**, *17*, 195–214.
19. Hopmans, P.; Flinn, D.W.; Farrel, P.W. Nitrogen mineralization in a sandy soil under native Eucalypt forest and exotic Pine plantations in relation to moisture content. *Commun. Soil Sci. Plant Anal.* **1980**, *1*, 71–79.
20. Keeney, D.R.; Bremner, J.M. Comparison and evaluation of laboratory methods of obtaining an index of soil nitrogen availability. *Agron. J.* **1966**, *58*, 498–503.
21. Corbeels, M.; Mc Murtrie, R.E.; Pepper, D.A.; Mendham, D.S.; Grove, T.S.; O'Connell, A.M. Long-term changes in productivity of eucalypt plantations under different harvest residue and nitrogen management practices: A modelling analysis. *For. Ecol. Manag.* **2005**, *217*, 1–18.
22. Alvares, C.A.; Stape, J.L.; Sentelhas, P.C.; Gonçalves, J.L.M.; Sparovek, G. Köppen's climate classification map for Brazil. *Meteorologische Zeitschrift* **2013**, *22*, 711–728.
23. Embrapa. *Sistema Brasileiro de Classificação de Solos*, 3rd ed.; Embrapa: Brasília, Brazil, 2013; p. 353.
24. Gonçalves, J.L.M.; Alvares, C.A.; Gonçalves, T.D.; Moreira, R.M.; Mendes, J.C.T.; Gava, J.L. Mapeamento de solos e da produtividade de plantações de Eucalyptus grandis em Itatinga, SP, com uso de sistema de informação geográfica. *Sci. For.* **2012**, *94*, 187–201.
25. Alvares, C.A.; Gonçalves, J.L.M.; Vieira, S.R.; Silva, C.R.; Franciscatte, W. Spatial variability of physical and chemical attributes of some forest soils in southeastern of Brazil. *Sci. Agri.* **2011**, *68*, 697–705.
26. Alvares, C.A.; Stape, J.L.; Sentelhas, P.C.; Gonçalves, J.L.M. Modeling monthly mean air temperature for Brazil. *Theor. Appl. Climatol.* **2013**, *113*, 407–427.
27. IPT—Instituto de Pesquisas Tecnológicas do Estado de São Paulo. *Mapa geológico do Estado de São Paulo*; escale 1:500.000. Secretaria da Indústria, Comércio, Ciência e Tecnologia: São Paulo, Brazil, 1981.
28. EMBRAPA. Manual de métodos de análise de solo; Embrapa: Rio de Janeiro, Brazil, 1999; p. 212.
29. Van Raij, B.; Andrade, J.C.; Cantarella, H.; Quaggio, J.A. *Análise química para avaliação da fertilidade de solos tropicais*; Instituto Agronômico: Campinas, Brazil, 2001; p. 285.
30. Bremner, J.M. Organic nitrogen in soils. In *Soil Nitrogen*; Bartholomew, W.V., Clark, F.E., Eds.; American Society of Agronomy: Madson, WI, USA, 1965; pp. 93–149.

31. Raison, R.J.; Connell, M.J.; Khanna, P.K. Methodology for studying fluxes of soil mineral-N *in situ*. *Soil Biol. Biochem.* **1987**, *19*, 521–530.
32. Arnold, J.; Corre, M.D.; Veldkamp, E. Cold storage and laboratory incubation of intact soil cores do not reflect in-situ nitrogen cycling rates of tropical Forest soils. *Soil Biol. Biochem.* **2008**, *40*, 2480–2483.
33. Kamogawa, M.Y.; Teixeira, M.A. Auto-amostrador de baixo custo para análise por injeção em fluxo. *Quim. Nova* **2009**, *32*, 1644–1646.
34. Jenkinson, D.S.; Ladd, J.N. Microbial biomass in soil: Measuremente and turnover. In *Soil Biochemistry*; Paul, E.A., Ladd, J.N., Eds.; Marcel Deker: New York, NY, USA, 1981; pp. 425–471.
35. Van Veen, J.A.; Ladd, J.N.; Martin, J.K.; Amato, M. Turnover of carbon, nitrogen and phosphorus through the microbial biomass in soils incubated with ^{14}C-, ^{15}N- and ^{32}-P-labelled bacterial cells. *Soil Biol. Biochem.* **1987**, *19*, 559–565.
36. Eaton, W.D. Microbial and nutrient activity in soils from three different subtropical forest habitats in Belize, Central America, before and during the transition from dry to wet season. *Appl. Soil Ecol.* **2001**, *16*, 219–227.
37. Carlyle, J.C. Nitrogen cycling in forested ecosystems. *For. Abstr.* **1986**, *47*, 307–336.
38. Smethurst, P.J.; Nambiar, E.K. Distribution of carbon and nutrients and fluxes of mineral nitrogen after-clearfelling a *Pinus radiata* plantation. *Can. J. For. Res.* **1990**, *20*, 1490–1497.
39. Gonçalves, J.L.M.; Carlyle, J.C. Modelling the influence of moisture and temperature on net nitrogen mineralization in a forested sandy soil. *Soil Biol. Biochem.* **1994**, *26*, 1557–1564.
40. Garnett, T.P.; Shabala, S.N.; Smethurst, P.J.; Newman, I.A. Kinetics of ammonium and nitrate uptake by eucalypt roots and associated proton fluxes measured using ion selective microelectrodes. *Funct. Plant Biol.* **2003**, *30*, 1165–1176.
41. Vale, F.R.; Novais, R.F.; Barros, N.F. Efeito do alumínio sobre a cinética de absorção de amônio e nitrato em raízes intactas de *Eucalyptus alba*. *Rev. Árvore* **1984**, *8*, 123–132.
42. Adams, M.A.; Attiwill, P.M. Nutrient cycling and nitrogen mineralization in eucalypt forests of souther-eastern Australia- II. Indices of nitrogen mineralization. *Plant Soil* **1986**, *92*, 341–362.
43. Khanna, P.K. Measurement of changes in the nitrogen status of forest soils due to management practices. In Proceedings of the International Symposium on Forest Soils, Harbin, China, 22–27 July 1990; p.10.
44. Theodorou, C.; Bowen, G.D. Nitrogen transformations in first- and second-rotation *Pinus radiata* forest soil. *Austral. For. Res.* **1983**, *13*, 103–112.
45. Theodorou, C.; Bowen, G.D. Effects of temperature, moisture and litter on nitrogen mineralization in *Pinus radiata* forest soils. *Austral. For. Res.* **1983**, *13*, 113–119.
46. Pottker, D.; Tedesco, M.J. Efeito do tipo e tempo de incubação sobre a mineralização da matéria orgânica e nitrogênio total em solos do Rio Grande do Sul. *Rev. Bras. Cienc Solo* **1979**, *3*, 20–24.
47. Noble, A.D.; Herbert, M.A. Influence of soil organic matter content on the responsiveness of Eucalyptus grandis to nitrogen fertiliser. *S. Afr. For. J.* **1991**, *156*, 23–27.
48. Maquere, V.; Laclau, J.P.; Bernoux, M.; Saint-Andre, L.; Gonçalves, J.L.M.; Cerri, C.C.; Piccolo, M.C.; Ranger, J. Influence of land use (savanna, pasture, *Eucalyptus* plantations) on soil carbon and nitrogen stocks in Brazil. *Eur. J. Soil Sci.* **2008**, *59*, 863–877.

49. Montero, L.L. Carbono em solos de cerrado: efeitos do uso florestal. Tese (Doutorado em Ciências, na área de Ecologia)—Instituto de Biociências, Universidade de São Paulo, São Paulo, Brazil, 2008.
50. Lima, A.M.N.; Silva, I.R.; Neves, J.C.L.; Novais, R.F.; Barros, N.F.; Mendonça, E.D.; Smyth, T.J.; Moreira, M.S.; Leite, F.P. Soil organic carbon dynamics following afforestation of degraded pastures with eucalyptus in southeastern Brazil. *For. Ecol. Manag.* **2006**, *235*, 219–231.
51. Vitousek, P.M. Litterfall, nutrient cycling and nutrient limitation in tropical forest. *Ecology* **1984**, *65*, 285–298.
52. Attiwill, P.M.; Leeper, G.W. *Forest Soils and Nutrient Cycles*. Melbourne University: Carlton, Australia, 1987; p. 202.
53. Smethurst, P.J.; Nambiar, E.K.S. Changes in soil carbon and nitrogen during the establishment of a second crop of *Pinus radiata*. *For. Ecol. Manag.* **1995**, *73*, 145–155.
54. Barros, N.F.; Neves, J.C.L.; Novais, R.F. Recomendação de fertilizantes minerais em plantios de eucalipto. In *Nutrição e fertilização florestal*; Gonçalves, J.L.M., Benedetti, V., Eds.; IPEF: Piracicaba, Brazil, 2000; pp. 269–286.
55. Ryan, M.G.; Stape, J.L.; Binkley, D.; Fonseca, S.; Loos, R.A.; Takahashi, E.N.; Silva, C.R.; Silva, S.R.; Hakamada, R.E.; Ferreira, J.M.; *et al.* Factors controlling Eucalyptus productivity: How water availability and stand structure alter production and carbon allocation. *For. Ecol. Manag.* **2010**, *259*, 1695–1703.

4

Residual Long-Term Effects of Forest Fertilization on Tree Growth and Nitrogen Turnover in Boreal Forest

Fredrik From [1,*], Joachim Strengbom [2] and Annika Nordin [1]

[1] Department of Forest Genetics and Plant Physiology, Umeå Plant Science Centre, Swedish University of Agricultural Sciences, Skogsmarksgränd 1, 90183 Umeå, Sweden; E-Mail: annika.nordin@slu.se

[2] Department of Ecology, Swedish University of Agricultural Sciences, Box 7044, 75007 Uppsala, Sweden; E-Mail: joachim.strengbom@slu.se

* Author to whom correspondence should be addressed; E-Mail: fredrik.from@slu.se;

Academic Editor: Heinz Rennenberg

Abstract: The growth enhancing effects of forest fertilizer is considered to level off within 10 years of the application, and be restricted to one forest stand rotation. However, fertilizer induced changes in plant community composition has been shown to occur in the following stand rotation. To clarify whether effects of forest fertilization have residual long-term effects, extending into the next rotation, we compared tree growth, needle N concentrations and the availability of mobile soil N in young (10 years) *Pinus sylvestris* L. and *Picea abies* (L.) H. Karst. stands. The sites were fertilized with 150 $kg \cdot N \cdot ha^{-1}$ once or twice during the previous stand rotation, or unfertilized. Two fertilization events increased tree height by 24% compared to the controls. Needle N concentrations of the trees on previously fertilized sites were 15% higher than those of the controls. Soil N mineralization rates and the amounts of mobile soil NH_4-N and NO_3-N were higher on sites that were fertilized twice than on control sites. Our study demonstrates that operational forest fertilization can cause residual long-term effects on stand N dynamics, with subsequent effects on tree growth that may be more long-lasting than previously believed, *i.e.*, extending beyond one stand rotation.

Keywords: carry-over fertilization effects; forest growth; Picea abies; Pinus sylvestris; N mineralization; *in situ* incubation; ion-exchange resin (IER)

---

1. Introduction

Nitrogen (N) availability limits forest growth in most boreal and temperate forests [1,2], and fertilization with N is one of the most cost-effective silvicultural methods to increase boreal forest yield [3]. In fact, timber harvest profitability can increase by nearly 15% if forest stands are fertilized about 10 years before final felling [4]. Fertilization has been practiced commercially since the middle of the previous century in many parts of the boreal and temperate forest regions, for example, in Sweden [5,6], USA [7], Canada [8] and Finland [9]. During the 20th century approximately 10% of the productive forest land in Sweden, *i.e.*, circa (ca) 2 million hectares (ha), was fertilized at least once [3,6,10].

According to previous studies, forest fertilization with N additions corresponding to 150 $kg{\cdot}N{\cdot}ha^{-1}$ $application^{-1}$, for up to three times during a forest rotation period [11], will only affect the subjected forest stand for up to 10 years after the last fertilizer application [3,9,12–16]. This conclusion is mainly based on studies of growth effects within the tree rotation period fertilized. Also a recent study, particularly targeting carry-over effects of fertilization between forest stand rotations, showed that growth of five year old pine seedlings was not affected by fertilization of the site three to nine years before harvest of the previous stand rotation [17], *i.e.*, supporting the view that there are no, or only very restricted long-term effects of fertilization. In contrast, a study from North America showed that the height growth of young Douglas fir (*Pseudotsuga menziesii* [Mirb.] Franco) trees were greater on plots that had been fertilized in the previous stand rotation period than on plots never fertilized [18].

Fertilization effects on species composition of the forest floor vegetation within one forest rotation have been extensively documented, e.g., summarized in Nohrstedt [3] and Saarsalmi and Malkonen [9], but only a few studies have addressed carry-over effects between forest stand rotations. In a study summarizing carry-over fertilization effects on forest floor vegetation, Hedwall, *et al.* [19] reported a higher abundance of graminoids and a lower abundance of dwarf-shrubs on experimental plots fertilized in the previous stand rotation period than on control plots never fertilized. Strengbom and Nordin [20] found carry-over effects between stand rotations on the composition and abundance of forest floor vegetation of operational forest fertilization, *i.e.*, with 150 $kg{\cdot}N{\cdot}ha^{-1}$ at one or two occasions during the previous stand rotation.

When fertilizing forests, a large part of the N added is retained in the soil layer [21,22]. Strengbom and Nordin [23] suggested that the disturbance from harvest and site scarification leads to increased mineralization rates of immobilized soil N, explaining why effects of previous (more than 25 years ago) N fertilization was more evident in the early phase of the subsequent stand rotation than within the same stand rotation that were fertilized.

In this study, we revisited young forest stands with documented effects on species composition of the forest floor vegetation from fertilization of the previous stand rotation. We hypothesized that fertilization of the previous forest stand would cause (1) increased tree growth in the following rotation period; and (2) increased soil N mineralization rates, both in comparison to control stands that never had been fertilized. Thus, the aim of the study was to determine fertilization carry-over effects between tree rotations on tree height and diameter growth and on soil N turnover, *i.e.*, mineralization rates and amounts of mobile soil N.

2. Materials and Methods

2.1. Study Area

We studied forest sites each sized between 4.7 and 22.4 ha situated within an 8500 ha forest land area owned by the company Sveaskog. The forest land area is in the middle boreal zone [24] in central Sweden (63°00′ N, 16°40′ E). The studied forest sites, either had, or had not, been subjected to N fertilization before harvest and regeneration (Table 1). Control (C) sites (n = 7) were never fertilized; N1 sites (n = 7) were fertilized with 150 kg·N·ha^{-1} once (in 1985), while N2 sites (n = 7) were repeatedly fertilized with 150 kg·N·ha^{-1} (in 1977 and 1985), *i.e.*, our study included seven stands of each treatment. Nitrogen was added as granules of ammonium nitrate (NH_4NO_3) spread by tractor or aircraft.

Table 1. Number of sites per treatment, elevation (above sea level, a.s.l.), temperature sum (GDD), forest floor slope, the site index (SI), the average numbers of trees ha^{-1} and the average sample tree ages (2010). The sites were fertilized once (N1) with 150 kg·N·ha^{-1} 25 years before the present study, twice (N2) 33 and 25 years before the present study or never fertilized (C).

Sites (n)	Control	N1	N2	*F*-value	*p*-value
	7	7	7		
Elevation a.s.l. (m)	334 ± 21	358 ± 20	334 ± 17	0.54	0.59
Temperature sum (GDD)	937 ± 6	900 ± 22	923 ± 11	1.59	0.23
Slope (1 to 5)	2.00 ± 0.22	1.86 ± 0.26	1.72 ± 0.18	0.41	0.67
Site index (H_{100}, m)	20.6 ± 0.2	19.7 ± 0.3	20.4 ± 0.3	3.00	0.08
Trees (ha^{-1})	2721 ± 378	2982 ± 416	2532 ± 206	0.43	0.65
Tree age (years)	10.4 ± 0.2	10.4 ± 0.2	10.3 ± 0.2	0.80	0.47

Forest inventory records describing the forest sites before clear-cut kept by the land owner were used when selecting the sites. In particular care was taken to select sites with similar productivity (site) indices (Table 1). The other criteria for including sites in our study were temperature sum (growing degree days, GDD), *i.e.*, the annual temperature sum of all daily mean temperatures above 5 °C [25]; slope, *i.e.*, the forest floor incline, on a scale from 1 (<10% inclination) to 5 (>50%); current stand age; tree species; and soil conditions. The sites were spread out in the 8500 ha forest land area and spatially separated from each other (>1 km). The annual precipitation in the area is between 600 to 700 mm per year and the atmospheric background N deposition ranges from 2.5 to 3.5 kg ha^{-1} year^{-1} [26].

All stands were on mesic sites with moraine soil, geotechnical nomenclature coarse-grained till [27], with Udic moisture regime [28]. Before clear-felling all sites had a mixed tree species composition with *P. sylvestris* and *P. abies* as dominant trees and *Betula* spp. as subordinate trees. The relative abundance of the two dominant species prior to clear-felling ranged from 25% to 75% with no difference between fertilized and unfertilized stands (Mann–Whitney U-test: U = 83.5, p = 0.348) [20]. The forest field layer was dominated by dwarf shrubs such as *Vaccinium myrtillus* L. and *V. vitis-idaea* L., *i.e.*, the stands were classified as spruce forest of bilberry type [29]. All sites were harvested by clear-felling between 1997 and 2000 and were thereafter treated with soil scarification and regenerated by planting *P. sylvestris* and *P. abies* seedlings at a density of 2200 to 2300 stems ha^{-1}. Due natural regeneration, all forest stands at the time of our study had a mix of both coniferous species and a sub population of *Betula pendula* Roth, *Betula pubescence* Ehrh and *Populus tremula* L.

2.2. Data Collection

Data on tree growth and needle N concentration were collected in August 2010 from four 100 m^2 (r = 5.64 m) circular plots per forest site. The plots were distributed along transects that were approximately stand-centered, described in Strengbom and Nordin [20], to minimize the influence of surrounding stands. Tree height (cm), diameter (mm) at breast height (DBH, at 1.3 m above ground) and species were noted for all trees taller than 1.3 m using cross-callipering and a five meter long, telescopic measuring stick (Teleskopmeter, Skogma, Hammerdal, Sweden). Across all plots, 38% of the trees were *P. abies*, 29% were *P. sylvestris* and 33% *Betula* spp. Less than 1% were *P. tremula.* From each site, 16 undamaged sample-trees (the four tallest from each circular plot) were selected. *P. abies* (56%) and *P. sylvestris* (44%) were exclusively used as sample-trees. Sites were classified as dominated by *P. abies* or *P. sylvestris* (*P. abies* = four C, three N1 and four N2; *P. sylvestris* = three C, four N1 and three N2). There was no statistically significant difference in site species classification between treatments (Kruskal-Wallis one-way analysis of variance test = 0.364, $p = 0.834$). Sample-tree annual shoot growth was measured from first visible node and upwards toward full tree height. Tree age was determined from the number of nodes plus one per sample tree.

Current year needles were collected for C and N analyses in early august 2010. Needles were cut from the sample tree's top branches facing the centre of the circle. All needle samples were dried in 70 °C for 24 h and grinded to a fine powder using a bead mill (five minutes per sample), the powder was stored in clear glass vials. In late September, small amounts (ca 3 mg per sample tree) of the ground needle samples were sealed in small tin foil cups, one per sample tree, and sent to a lab (Umeå Plant Science Centre, UPSC) for C and N content analyses using an AutoAnalyzer 3 (SEAL Analytical, OmniProcess AB, Solna, Sweden). Before statistical analyses replicate samples from the same site were pooled.

The buried bag technique was used to investigate humus N mineralization. At each site seven samples from the organic mor-layer were collected between 1st and 3rd June in 2009 with a cylindrical (diameter 10 cm) soil corer. After gently removing large roots and living green material, half of the sample from each mor-layer core was placed in plastic bags in a cooler and transported to the lab for analyses. The second half was placed in plastic bags and buried in the mor-layer *in situ*, and were retrieved between the 28th and 30th September the same year, stored in a cooler and transported to the lab for analyses. The mor-layer samples were extracted in 1 m KCL and analyzed for NH_4 and NO_3 using a FIAstar 5012 Analyzer (Tecator, Höganäs, Sweden). Mineralization was calculated through differences in NH_4-N and NO_3-N content ($mg \cdot g^{-1}$ dry weight [DW] soil) between sample times, *i.e.*, early June and late September. All contaminated samples (*i.e.*, damaged bags) were excluded before estimating mineralization. In total 17 samples were excluded from analyses (2 from C sites, 9 from N1 sites and 6 from N2 sites). Before statistical analyses replicate samples of NH_4-N and NO_3-N from the same site were pooled to obtain one value of mineralization per site.

Resin ion-exchange capsules (PST-1, Universal Bioavailability Environment/Soil Test, MT, USA) were used to estimate the amount of soil mobile NH_4-N and NO_3-N (mg per capsule). Six capsules were buried just beneath the mor-layer at each site in early June in 2009 and retrieved in late September in the same year. Following retrieval the capsules were placed in plastic bags in a cooler and transported to the lab for analyses. The ion-exchange resins were brushed off to get them as clean as possible before analyses. The capsules were placed in 50 mL Falcon tubes and 7 mL of 1 m KCL was pipetted into the tubes that thereafter were oscillated for 30 min in room temperature, thereafter the 7 mL was poured into a second collecting falcon tube. This process was repeated three times ($3 \times 7 = 21$ mL), and the tubes were stored in a cooler between oscillations. Soil mobile NH_4-N and NO_3-N were analyzed using a FIAstar 5012 Analyzer (Tecator, Höganäs, Sweden). Before statistical analyses replicate samples of NH_4-N and NO_3-N from the same site were pooled to obtain one value of soil mobile N per site.

2.3. Experimental Design and Statistical Analyses

For all statistical analyses the experiment is regarded as a completely randomized design. To test for differences in annual shoot growth among treatments (C, N1 and N2) we used repeated measures ANOVA, with annual growth (cumulative height growth during 10 years) as response and treatment and time as dependent variables. The degrees of freedom were corrected with the Greenhouse-Geisser elipson as the assumption of sphericity was not met, tested by the Mauchly's test of sphericity [30]. Sphericity is not met when the variances of the differences between all combinations of associated groups are not equal; correcting the degrees of freedom with the Greenhouse-Geisser epsilon produces a more accurate, upward adjusted p-value in the subsequent analysis. After adjusting the degrees of freedom, Tukey's post hoc HSD method was applied to detect significant ($\alpha = 0.05$) differences between treatments when the analysis of variance showed significant main N treatment effects. For the repeated measures ANOVA we used the software package IBM SPSS (v. 20).

Differences in tree height (in 2010), trees per ha, sample tree age, needle N concentration (%-DW), sample tree DBH, total soil mobile N (NH_4-N + NO_3-N, mg per ion capsule) and total mineralization rates (ammonification plus nitrification) between the treatments were analyzed with one-way ANOVAs followed by Tukey's post hoc test for pairwise comparison when the analysis of variance showed significant ($\alpha = 0.05$) main N treatment effects. Response variables were checked for normal distribution and homoscedasticity, and when these requirements were not met (for mineralization rate, amount of mobile soil N and needle N), data was transformed with the natural logarithm (LN). For the one-way ANOVAs we used the software package Minitab (v. 16). All data showed in figures and tables are original, untransformed mean values ± 1 standard error (SE).

3. Results

On the sites that were fertilized twice (N2) during the preceding stand rotation (fertilizer applied 25 and 33 years before our study), the annual shoot height growth over time was greater ($F_{2.857,25.715} = 5.09$, $p = 0.018$; Table 2) than those of trees on N1 sites and on control sites never fertilized (Figure 1). Annual shoot height growth over time on N1 plots was, however, not significantly greater than that on control plots ($F_{2.857,25.715} = 5.09$, $p = 0.255$). In 2010 the 10 year old trees on N2 sites were on average 24% higher than trees on control sites ($F_{2,18} = 4.50$, $p = 0.026$, Table 3). Average tree height on N1 sites was not different from that on control sites ($F_{2,18} = 4.50$, $p = 0.214$). Tree diameter (DBH) was not affected by the previous fertilization treatments ($F_{2,18} = 0.75$, $p = 0.486$, Table 3).

Table 2. Results from repeated measures ANOVA testing the effects of fertilization of the previous stand rotation on annual tree shoot height growth of the current stand rotation.

Within Subjects	Degrees of Freedom	*F*-value	*p*-value
Year	1.43	948.88	<0.001
Year · Treatment	2.86	3.33	0.037
Error (year)	25.72	1217.1	
Between subjects			
Treatment	2	5.09	0.018
Error	18		

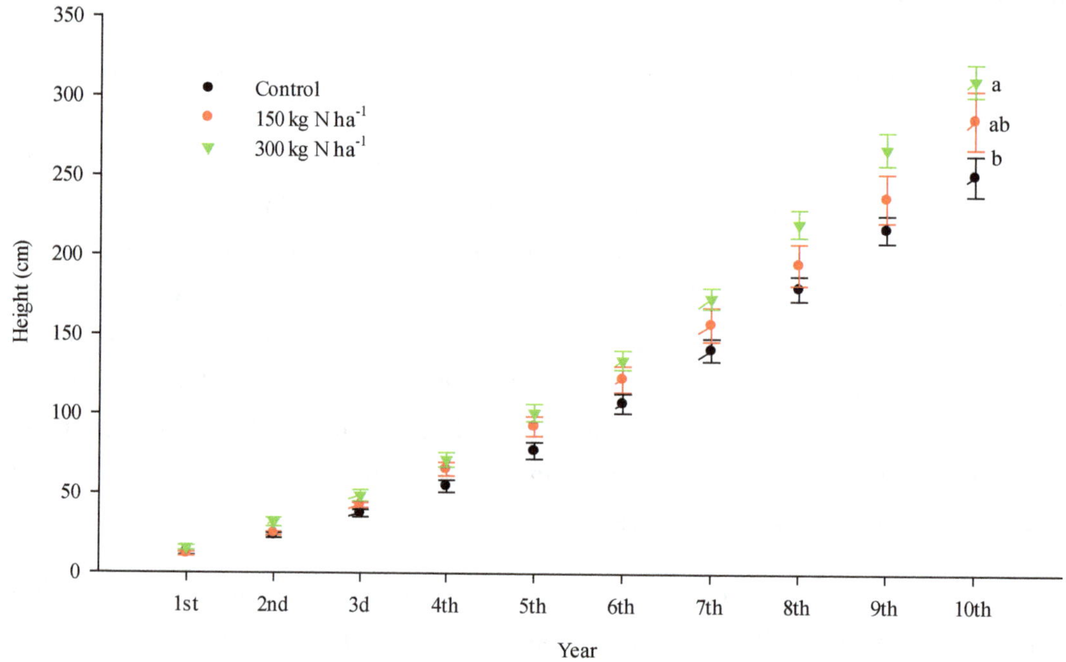

Figure 1. The mean (± 1 SE) annual tree height on sites with no fertilization (C), on sites fertilized once with 150 kg·N·ha^{-1} during the previous stand rotation (N1) and on sites fertilized twice with 150 kg·N·ha^{-1} during the previous stand rotation (N2). Different lower case letters (a or b) indicate a significant difference ($p < 0.05$) between sites analyzed with a repeated measures ANOVA followed by a Tukey's post hoc HSD test.

Trees on control sites had lower needle N concentration than trees on both N1 and N2 sites ($F_{2,17} = 5.04$, $p = 0.018$; Table 3). On N1 and N2 sites the needle N concentration was approximately the same, about 16% and 14% higher on N2 and N1 sites respectively compared to the control. There were no differences in needle carbon concentration (data not shown) between sites with the different N treatments.

Although the variation among sites with the same treatment was large, soil N mineralization rates and soil mobile N differed between the treatments (Table 3). Soil N mineralization rates were nearly four times higher on N2 sites than on control sites ($F_{2,18} = 5.22$, $p = 0.016$), whereas there were no differences in N mineralization rates between the control and the N1 sites. The total amount of soil mobile N that were captured on ion exchange resins placed in the soil was 74% higher on N2 sites than on control sites ($F_{2,18} = 3.71$, $p = 0.045$; Table 3).

Table 3. The mean (± 1 SE) tree needle N concentration (%-DW), diameter at breast height (DBH, mm), tree height in 2010, the total mineralization rates (NH_4-N and NO_3-N mg g^{-1} DW soil) and the total soil mobile N (NH_4-N and NO_3-N) concentration. The 10 year old trees grew on sites never fertilized (C), fertilized once with 150 kg·N·ha^{-1} during the previous stand rotation (N1) and fertilized twice with 150 kg·N·ha^{-1} during the previous stand rotation (N2). Different lower case letters (a or b) besides means in each column indicate significant post hoc differences between treatments (Tukey's HSD test).

Treatment	Needle N Concentration (%-DW)	DBH (mm, 2010)	Tree Height (cm, 2010)	Total Mineralization (N mg g^{-1} DW soil)	Total Soil Mobile N (mg $capsule^{-1}$)
C	1.19 ± 0.05 a	31 ± 5 a	251 ± 12 a	0.083 ± 0.012 a	0.057 ± 0.023 a
N1	1.38 ± 0.05 b	34 ± 5 a	286 ± 13 ab	0.090 ± 0.021 a	0.072 ± 0.022 ab
N2	1.35 ± 0.04 b	39 ± 4 a	311 ± 17 b	0.293 ± 0.109 b	0.099 ± 0.020 b
F-value	5.04	0.75	4.50	5.22	3.71
P-value	0.018	0.486	0.026	0.016	0.045

4. Discussion

According to common forestry practice the growth enhancing effect of N fertilization is expected to diminish within 10 years following treatment [3,9,31]. However, 25 years following the latest fertilization event and 10 years after harvest, scarification and tree planting, we detected positive effects of the previous N fertilization on tree growth in the subsequent young forest stands (Figure 1, Table 2). Our study revealed a fertilizer induced increment in tree growth and needle N concentration. A positive relationship between tree growth and needle N concentration is reported in previous studies, e.g., Bauer, *et al.* [32] and Iivonen, *et al.* [33], and needle N concentration is in turn indicative of the plant available N pool [34]. The higher needle N concentration, in combination with the higher soil N mineralization rates and higher amount of soil mobile N on previously fertilized sites, indicate that N dynamics in the previously fertilized sites remained altered despite that more than 25 years had passed since the last fertilization event.

When fertilizing forests, a large proportion of the N added is normally retained in the soil layer [21,22,35], and disturbances such as clear-felling and soil scarification have been shown to increase soil N mineralization rates and the amount of available soil nutrients [36–38], generally due to increased soil microbial activity. Previously added N retained in the soil layer may thus be accessible to above ground vegetation as N mineralization rates increases after clear-felling and soil scarification. Previous studies of the same sites as used in the present study revealed fertilizer induced shifts in species composition of the forest floor vegetation [20,23]. In combination with the results from the present study this clearly demonstrates that carry-over effects from previous fertilization events influence several components of the ecosystem, resulting in a long-lasting effect on site productivity.

Our results are in contrast to several studies of long-term effects of forest fertilization. For example, Pettersson and Högbom [39] studied old fertilization experiments in boreal forests where 100 to 240 kg of N had been applied one to four times, 14 to 28 years before their study. They found that the previous N addition did not increase *P. abies* or *P. sylvestris* growth in the long term. The majority of the previous studies have, however, addressed the fertilizer effects on site productivity in the forest stand subjected to fertilization, *i.e.*, within the same stand rotation. There are only a few studies targeting carry-over effects of N fertilization across stand rotation periods. In accordance with our results Högbom *et al.* [40,41] found that sites fertilized during the preceding stand rotation had an increased amount of inorganic soil N as well as increased soil N mineralization rates in comparison to unfertilized sites. Worth noting is that this study used N doses largely exceeding the doses used in our study (up to 1800 $kg \cdot N \cdot ha^{-1}$). Generally much higher amounts of N fertilizer than used in operational forestry practice, appears to be a common feature of most previous studies that have addressed carry-over effects between stand rotations, e.g., Smolander, Priha, Paavolainen, Steer and Malkonen [14], Pettersson and Högbom [39] and Olsson and Kellner [42]. Hence, an important contribution from our study is that N doses used in operational forestry practice can also result in a significant positive carry-over effect on tree growth. Moreover, the only other study we are aware of that targets carry-over effects across stand rotation periods from forest fertilization conducted according to the standard practice, report results that contrast ours, *i.e.*, no positive effect on *P. sylvestris* seedlings growth during the initial five year period following plantation [17]. Although there might be several potential explanations to the contrasting results, our study indicates that it may take more than five years to distinguish a consistent effect of previous fertilization on tree height growth. In accordance with our results, a North American study showed that *Pseudotsuga menziesii* height growth in the first couple of years after seedling establishment was unaffected by fertilization of the preceding forest stand, but that the positive response from past fertilization gradually increased over the years [18]. In their study, a significant difference with approximately 15% greater mean height in fertilized stands than in unfertilized stands became evident when the trees were seven to nine years old. In light of current and previous studies it could be argued that growth enhancement is cumulative, *i.e.*, the growth that the fertilization treatments trigger is built up gradually over time, to eventually entail significant differences in tree height between treatments.

Our study system is situated in a forest land area managed according to common operational practices. Since we lack information on initial site productivity conditions and it is not possible to find detailed information on the criteria used to select sites to be fertilized, we cannot exclude the possibility that initial differences may have influenced our final results, *i.e.*, that N2 sites were more fertile than C sites also before fertilization. However, all available information suggests that the sites selected for fertilization were not biased towards more productive sites. For example, the similar site productivity indices, expressed as estimated average tree height at tree age 100 years (20.6 on C sites and 20.4 on N2 sites) [43,44], and the overall similarity in site characteristics between the sites (see Table 1) supports the information provided by the forest companies employees that the sites selected for fertilization in 1977 and 1985 were, from a productivity point of view, randomly distributed over the forest land area. In addition, if results as those obtained in our study, with N1 sites consistently showing intermediate responses compared to the unfertilized controls and N2 sites (Table 3), were to be generated by pre-fertilization differences in productivity, sites selected for being fertilized would always have had to been more productive than unfertilized sites, and N2 sites would always have had to been more productive than N1 sites. Besides that it appears unlikely that sites by coincidence ended up being selected in such a way, we cannot see any logic reasons to why sites intentionally should be selected in such a biased way. Thus, we are confident that our results indicating residual effects of past fertilization are valid and not generated by initial differences in site productivity.

5. Conclusions

Our study highlighted that forest fertilization according to the standard practice may cause residual long-term effects on tree growth, needle N concentration, soil mineralization rates and amounts of mobile soil N in the stand rotation following the one fertilized. The findings presented herein emphasize that the effects from operational N fertilization on tree growth and biogeochemical ecosystem features can be of a more long-lasting character than previously thought. Whether this increase in site productivity will persist or in time level off is unclear, and our findings advocate that long-term effects from forest fertilization need to be studied further. Further research is also needed to elucidate the mechanisms on how fertilizer N is immobilized following fertilization, and how it is subsequently released following the disturbance caused by tree harvest and soil scarification to thereafter enhance growth of the young forest stand.

Acknowledgments

This work was financially supported by Future Forests and grants from the Swedish Research Council for Environment, Agricultural Sciences and Spatial Planning (FORMAS) to A. Nordin. We would like to thank Ann Sehlstedt and Sonja Wahlberg for assisting with the field work. We would also like to thank the two anonymous reviewers for their valuable suggestions.

Author Contributions

Annika Nordin and Joachim Strengbom conceived and designed the experiment. Fredrik From carried out the field work, data analysis and prepared the manuscript. Annika Nordin and Joachim Strengbom contributed to the data analysis and prepared the manuscript.

References

1. Tamm, C.O. Nitrogen in terrestrial ecosystems: Questions of productivity, vegetational changes, and ecosystem stability. In *Ecological Studies Analysis and Synthesis*; Springer-Verlag Berlin Heidelberg: Berlin, Germany, 1991; p. 115.
2. Vitousek, P.M.; Howarth, R.W. Nitrogen limitation on land and in the sea: How can it occur? *Biogeochemistry* **1991**, *13*, 87–115.
3. Nohrstedt, H.O. Response of coniferous forest ecosystems on mineral soils to nutrient additions: A review of Swedish experiences. *Scand. J. For. Res.* **2001**, *16*, 555–573.
4. Simonsen, R.; Rosvall, O.; Gong, P.C.; Wibe, S. Profitability of measures to increase forest growth. *For. Policy Econ.* **2010**, *12*, 473–482.
5. Kardell, L. Skogsgödslingen I backspegeln. In *Debatten om Storskogsbrukets Kvävegödsling I Sverige ca 1960–2009*; Future Forests Working Report; Future Forests: Umeå, Sweden, 2010.

6. Lindkvist, A.; Kardell, O.; Nordlund, C. Intensive forestry as progress or decay? An analysis of the debate about forest fertilization in Sweden, 1960–2010. *Forests* **2011**, *2*, 112–146.
7. Albaugh, T.J.; Allen, H.L.; Fox, T.R. Historical patterns of forest fertilization in the southeastern United States from 1969 to 2004. *South. J. Appl. For.* **2007**, *31*, 129–137.
8. Brockley, R.P. Fertilization of lodgepole pine in western Canada. In Proceedings of the Enhanced Forest Management, Fertilization & Economics Conference, Edmonton, AB, Canada, 1–2 March 2001; pp 44–55.
9. Saarsalmi, A.; Malkonen, E. Forest fertilization research in Finland: A literature review. *Scand. J. For. Res.* **2001**, *16*, 514–535.
10. Näslund, B.Å.; Stendahl, J.; Samuelsson, H.; Karlsson, L.; Kock-Hansson, G.; Svensson, H.; Engvall, C. Kvävegödsling på skogsmark. *Underlag för Skogsstyrelsen Föreskrifter och Allmänna Råd om Kvävegödsling*; Näslund, B.Å., Ed.; Swedish Forest Agency: Jönköping, Sweden, 2013; p. 48. Available online: http://www.regelradet.se/wp-content/files_mf/13702638072013_178_Rapport_kvavegodsling.pdf (accessed on 7 April 2015).
11. Enander, G.; Samuelsson, H. Skogsstyrelsens Allmänna Råd Till Ledning för Hänsyn Enligt 30 § Skogsvårdslagen (1979:429) vid Användning av Kvävegödselmedel på Skogsmark; Skogssstyrelsen. SKSFS **2007**, *3*, 5.
12. Pettersson, F. *New Predictive Functions Reveal Same Growth Response to N Fertilization as 30 Years Ago*; SkogsForsk (Forestry Research Institute of Sweden): Uppsala, Sweden, 1994; p. 6.
13. Priha, O.; Smolander, A. Nitrification, denitrification and microbial biomass N in soil from 2 N-fertilized and limed norway spruce forests. *Soil Biol. Biochem.* **1995**, *27*, 305–310.
14. Smolander, A.; Priha, O.; Paavolainen, L.; Steer, J.; Malkonen, E. Nitrogen and carbon transformations before and after clear-cutting in repeatedly N-fertilized and limed forest soil. *Soil Biol. Biochem.* **1998**, *30*, 477–490.
15. Högbom, L.; Jacobsson, S. Kväve 2002—En konsekvensbeskrivning av skogsmarksgödsling i sverige. *Skogforsk Redogörelse* **2002**, *6*. Available online: http://shop.skogsstyrelsen.se/shop/9098/art64/4645964-efc63c-1547.pdf (accessed on 7 April 2015).
16. Peterson, C.E.; Gessel, S.P. *Forest Fertilization in the Pacific Northwest: Results of the Regional Forest Nutrition Research Project*; General Technical Report; Pacific Northwest Forest and Range Experiment Station, USDA Forest Service: Washington, DC, USA, 1983; pp. 365–369.
17. Johansson, K.; Ring, E.; Hogbom, L. Effects of pre-harvest fertilization and subsequent soil scarification on the growth of planted Pinus Sylvestris seedlings and ground vegetation after clear-felling. *Silva Fennica* **2013**, *47*. doi:10.14214/sf.1016.
18. Footen, P.W.; Harrison, R.B.; Strahm, B.D. Long-term effects of nitrogen fertilization on the productivity of subsequent stands of douglas-fir in the pacific northwest. *For. Ecol. Manage.* **2009**, *258*, 2194–2198.
19. Hedwall, P.O.; Nordin, A.; Strengbom, J.; Brunet, J.; Olsson, B. Does background nitrogen deposition affect the response of boreal vegetation to fertilization? *Oecologia* **2013**, *173*, 615–624.
20. Strengbom, J.; Nordin, A. Commercial forest fertilization causes long-term residual effects in ground vegetation of boreal forests. *For. Ecol. Manag.* **2008**, *256*, 2175–2181.
21. Melin, J.; Nommik, H. Fertilizer nitrogen distribution in a *Pinus slvestris*/*Picea abies* ecosystem, central Sweden. *Scand. J. For. Res.* **1988**, *3*, 3–15.

22. Nohrstedt, H.O. Effects of repeated nitrogen fertilization with different doses on soil properties in a *Pinus-sylvestris* stand. *Scand. J. For. Res.* **1990**, *5*, 3–16.
23. Strengbom, J.; Nordin, A. Physical disturbance determines effects from nitrogen addition on ground vegetation in boreal coniferous forests. *J. Veg. Sci.* **2012**, *23*, 361–371.
24. Ahti, T.; Hämet-Ahti, L.; Jalas, J. Vegetation zones and their sections in northwestern Europe. *Ann. Bot. Fenn.* **1968**, *5*, 169–211.
25. Womach, J.; Becker, G.S.; Blodgett, J.; Buck, G.; Canada, C.; Chite, R.; Cody, B.; Copeland, C.; Corn, L.; Cowan, T.; *et al. Agriculture: A Glossary of Terms, Programs, and Laws*, 2005th ed.; BiblioGov: Columbus, OH, USA, 2005; p. 282.
26. Phil-Karlsson, G.; Akselsson, C.; Ferm, M.; Hellsten, S.; Hultberg, H.; Karlsson, P.E. *Totaldeposition av Kväve Till Skog*; Svensk Miljöinstitutet AB: Stockholm, Sweden, 2011.
27. Driessen, P.; Deckers, J.; Spaargaren, O.; Nachtergaele, F. Lecture notes on the major soils of the world. Diagnostic horizons, properties and materials. In *World Reference Base for Soil Resources (WRB)*; Driessen, P.M., Deckers, J., Spaargaren, O., Eds.; Food and Agriculture Organization: Rome, Italy, 2000; pp. 121–124.
28. Chesworth, W. Moisture regimes. In *Encyclopedia of Soil Science*; Chesworth, W., Ed.; Springer: Dordrecht, Netherlands, 2008; p. 485.
29. Påhlsson, L. Vegetationstyper I norden. In *Nordiska Ministerrådet Copenhagen*; Nordic Council of Ministers: Copenhagen, Denmark, 1995; p. 145–146.
30. Mauchly, J.W. Significance test for sphericity of a normal *n*-variate distribution. *Ann. Math. Stat.* **1940**, *11*, 204–209.
31. Nason, G.; Myrold, D.D. *Nitrogen Fertilizers: Fates and Environmental Effects in Forests*; Institute of Forest Resources Contrib. 73. College of Forest Resources, University of Washington: Seattle, WA, USA, 1992; pp. 67–81. Available online: http://www.cfr.washington.edu/research.smc/rfnrp /2FFC_Chap6.pdf (accessed on 7 April 2015).
32. Bauer, G.; Schulze, E.D.; Mund, M. Nutrient contents and concentrations in relation to growth of *Picea abies* and *Fagus sylvatica* along a European transect. *Tree Physiol.* **1997**, *17*, 777–786.
33. Iivonen, S.; Kaakinen, S.; Jolkkonen, A.; Vapaavuori, E.; Linder, S. Influence of long-term nutrient optimization on biomass, carbon, and nitrogen acquisition and allocation in Norway spruce. *Can. J. For. Res.* **2006**, *36*, 1563–1571.
34. Binkley, D.; Reid, P. Long-term increase of nitrogen availability from fertilization of douglas-fir. *Can. J. For. Res.* **1985**, *15*, 723–724.
35. Melin, J.; Nommik, H.; Lohm, U.; Flowerellis, J. Fertilizer nitrogen budget in a Scots pine ecosystem attained by using root-isolated plots and N^{15} tracer technique. *Plant Soil* **1983**, *74*, 249–263.
36. Lundmark-Thelin, A.; Johansson, M.B. Influence of mechanical site preparation on decomposition and nutrient dynamics of Norway spruce (*Picea abies* (L.) Karst) needle litter and slash needles. *For. Ecol. Manag.* **1997**, *96*, 101–110.
37. Vitousek, P.M.; Matson, P.A. Disturbance, nitrogen availability, and nitrogen losses in an intensively managed loblolly pine plantation. *Ecology* **1985**, *66*, 1360–1376.
38. Rosén, K.; Aronson, J.A.; Eriksson, H.M. Effects of clear-cutting on streamwater quality in forest catchments in central Sweden. *For. Ecol. Manag.* **1996**, *83*, 237–244.

39. Pettersson, F.; Högbom, L. Long-term growth effects following forest nitrogen fertilization in *Pinus sylvestris* and *Picea abies* stands in Sweden. *Scand. J. For. Res.* **2004**, *19*, 339–347.
40. Hogbom, L.; Nohrstedt, H.O.; Lundstrom, H.; Nordlund, S. Soil conditions and regeneration after clear felling of a *Pinus sylvestris* l. Stand in a nitrogen experiment, central Sweden. *Plant Soil* **2001**, *233*, 241–250.
41. Högbom, L.; Nohrstedt, H.Ö.; Lundström, H.; Nordlund, S. *Kvävegödsling Kan ge Varaktiga Effekter i Marken (Nitrogen Fertilization Can Have Long Term Effects on Soils)*; Skogforsk: Uppsala, Sweden, 2000; p. 4. (In Swedish)
42. Olsson, B.A.; Kellner, O. Long-term effects of nitrogen fertilization on ground vegetation in coniferous forests. *For. Ecol. Manag.* **2006**, *237*, 458–470.
43. Hägglund, B. *Forecasting Growth And Yield In Established Forests. An Outline And Analysis Of The Outcome Of A Subprogram Within The Hugin Project*; Swedish University of Agricultural Sciences: Umeå, Sweden, 1981; p. 145.
44. Hägglund, B. *Samband Mellan Ståndortsindex h100 Och Bonitet För Tall Och Gran i Sverige*; Swedish University of Agricultural Sciences: Umeå, Sweden, 1981; p. 90.

5

Importance of Arboreal Cyanolichen Abundance to Nitrogen Cycling in Sub-Boreal Spruce and Fir Forests of Central British Columbia, Canada

Ania Kobylinski [1] and Arthur L. Fredeen [2,*]

[1] Natural Resources and Environmental Studies (NRES) Graduate Program, University of Northern British Columbia (UNBC), 3333 University Way, Prince George, BC V2N 4Z9, Canada; E-Mail: aniakobylinski@gmail.com

[2] NRES Institute; Ecosystem Science and Management Program, UNBC, 3333 University Way, Prince George, BC V2N 4Z9, Canada

* Author to whom correspondence should be addressed; E-Mail: art.fredeen@unbc.ca;

Academic Editor: Heinz Rennenberg

Abstract: The importance of N_2-fixing arboreal cyanolichens to the nitrogen (N)-balance of sub-boreal interior hybrid spruce (*Picea glauca* × *engelmannii*) and subalpine fir (*Abies lasiocarpa*) forests was examined at field sites in central BC, Canada. Host trees were accessed by a single-rope climbing technique and foliage as well as arboreal macrolichen functional groups were sampled by branch height in eight random sample trees from each of two high (High Cyano) and two low (Low Cyano) cyanolichen abundance sites for a total of 32 sample trees. Natural abundances of stable isotopes of N (^{15}N, ^{14}N) and carbon (^{13}C, ^{12}C) were determined for aggregate host tree and epiphytic lichen samples, as well as representative samples of upper organic and soil horizons (Ae and Bf) from beneath host trees. As expected, N_2-fixing cyanolichens had 2–6-fold greater N-contents than chlorolichens and a $\delta^{15}N$ close to atmospheric N_2, while foliage and chlorolichens were more depleted in ^{15}N. By contrast, soils at all trees and sites were ^{15}N-enriched (positive $\delta^{15}N$), with declining (not significant) $\delta^{15}N$ with increased tree-level cyanolichen abundance. Lichen functional groups and tree foliage fell into three distinct groups with respect to $\delta^{13}C$; the tripartite cyanolichen *Lobaria pulmonaria* (lightest), host-tree needles (intermediate), and bipartite cyanolichens, hair (*Alectoria* and *Bryoria* spp.) and chlorolichens (heaviest). Branch height of host trees was an effective predictor of needle

$\delta^{13}C$. Our results showed a modest positive correlation between host tree foliage N and cyanolichen abundance, supporting our initial hypothesis that higher cyanolichen abundances would elevate host tree foliar N. Further study is required to determine if high cyanolichen abundance enhances host tree and/or stand-level productivity in sub-boreal forests of central BC, Canada.

Keywords: sub-boreal forest nitrogen; $\delta^{15}N$; $\delta^{13}C$; arboreal lichens; cyanolichens; lichen epiphyte nitrogen; *Lobaria pulmonaria*

---

1. Introduction

Nitrogen (N) is commonly limiting to conifer forests in much of the Pacific Northwest, USA [1] as well as the central interior of BC, Canada [2]. Biological N_2-fixing species have been shown to contribute and mediate important inputs of N to these forests [3]. In central British Columbia, mature spruce and fir forests, particularly in the wetter ecological zones, can have well-developed and vertically stratified communities of lichen epiphyte, with N_2-fixing cyanolichens predominating (when present) in the lower canopy branches [4,5]. This stratification and interaction between upper and lower canopy zones can be further altered by vertical gradients in lichen epiphyte guild to create significant variation in N content [4] and $\delta^{15}N$ [6] across a vertical profile of a host tree. Although the presence and guild of epiphytic lichens have been shown to enhance certain aspects of N-cycling, empirical evidence for their forest level significance has been equivocal. For example, the removal of lichens from an oak forest system had no effect on tree growth [7]. Thus, while atmospheric N contributions from cyanolichens are observed in sub-boreal spruce and fir ecosystems [8], it is difficult to predict the nature of combined cyanolichen and host conifer tree ecological interactions (*i.e.*, mutualistic to competitive) over tree to forest spatial and as well as temporal scales.

Stable isotopes have been used experimentally to infer many properties about elemental cycles of important nutrients (most notably C and N) at a variety of temporal and spatial scales and ecological contexts. The ratio of the heavier less common stable isotope to that of the lighter abundant isotope (e.g., ^{15}N:^{14}N and ^{13}C:^{12}C) can be expressed as atom %, but are typically represented in delta (δ) notation, which is not the absolute isotope ratio, but the difference between the sample measurement and an internationally accepted reference standard [9] in parts per thousand or per mil (‰) [10]. In theory, natural abundances of isotopes can be used to make inferences about the contributions of N_2-fixing species to forest trees by exploiting the naturally occurring differences in ^{15}N:^{14}N ratios between plant-available mineral N sources in the soil and that of atmospheric N_2 utilized by N_2-fixing species [11]. Non-N_2-fixing plants receive their entire N supply from soil N pools and can be expected to be isotopically heavier or lighter than N_2-fixing plants and lichens in concordance with the soil $\delta^{15}N$. In either case, natural abundances of stable N isotopes have frequently been used to estimate direct or indirect contributions of N_2-fixation to plant N content [12].

Most terrestrial materials have $\delta^{15}N$ compositions ranging between −20‰ and +30‰ [13], but lichens have a more restricted range of ^{15}N compositions (−21.5‰ [14]) to (+18‰ [15]), and plants even moreso (e.g., −5‰ to +2.9‰ [16]). The natural abundances and ratios of stable isotope compositions are potentially useful to the researchers quantifying these processes in that they can shift due to isotopic fractionation.

Isotopic fractionation can occur from a variety of processes. For example, N_2-fixation in organisms such as epiphytic cyanolichens does not discriminate between the ^{15}N and ^{14}N and therefore would normally represent N isotope ratios close to the atmospheric $\delta^{15}N$ standard of 0‰. By contrast, non-N_2-fixing elements and processes of forest systems such as ectomycorrhizae (ECM), mineralization of organic nitrogen, nitrification, microbial assimilation of inorganic N, and denitrification can fractionate stable isotopes of N belowground. Nutrient uptake in temperate and boreal trees is predominantly dependent on ECM hyphae growing into the soil from the mycorrhizal root tips [17]. Högberg *et al.* [18] found that ECM roots of Norway spruce (*Picea abies*) and beech (*Fagus sylvatica*) were 2‰ enriched in ^{15}N relative to non-mycorrhizal roots. The study also found ECM fungi were enriched in ^{15}N compared to their host plants, further suggesting that ^{15}N ECM discrimination was fungal in origin. ECM fungi were also more enriched in ^{13}C relative to total soil C [19], demonstrating ^{13}C ECM discrimination. Biologically mediated reactions that control elemental dynamics in soils can result in ^{15}N and ^{13}C enrichment or depletion, making inferences on forest ecosystem N- and C-cycling based on natural abundances of stable isotopes challenging.

Bulk soil $\delta^{15}N$ values are generally higher than atmospheric N (positive, ~5–10‰), a result of faster losses of the lighter isotope in soil N during decomposition [20], usually with increasing (more positive) deltas with increasing depth of organic ('L', 'F', and 'H') and mineral horizons [21]. Thus, fractionation of isotopes during litter decomposition in forests causes surface soils to have lower $\delta^{15}N$ values than deeper soil horizons [22]. Even slight fractionations occurring over decades of transfers of N from mineral soil to forest biomass can be sufficient enough to increase $\delta^{15}N$ of soil organic matter by ~6–8‰ [23]. Surface soils located beneath trees have also been found to have lower $\delta^{15}N$ values than those in open areas as a result of litter deposition [24], making decomposing epiphytic lichens an important factor influencing soil N isotope composition.

Nitrogen-fixing plants in western North American forest systems typically have more positive $\delta^{15}N$ values than non-N_2-fixing plants, and/or are closer to the of atmospheric N_2 [25,26]. The differences between these plant $\delta^{15}N$ values provide the basis of the ^{15}N natural abundance technique for estimating fixed-N contributions to terrestrial ecosystems [24]. However, attempting to trace fixed-N through ecosystems can be complicated by a myriad of fractionations, which are caused by numerous and often serial pathways of mineralization, nitrification, immobilization, and denitrification within the soil, plant root and mycorrhizal fungus mediation of soil N uptake [27].

Nutrient cycling processes in the soil are known to vary with plant community composition [28]. The processes associated with litter decomposition and enzymatic transformations of organic substrates in particular can vary because of differences in a variety of factors such as chemistry of the litter material, soil biota and soil chemistry associated with different plant species [29]. Both litter accumulation and stem flow can deliver N to epiphytes from various N pools [6], making the abundance and branch position of lichens on a tree significant. Lichens may play an important role in nutrient cycling in forest ecosystems, but the relative impact of lichens compared to other ecosystem components is not well understood. The effect of lichens on forest N-cycling could range from large in ecosystems where epiphytic lichens, especially cyanolichens are abundant, to insignificant where lichens are only present in low amounts [30].

In this study, we examine the importance of the vertical distributions of arboreal lichen biomasses [5] to the N-status of wet sub-boreal spruce and fir forest ecosystems of the central interior of BC, Canada. To do this, we measured the ^{15}N:^{14}N, %N of host tree (*i.e.*, conifer) foliage, host tree epiphytic lichen functional groups and organic and mineral soil horizons under the tree crowns at sites containing variable amounts of cyanolichen tripartite (*i.e.*, *L. pulmonaria)* and bipartite lichens from relatively low to high levels. We hypothesized that sub-boreal forest trees with high cyanolichen abundance should have higher needle %N reflecting the greater inputs of biologically fixed-N into these systems [8,31]. Knowledge obtained from this study should provide valuable information on the functional importance of epiphytic (N_2-fixing) cyanolichens to the N-status of wet sub-boreal conifer ecosystems of central BC.

2. Experimental Section

2.1. Study Area

The study area was described previously by Kobylinski and Fredeen [5]. In brief, this field study was carried out on the north side of the Fraser River near the town of Upper Fraser, BC, located approximately 70 km NE of Prince George, BC. All study sites were characterized by having relatively cool and moist summers and cold, snowy winters. Four mature sub-boreal forest sites were chosen based on having trees with predominantly high (High Cyano) or low (Low Cyano) epiphytic cyanolichen abundance and diversity, denoted as 'Upper Fraser' and 'Herrick' in previous work [5,32], respectively.

The two dominant tree species representing the vast majority of the canopy trees at all four sites were interior hybrid spruce (*Picea engelmannii* Parry ex Engelm. x *glauca* (Moench)) and subalpine fir (*Abies lasiocarpa* (Hook.) Nutt.) High Cyano sites were in "sub-boreal spruce" (SBS) 'wk1' and the 'vk' subzones (BC Biogeoclimatic Ecosystem Classification scheme) with a mean elevation of 680 m above sea level (a.s.l.), mean summer temperature of 11.8 ± 5.3 °C and relative humidity of 78% [8]. Low Cyano sites had lower average cyanolichen abundances and diversities and were located within 10–20 km of the High Cyano sites at a slightly higher mean elevation of 850 m a.s.l. and similar mean summer temperature and relative humidity of 10.8 ± 5.3 °C and 78%, respectively [8]. Soils at all sites were Orthic Humo-Ferric Podzols formed from sandy-colluvial materials at the High Cyano sites and from sandy–skeletal glaciofluvial materials at the Low Cyano sites. Average precipitation in the ecotonal study area is approximately 897 mm per year [8].

2.2. Canopy Access and Biomass Sampling

Sample tree selection, canopy access and biomass sampling procedures were previously described in Kobylinski and Fredeen [5]. In brief, trees were selected randomly from High Cyano and Low Cyano sites, with the exception of the fact that trees adjacent to Sitka alder (*Alnus viridis*) were excluded from the study given that annual N-inputs from Sitka alder can be substantial [26] and could have confounded ^{15}N natural abundance interpretation. All study trees were canopy trees and in excess of 22 m in height and 20 cm in DBH. Needles and lichens were sampled from each canopy height zone at the highest accessible point of each of the 32 study trees. Access into canopies was achieved through a single rope technique [33,34]. Selected trees were rigged, climbed and assessed vertically for epiphytic lichen biomass.

Lichen and needle sampling was described previously [5]. Briefly, epiphytic lichen functional group or species and host tree needle samples were collected over two summers: June to August 2008 and May to September 2009 at various heights within sample tree canopies. All cyanolichens in these mature forest tree sites were almost exclusively arboreal and on branches, with negligible cyanolichen occurring on the forest floor [35]. Bryophytes were essentially absent from all canopies, as is normal for these forest types [35]. Lichen biomasses were separated into five primary categories or functional groups based on biomass dominance by single species or functional group properties within different vertical canopy zones: *Alectoria sarmentosa* (Ach.) Ach., *Bryoria* Brodo & D. Hawksw. species, foliose chlorolichens, bipartite cyanolichens, and *Lobaria pulmonaria* (L.) Hoffm., the only tripartite cyanolichen at the site. The needle cohort was taken exclusively from the previous year's foliage to keep the age of the sampled needles constant. A processed and homogeneous sample for each functional group or needle was sent for isotopic analysis from the upper, middle and lower canopy height zones provided they were present in that zone.

2.3. Composite Soil Sampling

Soil samples were collected from the base of each sample tree between June and August 2009. We extracted soil cores using a soil auger (7.5 cm in diameter) by rotating the auger while applying downward force and lifting out the full blades. Soil core samples were extracted from a 1-m radius around each sample tree base in each of the cardinal directions. Samples were separated into three distinct layers: F-folic layer composed of organic matter rich in mycelia (~5 cm ± 2 cm deep), Ae-grayish surface soil layer (~2–10 cm thick); and Bf—yellowish brown to reddish brown subsoil layer (~10–20 cm). All layers were distinct and could be easily separated from each other; the soil auger was cleaned between cores. The three layers of the four subsamples were air dried for two days at room temperature (~22–25 °C), roots were removed and soil sieved through a 2-mm sieve. The F-layer samples were separated twice more (0.85 mm and 0.3 mm sieves) to remove small pieces of woody debris before composite samples were prepared. Composite samples were also made from the four Ae and Bf subsamples from each site. A dry weight of 2 g from each of the three layers of the four samples was mixed into one composite sample per soil layer (8 g dry weight), for a total of 96 soil samples.

2.4. Stable Isotope Analyses

Sample preparation of lichen, needle and soil samples took place at the University of Northern British Columbia (UNBC) before samples were shipped to the Stable Isotope lab at the University of Saskatchewan (U of S) in Saskatoon, Canada for analysis. Lichen and needle samples were oven dried for three days at 55 °C in paper bags. Samples were ball-milled (Retsch MM301, Hann, Germany) to a particle size of less than 250 μm and any fibrous matter or visible granules removed to improve precision of isotopic analysis. Ball-mill chambers were cleaned with deionized water and dried between samples. Samples (2 ± 0.2 mg) were stored in scintillation vials and encapsulated in 8 × 5-mm tin capsules (Catalog # D1008, Elemental Microanalysis Limited, Okehampton, UK) before combustion.

Soil, needle and lichen samples were analyzed using an elemental analyzer (Costech ECS4010, Valencia, CA, USA) coupled to a Delta V Advantage isotope ratio mass spectrometer with continuous flow (Conflo IV, Thermoscientific, Waltham, MA, USA) interface. The continuous flow gas isotope ratio mass spectrometer was used to measure % N, % C and the relative abundance of stable isotopes (^{15}N, ^{14}N, ^{13}C, and ^{12}C) in needle, lichen, and soil samples. Isotope ratio calibrations were performed with IAEA-N1, IAEA-N2, IAEA-NO-3, IAEA-CH6, and USGS-24 standards (International Atomic Energy Agency, 1995).

2.5. Natural Abundance Method

The δ values for ^{15}N and ^{13}C were calculated using equations 1 and 2, respectively.

$$(\delta\ ^{15}N\ (‰\ \text{relative to atmospheric}\ N_2) = (R_{sample}/R_{standard} - 1) \times 1000) \quad (1)$$

$$(\delta\ ^{13}C\ [‰\ \text{relative to V-PDB}] = (R_{sample}/R_{standard} - 1) \times 1000) \quad (2)$$

The value *R* is the ratio of heavy isotope (^{15}N and ^{13}C) to their lighter more abundant isotope.

2.6. Statistical Analyses

Data were analyzed using Sigma Plot 11 (Systat Software Inc., San Jose, CA, USA). Ordinary least square (OLS) multiple linear regression analyses were performed to address whether there were statistically significant differences in: (1) $\delta^{15}N$ and $\delta^{13}C$ between lichen functional groups and needles controlling for tree branch height, tree species, and lichen abundance; (2) $\delta^{15}N$ and $\delta^{13}C$ by tree branch height controlling for lichen functional groups and needles; (3) $\delta^{15}N$ and $\delta^{13}C$ between High and Low Cyano sites controlling for lichen functional groups, needles and soil; (4) $\delta^{15}N$ and $\delta^{13}C$ between tree species (spruce and fir) controlling for the lichen functional groups, needles and soil; and (5) foliar (needle) %N at sites with higher abundances of cyanolichen. The unstandardized coefficient (B) for OLS linear regression analyses provide an estimate of the change in the dependent variable associated with a one-unit change in the identified independent variable controlling for all other independent variables.

Six OLS regression models were created to predict: (i) $\delta^{15}N$ in lichen functional groups; (ii) $\delta^{15}N$ in needles; (iii) $\delta^{13}C$ in lichen functional groups; iv) $\delta^{13}C$ in needles; (v) $\delta^{15}N$ in soil; and (vi) $\delta^{13}C$ in soil. Correlation analyses were conducted between all the variables in each model to check for potential collinearity. Additional correlations were also conducted to look at the relationships between needle $\delta^{15}N$ and cyanolichen abundance, % N and cyanolichen abundance, and soil $\delta^{15}N$ and cyanolichen abundance.

Kolmogorov-Smirnov and Shapiro-Wilk tests were used to determine normality of data. Both tests showed lichen and needle $\delta^{15}N$ (KS = 0.191, $p < 0.001$ and W = 0.886, $p < 0.001$) and $\delta^{13}C$ (KS = 0.122, $p < 0.001$ and W = 0.949, $p < 0.001$) were not normally distributed. However, because of bimodal distributions of the dependent variable, data transformations were not used. The distribution of lichen and needle $\delta^{15}N$ and $\delta^{13}C$ (but not soil) showed some signs of heteroscedasticity and non-linearity across all independent variables.

3. Results

Nitrogen contents varied by over six-fold among lichen epiphytes sampled. At the extremes, *Alectoria sarmentosa* had the lowest % N (0.50 ± 0.13) and bipartite cyanolichens had the greatest %N (3.24 ± 0.52) on a dry weight basis (Figure 1). *Bryoria* spp. and foliose chlorolichens had very similar %N contents of 0.77 ± 0.18 and 0.73 ± 0.19, respectively. Needles had a higher %N content (1.11 ± 0.14) than hair lichens and foliose chlorolichens but not as high as *L. pulmonaria* (2.26 ± 0.30). The differences in mean %N between functional groups were significant ($F_{5,404} = 950.3$, $p < 0.001$: Tukey test) except for *Bryoria* spp. and foliose chlorolichens (Figure 1).

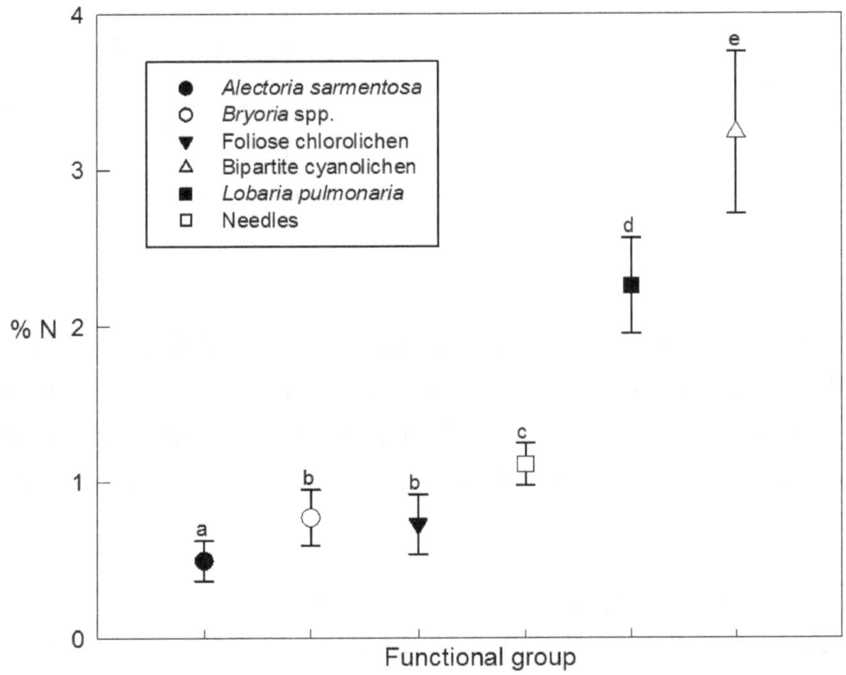

Figure 1. Mean %N (± SD) of lichen biomass in functional groups of lichen and foliage (needle) from 32 spruce and fir study trees with varying amounts of tree-level cyanolichen abundance. Different letters above error bars indicate significant ($p < 0.001$) differences between lichen functional groups (Tukey multiple mean comparison test).

All lichen functional groups or species and conifer foliage (needle) had mostly unique $\delta^{15}N$ signatures relative to their %N (Figure 2). *Bryoria* spp. hair lichens at the top of the canopy had the lowest $\delta^{15}N$, while the hair lichen *Alectoria sarmentosa* found lower in the canopy had slightly lower %N and higher $\delta^{15}N$ than *Bryoria.* Host tree needles and foliose chlorolichens had $\delta^{15}N$ values that were intermediate between hair lichens and cyanolichens but similar %N. While *L. pulmonaria* and bipartite cyanolichens both had $\delta^{15}N$ values close to atmospheric N_2 ($\delta^{15}N = 0$), *L. pulmonaria* had 1.5-times less N on average (Figure 2). Mean $\delta^{15}N$ isotope values were slightly higher at Low Cyano sites for all lichen functional groups and host tree needles, but there was no significant difference in the $\delta^{15}N$ values or %N of lichens or needles on trees from High Cyano (black symbols) *versus* Low Cyano (white symbols) sites (Figure 2).

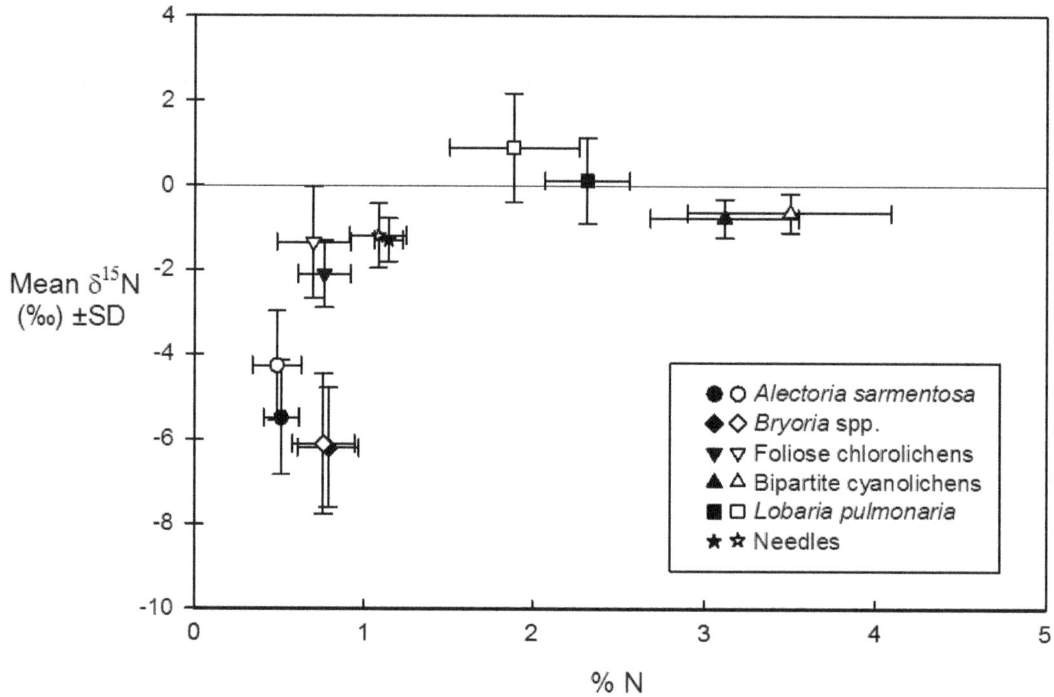

Figure 2. Mean (± SD) % N and δ^{15}N signatures of five epiphytic lichen and host tree foliage (needle) biomasses across four sites and two host tree species from sub-boreal spruce and fir forests in central interior BC. Half of the host trees were from high (High Cyano: black symbols) and the other half from low (Low Cyano: white symbols) cyanolichen abundance sites.

When we compared δ^{15}N of living conifer foliage (needle) on each of the 32 study trees to their actual cyanolichen abundance, there was a small, but ultimately not significant ($r = -0.16$, $p = 0.121$) negative correlation between mean needle δ^{15}N and increasing cyanolichen abundance (Figure 3a). Host tree foliage (needles) %N showed a significant but weak positive correlation to actual tree-level cyanolichen abundance ($r = 0.24$, $p = 0.019$; Figure 3b).

When lichen species, tree branch height, and tree species were controlled for, cyanolichen abundance in the canopy was a statistically significant but ineffective predictor of δ^{15}N (B < 0.001, $p < 0.001$, Table 1). The regression coefficients of the dependent variable δ^{15}N suggest that lichen species and cyanolichen abundance, but not tree species or tree branch height, were significant predictors of δ^{15}N in lichen functional groups. The regression analyses also show that *A. sarmentosa* (B = −3.657, $p < 0.001$), *Bryoria* spp. (B = −4.838, $p < 0.001$) and foliose chlorolichen (B = −0.482, $p = 0.002$) tree-level abundances were all negatively related to their δ^{15}N signatures. Conversely, bipartite cyanolichens (B = 0.561, $p < 0.001$) and *L. pulmonaria* (B = 1.608, $p < 0.001$) tree-level abundances were positively related to their δ^{15}N signatures. Both tree species (B = −0.166, $p = 0.095$) and tree branch height (B = 0.010, p =0.238) were not significantly related to δ^{15}N levels in lichen functional groups (data not shown) and insignificant or with a very small coefficient for δ^{15}N in needles (B = 0.028, $p = 0.002$). Similarly, tree species (B = −0.146, $p = 0.267$) and cyanolichen abundance (data not shown) were not significant predictors of δ^{15}N in lichen or host-tree foliage.

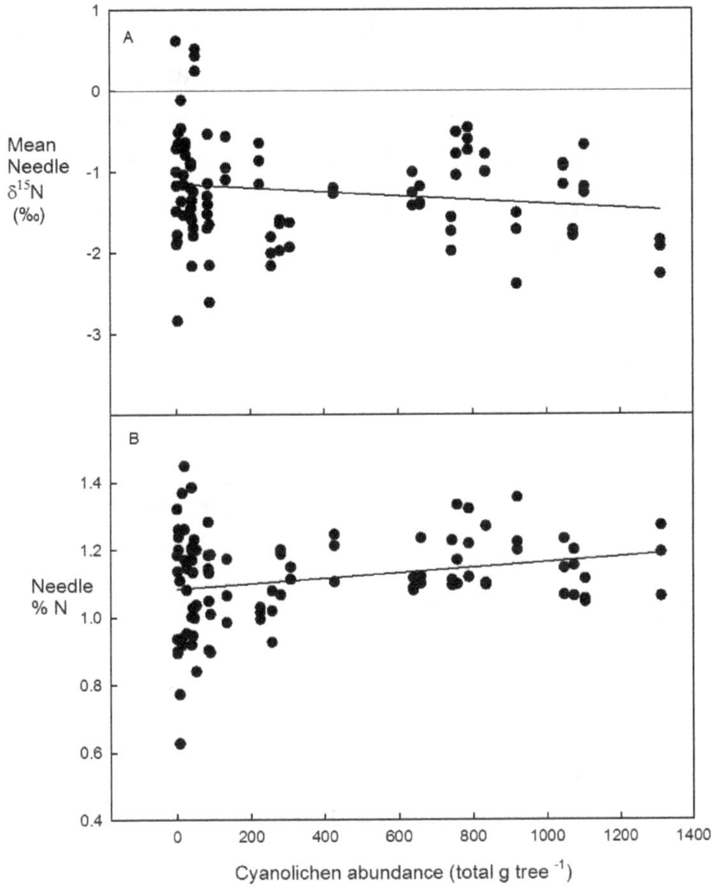

Figure 3. Mean **(a)** δ^{15}N and **(b)** %N of host tree foliage (needle) from low to high tree-level cyanolichen abundance (total g tree^{-1}).

Table 1. Linear regression (OLS) coefficients with standard error for the regression model predicting δ^{15}N of lichen functional groups (F = 220.501, $p < 0.001$, $R^2 = 0.817$) and host tree needles (F = 4.363, $p = 0.006$, $R^2 = 0.126$) based on tree species (spruce and fir), cyanolichen abundance, tree branch height, and lichen species or functional group.

	Lichen Functional Groups (Model i)		**Needles (Model ii)**	
	Unstandardized coefficient (B)	Std Error	Unstandardized coefficient (B)	Std Error
Intercept	−1.126 ***	0.165	−1.499 ***	0.171
Tree species (Spruce)	−0.166	0.099	−0.146	0.131
Tree branch height	0.010	0.008	0.028 **	0.009
Alectoria sarmentosa	−3.657 ***	0.168		
Bryoria spp.	−4.838 ***	0.183		
Foliose chlorolichen	−0.482 **	0.156		
Bipartite cyanolichen	0.561 ***	0.137		
Lobaria pulmonaria	1.608 ***	0.165		

*** $p < 0.001$; ** $p < 0.01$; * $p < 0.05$.

Soil horizon (F, Ae and Bf) δ^{15}N values were consistently higher at Low Cyano (means: F = 1.340, Ae = 5.130, Bf = 6.262) than at High Cyano (F = 0.782, Ae = 4.765, Bf = 5.383) sites, and decreased with increasing cyanolichen abundance in all three horizons (Figure 4A–4C), respectively. The abundances of cyanolichens and $\delta^{15}N$ were negatively correlated in all three soil horizons, with correlations being nearly significant for the uppermost horizons (F: $r = -0.341$, $p = 0.056$; Ae: $r = -0.263$, $p = 0.146$), and significant for the Bf horizon ($r = -0.356$, $p = 0.045$). Total %N in the F-horizon decreased with increasing tree-level cyanolichen abundance (Figure 4D), but was relatively unaffected in Ae and Bf horizons (Figure 4E and 4F).

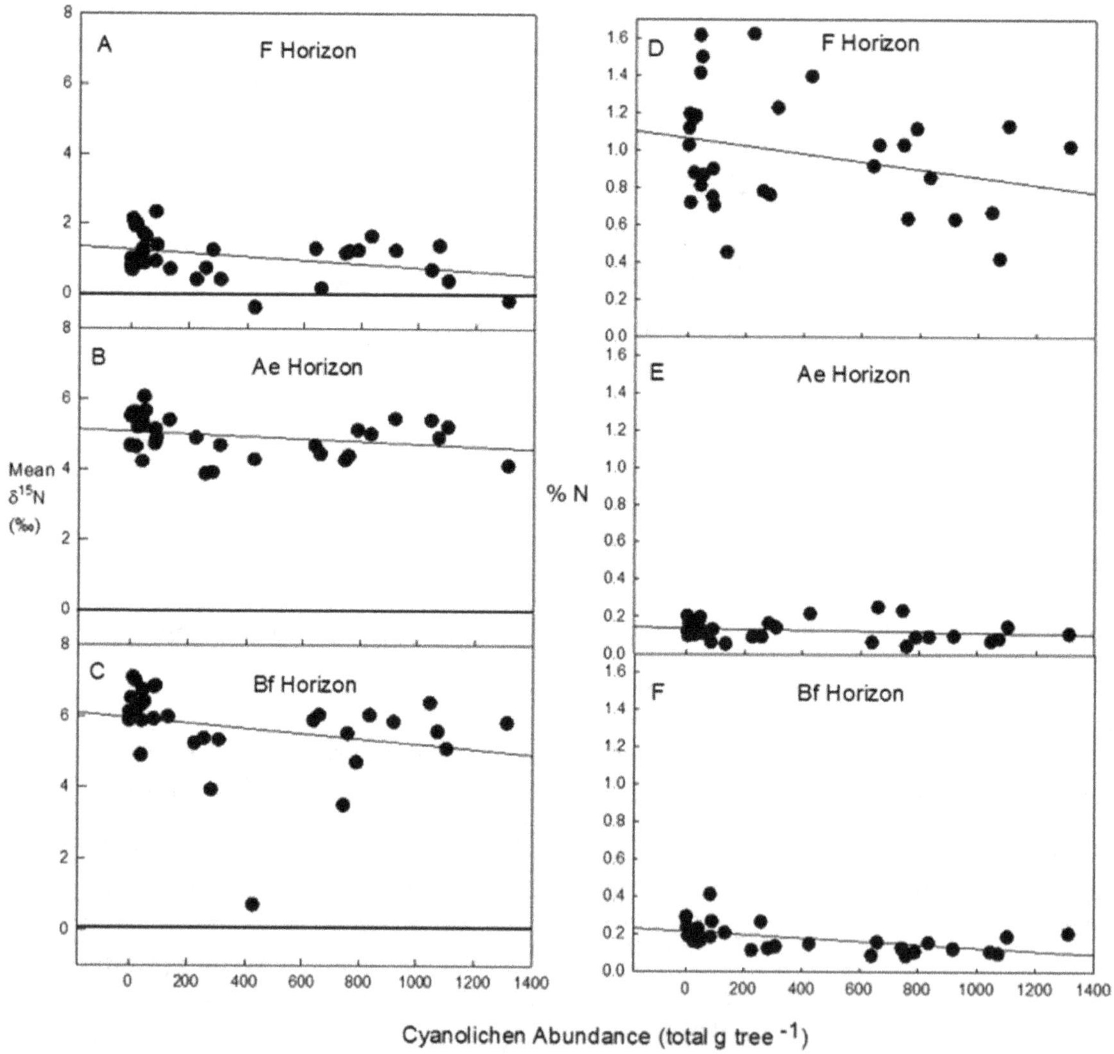

Figure 4. Mean $\delta^{15}N$ and %N of (A and B) the surface organic F horizons, (B and E) upper-most mineral (Ae) horizons and (C and F) lower mineral (Bf) horizons, respectively, beneath host trees with varying levels of cyanolichen abundance (total g tree^{-1}).

Tree species and both upper mineral soil horizons were all significant predictors and all contributed to the overall relationship of both soil $\delta^{15}N$ and $\delta^{13}C$ (Table 2). Cyanolichen abundance was a significant predictor for both soil $\delta^{15}N$ and $\delta^{13}C$, but coefficients were <0.01 and therefore not relevant (data not shown). There was also a statistically significant difference in soil $\delta^{15}N$ (B = −0.525, $p < 0.001$) and soil $\delta^{13}C$ (B = −0.221, $p = 0.009$) between host tree species (spruce and fir).

Table 2. Linear regression (OLS) coefficients with standard error for the regression model predicting $\delta^{15}N$ (F = 314.044, $p < 0.001$, $R^2 = 0.932$) and $\delta^{13}C$ (F = 117.897, $p < 0.001$, $R^2 = 0.838$) of soil based on tree species, and Ae and Bf soil horizons.

	$\delta^{15}N$		$\delta^{13}C$	
	Unstandardized coefficient (B)	Std. Error	Unstandardized coefficient (B)	Std. Error
Intercept	1.586 ***	0.140	−26.923 ***	0.094
Tree species	−0.525 ***	0.123	−0.221 **	0.062
Mineral surface soil (Ae)	3.887 ***	0.145	1.523 ***	0.097
Mineral subsoil (Bf)	4.777 ***	0.145	1.961 ***	0.097

*** $p < 0.001$; ** $p < 0.01$; * $p < 0.05$.

When controlling for all independent variables, tree species and branch height were significant predictors of $\delta^{13}C$ of host-tree foliage (needles) and to lesser extents for lichen epiphytes (Table 3 and Figure 5A). Trends of increasing (less negative) $\delta^{13}C$ values with increasing branch height were evident for both *L. pulmonaria* and tree foliage (Figure 5A). There was evident clumping of $\delta^{13}C$ into three distinct groups (Figures 5A and 6). Specifically, $\delta^{13}C$ values of the lone tripartite cyanolichen (*L. pulmonaria*) were lower than host tree needles, which were in turn lower than hair lichens (*A. sarmentosa* and *Bryoria* spp.), bipartite cyanolichens and foliose chlorolichens.

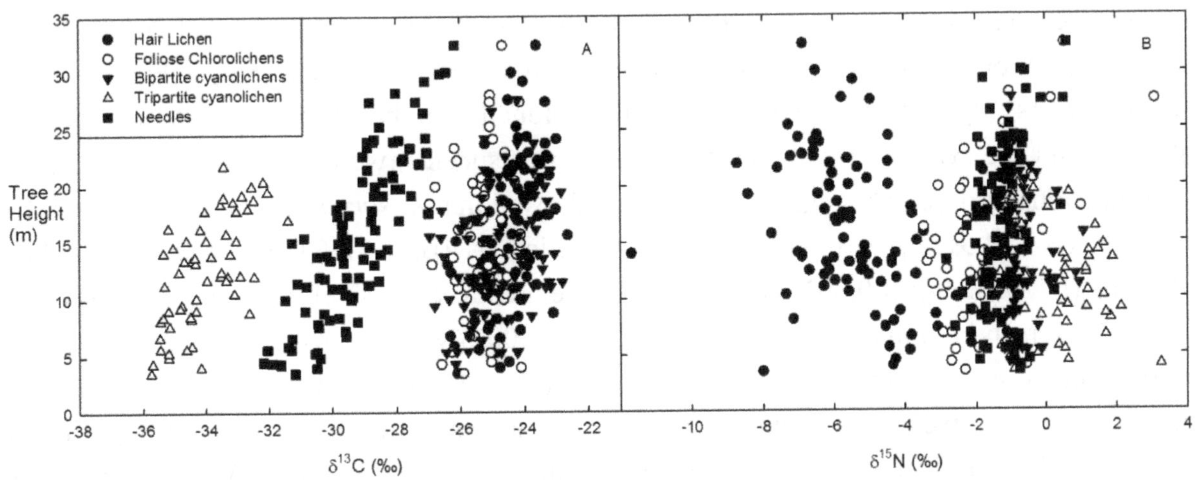

Figure 5. Variation in the natural abundances of (**A**) $\delta^{13}C$ and (**B**) $\delta^{15}N$ of five epiphytic lichen functional group and host tree foliage (needle) biomasses according to tree branch height for all sites.

Table 3. Linear regression (OLS) coefficients with standard error for the regression model predicting δ^{13}C in lichen functional groups ($F = 711.517$, $p < 0.001$, $R^2 = 0.942$) and needles ($F = 79.545$, $p < 0.001$, $R^2 = 0.724$) based on tree species (spruce), tree branch height and lichen functional groups.

	Lichen Functional Groups (Model iii)		**Needles (Model iv)**	
	Unstandardized coefficient (B)	Std. Error	Unstandardized coefficient (B)	Std. Error
Intercept	−30.523 ***	0.154	−31.868 ***	0.204
Tree species (Spruce)	0.048	0.089	0.622 ***	0.156
Tree branch height	0.093 ***	0.008	0.148 ***	0.011
Alectoria sarmentosa	4.899 ***	0.151		
Bryoria spp.	4.513 ***	0.164		
Foliose chlorolichen	3.969 ***	0.140		
Bipartite cyanolichen	4.324 ***	0.166		
Lobaria pulmonaria	−4.358 ***	0.149		

*** $p < 0.001$; ** $p < 0.01$; * $p < 0.05$.

Hair lichens *A. sarmentosa* and *Bryoria* spp. had uniformly negative δ^{15}N values that were also more negative than all other functional groups (Figure 5B), and trending lower with increase in branch height. The δ^{15}N of other lichen functional groups and host-tree foliage did not appear to be influenced by branch height.

Host tree needles had δ^{13}C values intermediate to *L. pulmonaria* and all other lichen functional groups (Figure 6), but at distinctly higher C concentrations than all lichen groups (Figure 7). Lichen functional groups, cyanolichen abundance, tree branch height, and the interaction between cyanolichens and site cyanolichen abundance (but not tree species) were significant predictors of δ^{13}C in lichen functional groups. The regression analyses showed that *A. sarmentosa* ($B = 4.899$, $p < 0.001$), *Bryoria* spp. ($B = 4.513$, $p < 0.001$), foliose chlorolichens ($B = 3.969$, $p < 0.001$), and bipartite cyanolichens ($B = 4.324$, $p < 0.001$) were positively related to δ^{13}C levels in lichen functional groups (Table 2). Conversely, *L. pulmonaria* ($B = -4.358$, $p < 0.001$) was negatively related to δ^{13}C levels in lichen functional groups. Both cyanolichen abundance ($B = -0.001$, $p < 0.001$) and tree branch height ($B = 0.093$, $p < 0.001$) were significantly related to δ^{13}C levels in lichen functional groups. There was also a statistically significant interaction between cyanolichens and site cyanolichen abundance and δ^{13}C increased slightly as site-level cyanolichen abundance increased ($B = 0.001$, $p = 0.005$, Table 2). There was no significant difference in δ^{13}C between sample tree species spruce and fir in lichen functional groups ($B = 0.048$, $p = 0.594$, Table 2). Alternatively, both tree species ($B = 0.622$, $p < 0.001$) and tree branch height ($B = 0.148$, $p < 0.001$), but not cyanolichen abundance ($B < 0.001$, $p = 0.942$), were significant predictors of δ^{13}C in needles ($B = 0.028$, $p = 0.002$).

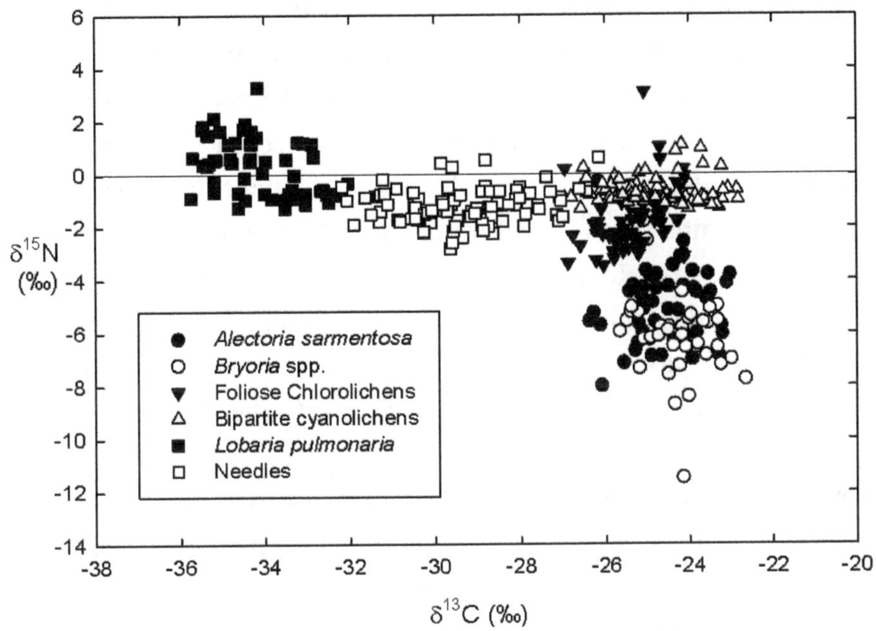

Figure 6. Dual natural abundance isotope plot of $\delta^{15}N$ *versus* $\delta^{13}C$ of the five epiphytic lichen functional group and host tree foliage (needle) biomasses across four sites and two host tree species.

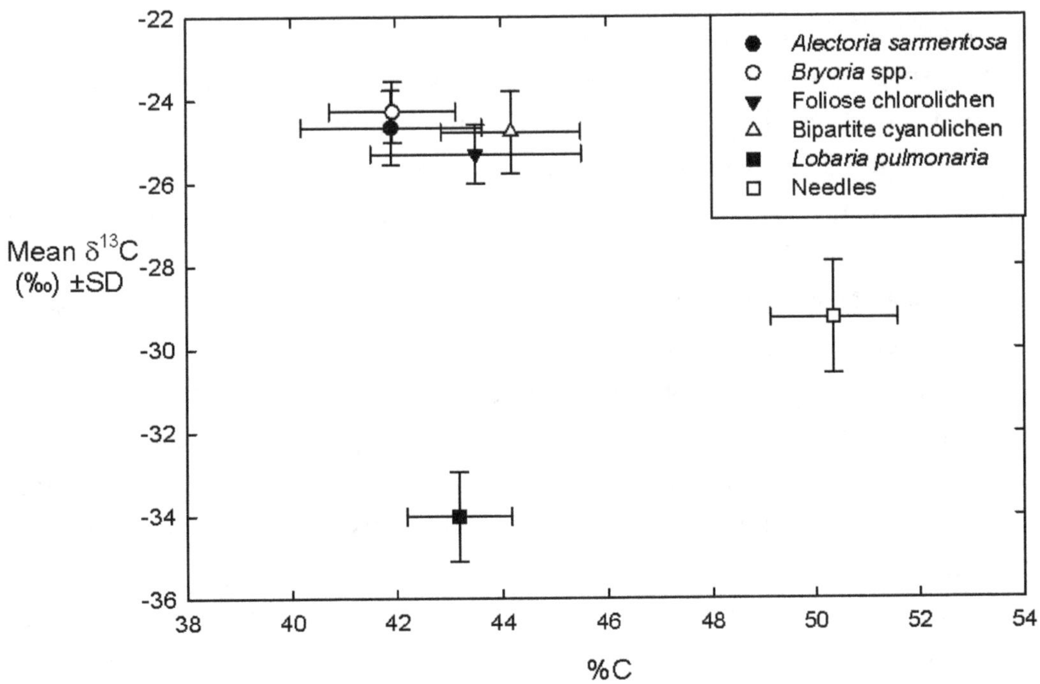

Figure 7. Mean (± SD) C concentrations *versus* $\delta^{13}C$ signals of five epiphytic lichen functional group and host tree foliage (needle) biomasses across four sites and two host trcc spccics.

4. Discussion

Nitrogen-fixing cyanolichen epiphytes can be both diverse and abundant in conifer forests of central BC, and have been conjectured to be an important component of the N-cycle in these forest locations [4,8,32,36]. Although lichen N pool sizes and decomposition rates in these studies were indicative of enhanced mineral N flux rates, as observed in other forests types [7], direct links between cyanolichen abundance and improved host tree N status have been elusive. In our study, gradients in cyanolichen abundance across host trees (interior hybrid spruce and subalpine fir) in two generally high cyanolichen abundance sites and two generally low cyanolichen abundance sites [8,32] provided us with an opportunity to examine the potential for enhanced cyanolichen N-inputs into these forests in central BC. We hypothesized that foliar (needle) N contents would be positively correlated with cyanolichen abundance, and in fact this was found to be the case, albeit the relationship was not strong (Figure 3b). We further sought to more directly link increased host tree foliar N with cyanolichen N using the natural abundances of ^{15}N and ^{14}N of forest tree, lichen and soil components of these systems.

Our results corroborate previous findings of $\delta^{15}N$ values in lichen functional groups and or species. It is already known that cyanolichens that fix N_2 typically have $\delta^{15}N$ values close to atmospheric N_2, *i.e.*, 0‰ [37,38]. We also found $\delta^{15}N$ values close to 0‰ for both bipartite and tripartite (*L. pulmonaria*) cyanolichens in our study. By contrast, other non-N_2-fixing arboreal macrolichens (hair and chlorolichen) as well as host-tree foliage all had $\delta^{15}N$ values that were negative and less enriched in ^{15}N than either cyanolichen functional group. Hair lichens had the lowest $\delta^{15}N$, while chlorolichens and host tree foliage were intermediate. Interestingly, *A. sarmentosa*, occurring lower in the canopy than *Bryoria* spp., and therefore overlapping with cyanolichen canopy zones [5], had $\delta^{15}N$ closer to atmospheric $\delta^{15}N$ (*i.e.*, 0) (Figure 2). This is consistent with *A. sarmentosa* receiving more leached-N from cyanolichens than *Bryoria* spp., and supported by the fact that an increase in the $\delta^{15}N$ of *A. sarmentosa* was observed in High Cyano sites, while relatively little change in $\delta^{15}N$ was observed for *Bryoria* spp. In both hair lichens, relatively low N contents and low $\delta^{15}N$ (~−6 ‰ in Low Cyano sites) were consistent with low atmospheric N inputs in central BC [39] and the negative $\delta^{15}N$ of precipitation N measured in other studies [31].

Although biological N_2-fixation may represent a significant N input into ecosystems, both via leaching (less well documented) and decomposition, it may be difficult to identify because of its small relative isotopic effect against the background of ecosystem component $\delta^{15}N$ signatures [12]. Nitrogen isotope signatures of epiphytes varied across functional groups and with canopy position, but the reason for lichen $\delta^{15}N$ variability with canopy position in most studies remains unexplained. The $\delta^{15}N$ of hair lichens, with single sources of N (*i.e.*, atmospheric fixed-N), are relatively easy to explain relative to other lichens (cyanolichen and chlorolichen) lower in the canopy where inputs and outputs are more complex. At least 10 processes have been identified that can alter $\delta^{15}N$ values, none of which can currently be separated out in field studies [9]. Key explanations for $\delta^{15}N$ variability in lichens are: (i) a preference for uptake of the lighter ^{14}N isotope which can then lead to $\delta^{15}N$ enrichment of outputs [40]; (ii) fractionation of N isotopes in gaseous phase ammonia is greater than in the liquid phase of nitrate; and (iii) transfer of organic N which can result in increased ^{15}N depletion of the photobiont and less depletion in the mycobiont [41]. In general, our results were in agreement with previous work showing that lichens with a green alga as their photobiont (all lichens except bipartite cyanolichens) showed greater relative ^{15}N depletion [42] (Figure 1).

Soil $\delta^{15}N$ values increased with soil depth in our study, consistent with results of Gebauer and Schulze [43]. In keeping with our expectations, we also observed decreasing trends in $\delta^{15}N$ with increasing cyanolichen abundance in all soil horizons (Figure 4A–4E) equating to less positive soil horizon $\delta^{15}N$ values, more proximal to the $\delta^{15}N$ of cyanolichen biomasses which were uniformly close to zero. There are many reasons why soil $\delta^{15}N$ wouldn't necessarily be closely correlated with tree $\delta^{15}N$. First, measured $\delta^{15}N$ of soil pools may not represent the true isotopic composition of N available to plant roots, since most of the N in soils is bound in forms that are not immediately available to plants [25,44]. Second, only a few % of total soil N becomes available in a year [18], and symbiotic fungi (mycorrhizae) can alter the $\delta^{15}N$ of the N transferred from soil to host plant. Nevertheless, a similar downward trend in tree foliage $\delta^{15}N$ at our sites with increasing abundance of cyanolichen (Figure 3) was consistent with the downward trend in soil $\delta^{15}N$ (Figure 4A–4C). Explaining the increase in %N of host tree foliage with increased cyanolichen abundance is difficult to reconcile with total soil N. If greater cyanolichen inputs of N increased soil N concentrations, then a straightforward mass action of greater plant N uptake could have explained enhanced foliar %N. However, soil N decreased in all soil horizons, though not significantly, with higher cyanolichen abundance (Figure 4D–4F). Since total soil N does not in any way equate with amounts and forms of available inorganic N, it is possible that more readily available N fractions were more available at high abundance cyanolichen sites, even though total %N was not. Further work on soil N at these sites would be required to test this hypothesis.

Foliar (needle) $\delta^{15}N$ values in our host trees ranged between 0.6 and −2.8‰ and did not differ significantly between High and Low Cyano sites or tree branch height. However, there was a trend of decreasing foliar $\delta^{15}N$ with increasing cyanolichen abundance (Figure 3a). Gebauer and Schulze [43] reported lower, but overlapping, $\delta^{15}N$ values for conifer needles (between −2.5 and −4.1‰), which varied according to stand and age, with foliage from the healthiest site having the lowest $\delta^{15}N$. It is unclear why the range of $\delta^{15}N$ values observed by Gebauer & Schulze were more negative than ours, but one possibility would be the presumed greater atmospheric N inputs in these European forests when compared to the N-limited forests of central BC. Gebauer and Schulze [43] also reported a similar trend of $\delta^{13}C$ values of needles, which ranged between −26.2 and −32.2‰ and did not differ between lichen abundance sites but did change with canopy height. Similar to our study, needle $\delta^{13}C$ values were not significantly different between abundance sites (Table 3), were more negative in the lower canopy and increasing with branch height (Figure 5).

Lichen $\delta^{13}C$ values were previously found to vary widely over a large range of habitats and species [45]. Our lichen functional groups and tree foliage (needles) fell into three distinct groups with respect to $\delta^{13}C$ values: the tripartite (and only cephalodic) cyanolichen *L. pulmonaria* with the lowest $\delta^{13}C$ values (−36 to −31), host tree foliage with intermediate $\delta^{13}C$ values (−33 to −26), and all other lichens including bipartite cyanolichens, hair lichens and chlorolichens with the least negative $\delta^{13}C$ values (−28 to −23). Our values are entirely consistent with those of Maguas *et al.* [46], namely that tripartite lichen associations typically have the lowest and bipartite lichens the highest $\delta^{13}C$. Carbon isotope discrimination in lichens [42] has been attributed to the species of photobiont present. Carbon concentrations also clearly distinguished between conifer foliage (mean %C of ~50.5%) and all lichen functional groups (means of %C ranging from ~42%–45%). This range of values are in close agreement with %C values observed previously for epiphytic lichens in sub-boreal BC [8] and conifer foliage [47], respectively.

Interestingly, we observed positive relationships between tree branch height and $\delta^{13}C$ values for both *L. pulmonaria* and foliage. Possible causes that may be responsible for the observed height-specific differences in these $\delta^{13}C$ values are CO_2 source, light level and factors influencing diffusion resistance such as water availability. Carbon dioxide source influences the $\delta^{13}C$ value of lichens [48,49], and lichens growing in the canopy close to the forest floor (*i.e.*, *L. pulmonaria*) would more readily assimilate CO_2 enriched by soil respiration [46,50,51]. Light levels also influence photosynthesis and can alter the CO_2 gradient inside the lichen thallus. Inorganic carbon acquisition by lichen photobionts fluctuate greatly with moisture and light availability [52], and lichen hydration is known to affect $\delta^{13}C$ values. Increased $\delta^{13}C$ values have been reported for lichens in drier habitats [45] and on thinner branches, both being directly and indirectly associated with desiccation stress, respectively [53]. Water stress would be expected to increase with branch height. Therefore, differences in $\delta^{13}C$ likely exist due to biological traits (e.g., foliage (stomatal) *versus* lichen, cephalodic *versus* non-cephalodic lichen), different microclimatic conditions in the canopy and/or the uptake of isotopically different CO_2 sources. Irrespective of mechanism and ecological significance, the $\delta^{13}C$—in concert with the $\delta^{15}N$—are clearly useful in chemically distinguishing among epiphyte functional group and host tree foliage.

5. Conclusions

Stable isotope natural abundance techniques have been important tools for ecophysiology and ecosystem research for many decades now [10,18], but have infrequently been applied to determining the contributions that arboreal cyanolichens make to forest N-cycling. Despite the narrow geographic and temporal scope of this study, the results from our study support previous work at our high and low cyanolichen abundance sites [8,32], suggesting that trees with a high abundance of arboreal cyanolichens provide greater N-inputs into N-limited conifer forests of central BC. High cyanolichen abundance, typically coupled with higher overall arboreal lichen diversity, is usually constrained to mature and old-growth stands in drier parts of the central interior of BC. However, the reasons for the variability in cyanolichen abundance at the tree and site level, and the contributions that cyanolichen abundances make to tree and forest growth remain uncertain and are the topics of current and future research at our forest research sites. Low-input, extensive forestry practiced in central British Columbia could benefit from this increased understanding.

Acknowledgements

The authors are grateful for Jocelyn Campbell's guidance and assistance with field site selection and project design. Kevin Jordan provided expert professional help and support in accessing canopies in year one of this study. Thanks are extended to Myles Stocki for stable isotope analyses and to the many graduate student field assistants for help in gathering samples in the canopy. This project was primarily funded by a Discovery Grant from the Natural Sciences and Engineering Research Council of Canada (NSERC 194405; ALF).

Author Contributions

Ania Kobylinski is the lead author and researcher. She climbed nearly half of the experimental trees, conducted the field sampling, prepared samples for analyses, and collated and analyzed data. This work comprised one chapter in her M.Sc. thesis. Art Fredeen conceived of the fundamental research questions, acquired the research funding, helped in decision-making regarding sampling, analyses and data processing, and helped in the editing and/or writing of the thesis chapter and manuscript.

References

1. Holub, S.M.; Lathja, K. The fate and retention of organic and inorganic ^{15}N-nitrogen in an old-growth forests soil in western Oregon. *Ecosystems* **2004**, *7*, 368–380.
2. Brockley, R.P.; Simpson D.G. Effects of Intensive Fertilization on the Foliar Nutrition and Growth of Young Lodgepole Pine and Spruce Forests in the Interior of British Columbia (E.P. 886.13): Establishment and Progress Report. Available online: https://www.for.gov.bc.ca/hfd/pubs/Docs/tr/tr018.pdf (accessed on 27 July 2015).
3. Coxson, D.S.; Nadkarni, N.M. Ecological Roles of Epiphytes in Nutrient Cycles of Forest Ecosystems. In *Forest Canopies*; Academic Press: San Diego, California, CA, USA, 1995.
4. Campbell, J.; Fredeen, A.L. Contrasting the abundance, nitrogen, and carbon of epiphytic macrolichen species between host trees and soil types in a sub-boreal forest. *Can. J. Bot.* **2007**, *85*, 31–42.
5. Kobylinski, A.; Fredeen, A.L. Vertical distribution and nitrogen content of epiphytic macrolichen functional groups in sub-boreal forests of central British Columbia. *For. Ecol. Manag.* **2014**, *329*, 118–128.
6. Bergstrom, D.M.; Tweedie, C.E. A conceptual model for integrative studies of epiphytes: Nitrogen utilisation, a case study. *Aust. J. Bot.* **1998**, *46*, 273–280.
7. Knops, J.M.H.; Nash, T.H.I.; Schlesinger, W.H. The influence of epiphytic lichens on the nutrient cycling of a blue oak woodland. *Ecol. Monogr.* **1997**, *66*, 75–82.
8. Campbell, J.; Fredeen, A.L.; Prescott, C.E. Decomposition and nutrient release from four epiphytic lichen litters in sub-boreal spruce forests. *Can. J. Bot.* **2010**, *40*, 1473–1484.
9. Robinson, D. δ^{15}N as an integrator of the nitrogen cycle. *Trends Ecol. Evol.* **2001**, *16*, 153–162.
10. Dawson, T.E.; Mambelli, S.; Plamboeck, A.H.; Templer, P.H.; Tu, K.P. Stable isotopes in plant ecology. *Annu. Rev. Ecol. Syst.* **2002**, *33*, 507–559.
11. Boddey, R.M.; Peoples, M.B.; Palmer, B.; Dart, P.J. Use of the ^{15}N natural abundance technique to quantify biological nitrogen fixation by woody perennials. *Nutr. Cycl. Agroecosyst.* **2000**, *57*, 235–270.
12. Vitousek, P.M.; Shearer, G.; Kohl, D.H. Foliar ^{15}N natural abundance in Hawaiian rainforest: Patterns and possible mechanisms. *Oecologia* **1989**, *78*, 383–388.
13. Junk, G.; Svec, H.J. The absolute abundance of the nitrogen isotopes in the atmosphere and compressed gas from various sources. *Geochim. Cosmochim. Acta* **1958**, *14*, 234–243.
14. Fogel, M.L.; Wooller, M.J.; Cheeseman, J.; Smallwood, B.J.; Roberts, Q.; Romero, I.;

Meyers, M.J. Unusually negative nitrogen isotopic compositions ($\delta^{15}N$) of mangroves and lichens in an oligotrophic, microbially-influenced ecosystem. *Biogeosciences* **2008**, *5*, 1693–1704.

15. Huiskes, A.H.L.; Boschker, H.T.S.; Lud, D.; Moerdijk-Poortvliet, T.C.W. Stable isotope ratios as a tool for assessing changes in carbon and nutrient sources in antarctic terrestrial ecosystems. *Plant Ecol.* **2006**, *182*, 79–86.
16. Amundson, R.; Austin, A.T.; Schurur, E.A.G.; Yoo, K.; Matzek, V.; Kendall, C.; Uebersax, A.; Brenner, D.; Baisden, W.T. Global patterns of the isotopic composition of soil and plant nitrogen. *Glob. Biogeological Cycles* **2003**, *17*, 1–10.
17. Wallander, H.; Göransson, H.; Rosengren, U. Production, standing biomass and natural abundance of 15N and 13C in ectomycorrhizal mycelia collected at different soil depths in two forest types. *Oecologia* **2004**, *139*, 89–97.
18. Högberg, P. Tansley Review No. 95 ^{15}N natural abundance in soil-plant systems. *New Phytol.* **1997**, *137*, 179–203.
19. Högberg, P.; Högberg, M.N.; Quist, M.E.; Ekblad, A.; Nasholm, T. Nitrogen isotope fractionation during nitrogen uptake by ectomycorrhizal and non-mycorrhizal Pinus sylvestris. *New Phytol.* **1999**, *142*, 569–576.
20. Peterson, B.J.; Fry, B. Stable isotopes in ecosystem studies. *Annu. Rev. Ecol. Syst.* **1987**, *18*, 293–320.
21. Mariotti, A.; Pierre, D.; Vedy, J.; Bruckert, S.; Guillemot, J. The abundance of natural nitrogen-15 in the organic matter of soils along an altitudinal gradient. *Catena* **1980**, *7*, 293–300.
22. Nadelhoffer, K.J.; Fry, B. Controls on natural nitrogen-15 and carbon-13 abundances in forest soil organic matter. *Soil Sci. Soc. Am. J.* **1988**, *52*, 1633–1640.
23. Billings, S.A.; Richter, D.D. Changes in stable isotopic signatures of soil nitrogen and carbon during 40 years of forest development. *Oecologia* **2006**, *148*, 325–333.
24. Shearer, G.B.; Kohl, D.H. N_2-fixation in field settings: Estimations based on natural ^{15}N abundance. *Aust. J. Plant Physiol.* **1986**, *13*, 699–756.
25. Binkley, D.; Hart, S.C. Components of nitrogen availability assessments in forest soils. *Adv. Soil Sci.* **1989**, *10*, 57–112.
26. Sanborn, P.; Preston, C.; Brockley, R. N_2 fixation by Sitka alder in a young lodgepole pine stand in central interior British Columbia, Canada. *For. Ecol. Manag.* **2002**, *167*, 223–231.
27. Lajtha, K.; Schlesinger, W.H. Plant response to variations in nitrogen availability in a desert shrubland community. *Biogeochemistry* **1986**, *2*, 29–37.
28. Ehrenfeld, J.G.; Ravit, B.; Elgerma, K. Feedback in the plant–soil system. *Annu. Rev. Ecol. Syst.* **2005**, *30*, 1–41.
29. Aerts, R. Climate, leaf litter chemistry and leaf litter decomposition in terrestrial ecosystems: A triangular relationship. *Oikos* **1997**, *79*, 439–449.
30. Pike, L.H. The importance of epiphytic lichens in mineral cycling. *Bryol.* **1978**, *81*, 247–257.
31. Freyer, H.D. Seasonal variation of $^{15}N/^{14}N$ ratios in atmospheric nitrate species. *Tellus* **1991**, *43*, 30–44.
32. Campbell, J.; Bradfield, G.E.; Prescott, C.E.; Fredeen, A.L. The influence of overstorey *Populus* on epiphytic lichens in subboreal spruce forests of British Columbia. *Can. J. For. Res.* **2010**, *40*, 143–154.
33. Denison, W.C. Life in tall trees. *Sci. Am.* **1973**, *228*, 74–80.

34. Perry, D.; Wiliams, J. The tropical rain forest canopy: A method providing total access. *Biotropica* **1981**, *13*, 283–285.
35. Botting, R.S.; Fredeen, A.L. Contrasting terrestrial moss, lichen and liverwort diversity and abundance between old-growth and young second-growth sub-boreal spruce forest in central British Columbia. *Can. J. Bot.* **2006**, *84*, 120–132.
36. Campbell, J.; Fredeen, A.L. *Lobaria pulmonaria* abundance as an indicator of macrolichen diversity in interior cedar hemlock forests of East-Central British Columbia. *Can. J. Bot.* **2004**, *82*, 970–982.
37. Macko, S.A.; Fogel, M.L.; Hare, P.E.; Hoering, T.C. Isotopic fractionation of nitrogen and carbon in the synthesis of amino acids by microorganisms. *Chem. Geol.* **1987**, *65*, 79–92.
38. Virginia, R.A.; Delwiche, C.C. Natural ^{15}N abundance of presumed N_2-fixing and non-N_2-fixing plants from selected ecosystems. *Oecologia* **1982**, *54*, 317–325.
39. Hope, G. The Soil Ecosystem of an ESSF Forest and Its Response to a Range of Harvesting Disturbances (Extension Note 53). Available online: http://www.for.gov.bc.ca/hfd/pubs/Docs/En /En53.pdf (accessed on 11 February 2014).
40. Wania, R.; Hietz, P.; Wanek, W. Natural ^{15}N abundance of epiphytes depends on the position within the forest canopy: Source signals and isotope fractionation. *Plant Cell Environ.* **2002**, *25*, 581–589.
41. Hobbie, E.A.; Hobbie, J.E. Natural abundance of ^{15}N in nitrogen-limited forests and tundra can estimate nitrogen cycling through mycorrhizal fungi: A review. *Ecosystems* **2008**, *11*, 815–830.
42. Riera, P. $\delta^{13}C$ and $\delta^{15}N$ comparisons among different co-occurring lichen species from littoral rocky substrata. *Lichenologist* **2005**, *37*, 93–95.
43. Gebauer, G.; Schulze, E.D. Carbon and nitrogen isotope ratios in different compartments of a healthy and a declining *Picea abies* forest in the Fichtelgebirge, NE Bavaria. *Oecologia* **1991**, *87*, 198–207.
44. Jansson, S.L. Tracer studies on nitrogen transformations in soil with special attention to mineralization-immobilization relationships. *Ann. R. Agri. Coll. Swed.* **1958**, *24*, 101–361.
45. Batts, J.E.; Calder, L.J.; Batts, B.D. Utilizing stable isotope abundances of lichens to monitor environmental change. *Chem. Geol.* **2004**, *204*, 345–368.
46. Máguas, C.; Griffiths, H.; Broadmeadow, M.S.J. Gas exchange and carbon isotope discrimination in lichens: Evidence for interactions between CO_2-concentrating mechanisms and diffusion limitation. *Planta* **1995**, *196*, 95–102.
47. Thomas, S.C.; Martin, A.R. Carbon content of tree tissues: A synthesis. *Forests* **2012**, *3*, 332–352.
48. Farquhar, G.D.; Ehleringer, J.R. Carbon isotope discrimination and photosynthesis. *Annu. Rev. Plant Physiol. Plant Mol. Biol.* **1989**, *40*, 503–537.
49. Lakatos, M.; Hartard, B.; Maguas, C. The stable isotopes of $\delta^{13}C$ and $\delta^{18}O$ of lichens can be used as tracers of microenvironmental carbon and water sources. In *Stable Isotopes as Indicators of Ecological Change (Terrestrial Ecology)*; Dawson, T.E., Siegwolf, R.T.W., Eds.; Elsevier: Oxford, UK, 2007; pp. 73–88.
50. Broadmeadow, M.S.J.; Griffiths, H.; Maxwell, C.; Borland, A.M. The carbon isotope ratio of plant organic material reflects temporal and spatial variations in CO_2 within tropical forest formations in Trinidad. *Oecologia* **1992**, *89*, 435–441.

51. Máguas, C.; Griffiths, H.; Ehleringer, J.R.; Serodio, J. Characterizations of photobiont associations in lichens using carbon isotope discrimination techniques. In *Stable Isotopes and Plant Carbon-Water Relations*; Academic Press London: New York, NY, USA, 1993; pp. 201–212.
52. Palmqvist, K. Tansley review No. 117 Carbon economy in lichens. *New Phytol.* **2000**, *148*, 11–36.
53. Hietz, P.; Wanek, W.; Wania, R.; Nadkarni, N.M. Nitrogen-15 natural abundance in a montane cloud forest canopy as an indicator of nitrogen cycling and epiphyte nutrition. *Oecologia* **2002**, *131*, 350–355.

6

Seasonal Pattern of Decomposition and N, P and C Dynamics in Leaf litter in a Mongolian Oak Forest and a Korean Pine Plantation

Jaeeun Sohng [1,4], Ah Reum Han [1], Mi-Ae Jeong [1,5], Yunmi Park [2], Byung Bae Park [3] and Pil Sun Park [1,*]

[1] Department of Forest Sciences, Seoul National University, Seoul 151-921, Korea; E-Mails: jaeeun.sohng@yale.edu (J.S.); phoya@naver.com (A.R.H); sangdi@snu.ac.kr (M.-A.J)

[2] Division of Special-purpose Trees, Korea Forest Research Institute, Suwon 441-847, Korea; E-Mail: pym5250@hanmail.net

[3] Department of Environment and Forest Resources, Chungnam National University, Daejeon 305-764, Korea; E-Mail: bbpark@cnu.ac.kr

[4] School of Forestry & Environmental Studies, Yale University, New Haven, CT 06511, USA

[5] Department of Landscape Architecture, Seoul National University, Seoul 151-921, Korea

* Author to whom correspondence should be addressed; E-Mail: pspark@snu.ac.kr;

External Editor: Heinz Rennenberg

Abstract: Distinct seasons and diverse tree species characterize temperate deciduous forests in NE Asia, but large areas of deciduous forests have been converted to conifer plantations. This study was conducted to understand the effects of seasons and tree species on leaf litter decomposition in a temperate forest. Using the litterbag method, the decomposition rate and nitrogen, phosphorous, and carbon dynamics of Mongolian oak (*Quercus mongolica*), Korean pine (*Pinus koraiensis*), and their mixed leaf litter were compared for 24 months in a Mongolian oak stand, an adjacent Korean pine plantation, and a Mongolian oak—Korean pine mixed stand. The decomposition rates of all the leaf litter types followed a pattern of distinct seasonal changes: most leaf litter decomposition occurred during the summer. Tree species was less influential on the leaf litter decomposition. The decomposition rates among different leaf litter types within the same stand were not significantly different, indicating no mixed litter effect. The immobilization of leaf litter N and P lasted for 14 months. Mongolian oak leaf litter and Korean pine leaf litter showed different N and P contents and dynamics during the decomposition, and soil P_2O_5 was highest in the Korean pine plantation, suggesting effects of plantation on soil nutrient budget.

Keywords: temperate deciduous forest; mixed leaf litter effect; soil nutrients; leaf litter nutrients; mineralization

---

1. Introduction

Leaf litter decomposition is a critical step in nutrient cycling and providing nutrients to plants [1]. Various factors control leaf litter decomposition, including climate [2], topography [3,4], chemical characteristics of leaf litter [5,6], and terrestrial microbiota [7]. Among those factors, climate, especially temperature and precipitation, is a dominant factor determining leaf litter decomposition patterns in regions experiencing distinct seasonal climate change [8].

Every ecosystem has its own physical characteristics and specific environment that foster the distinct features of its species composition. Each tree species affects leaf litter decomposition differently by providing leaf litters of different quality, which are closely related to the terrestrial microbial community and soil nutrient cycling [9,10]. In temperate natural forests, plant leaf litter usually decomposes in a mixed-species environment. Mixedwood forests are characterized by diverse soil conditions and biological activity within short distances, increasing the importance of small-spatial-scale research on soil nutrient pools, as well as integrating diverse chemical variables of single or mixed species into different temporal and spatial scales [6,11]. However, species composition is one of the factors most altered by human activities as natural mixedwood forests have been converted to single-species plantations. Stand conversion from natural forest to single species plantation changes leaf litter decomposition and eventually alters ecosystem functions [12].

Information on the nutrient cycling within plantations, such as patterns of N, P, and C release, immobilization and mineralization in a decomposing phase, and the planted species' interactions with existing species, is crucial to properly manage plantations and improve their vitality to be similar to natural states [13]. However, limited information is available on the effects of a plantation on nutrient cycling at a site and on plantation interactions with existing ecosystems [14]. Thus, an increasing number of studies investigate the effect of certain species on the decomposition of single or mixed leaf litters for better plantation management or restoration of natural stands [15,16]. Those studies simultaneously integrate features of species, mixed leaf litters, and forest floor conditions by forest type to explain the leaf litter decomposition process [17,18].

Korean pine (*Pinus koraiensis* Siebold et Zucc.) is a representative species in northern temperate old-growth forests and has been widely planted for wood and pine seeds in NE Asia [19,20]. The total plantation area of Korean pine in South Korea is *ca.* 214,357 ha, making Korean pine the third most planted species in Korea [20]. Most Korean pine plantations replace oak-dominant broadleaved stands and are still adjacent to broadleaved forests dominated by oak species such as the oriental cork oak (*Quercus variabilis* Blume), konara oak (*Quercus serrata* Thunb.), and Mongolian oak. The establishment of Korean pine plantations might alter leaf litter decomposition patterns and affect the nutrient budget of an area.

This study compared leaf litter decomposition characteristics and nutrient dynamics between an oak-dominant secondary deciduous stand and a Korean pine plantation to understand the plantation effect on nutrient cycling. We hypothesized that different leaf litter types would show different leaf litter decomposition rates and nitrogen (N), phosphorous (P), and carbon (C) mineralization patterns by forest type. We expected nutrient transport between the Mongolian oak leaf litter and the Korean pine leaf litter to induce complementary interactions, resulting in a synergistic effect that would give the mixed leaf litter the highest decomposition rate.

Efforts to introduce mixedwood forests that are more ecologically sound and resistant to disturbances than monoculture are increasing in NE Asia [21]. Understanding the leaf litter decomposition process in Korean pine plantations would show what happens when a plantation replaces a natural stand and how a plantation interacts with existing ecosystems.

2. Materials and Methods

2.1. Study Sites

The study was conducted at three types of stands in the Seoul National University Forest at Mt. Taehwa (38°17′16″–38°19′26″ N, 127°16′45″ E; 644 m asl; 796 ha), Gyeonggi Province, Korea. The area is a temperate zone with a continental monsoon climate. The annual mean precipitation in this area is 1371 mm with about sixty percent of the annual precipitation is concentrated between June and August [22]. Meteorological data from the weather station of the University Forest during 2009–2011 showed that the annual mean air temperature was 10.9 °C, with a maximum temperature of 33.5 °C in July and a minimum temperature of −21.3 °C in January. It was relatively dry in the winter, with precipitation of 90 mm in the form of snow from December 2010 to February 2011. Snow depth at the study sites ranged from 5 to 20 cm depending on the slope aspect. Soils in the forest are classified as coarse loamy, in the mesic family of Typic Dystrudepts according to the USDA soil classification system [23]. Soil parent material was an eluvium of metamorphic rock, such as biotite and granite gneiss. The depth of the leaf litter layer ranged 3–7 cm in native secondary forests and 2–6 cm in Korean pine plantations in both spring and autumn.

Mt. Taehwa belongs to a broadleaved deciduous forest zone; however, most of the lower slopes are covered by Korean pine and Japanese larch (*Larix kaempferi* (Lamb.) Carrière) plantations. The Korean pine plantations in Mt. Taehwa occupied 136 ha and most of them were established in late 1960s. Because the Korean pine plantations were established in a relatively short period of time and distributed on lower slopes that had similar soil condition, they had a similar stand structure. Secondary natural stands, occupying 497 ha, were distributed on the upper slopes and ridges and were dominated by oaks, such as Mongolian oak, oriental cork oak, and oriental white oak (*Quercus aliena* Blume).

Three types of stands were selected: a secondary natural stand dominated by Mongolian oak (hereafter called oak stand), an adjacent Korean pine plantation (5.5 ha) established in 1969, and a Korean pine—Mongolian oak mixed stand (hereafter called mixed stand). The mixed stand was located between the oak stand and the Korean pine plantation and was established as a Korean pine plantation. However,

Mongolian oak from the adjacent natural stands invaded the plantation and replaced part of the Korean pine. The Mongolian oak in the oak stand and the Korean pine in the Korean pine plantation occupied 79.1% and 97.6% of each stand's basal area, respectively. The respective basal areas within the mixed stand were 48.4% Korean pine and 39% Mongolian oak. All study sites were at an altitude of 229–285 m asl, and the slope steepness was 36–67%. The oak stand was on a NW slope and the mixed stand on a NE slope. The Korean pine plantation grew across different oriented slopes from 120–315°; however, it mostly faced south (Table 1).

In each type of stand, three 20 m × 20 m plots were randomly established. All trees ≥2.5 cm in diameter at breast height (DBH; 1.3 m above the ground) were measured within each 20 m × 20 m plot. For each tree, species names, DBH, tree condition (alive, dead, broken top, or leaning), crown width, and proportion of crown to tree height were recorded.

Table 1. Selected characteristics of the three study sites in the Seoul National University Forest at Mt. Taehwa, Gyeonggi Province.

Stand types	Oak stand	Mixed stand	Korean pine plantation
Altitude (m)	284–285	258–270	229–239
Slope steepness (%)	43–67	36–47	42–55
Aspect (°)	315–335	36–47	120–315
Soil texture	Loam	Loam	Loam
Sand (%)	38	36	36
Silt (%)	39	41	41
Clay (%)	23	23	23
Species (density (number of stems ha^{-1}), relative basal area (%))	*Quercus mongolica* (567, 79.1), *Pinus koraiensis* (308, 7.9), *Kalopanax septemlobus* (17, 2.9), Others (767, 10.1)	*P. koraiensis* (425, 48.4), *Q. mongolica* (256, 39), *Q. variabilis* (25, 3.7), Others (127, 8.9)	*P. koraiensis* (725, 97.6), *Q. variabilis* (17, 1.0), *Styrax obassia* (42, 0.7), Others (50, 0.7)

2.2. Soil Measurements, Sampling, and Analysis

A HOBO soil data logger (HOBO H21-002 Micro Station, On-set computer Corporation, USA) was installed in the center of each study site. The soil temperature (°C) and soil moisture content (m^3 m^{-3}) were measured from 28 March 2010 to 31 March 2011 at 30-minute intervals at 10 cm soil depth. The daily mean soil temperature and daily mean soil water content were calculated.

Soil samples were collected from the soil surface to a depth of 10 cm with a soil sampling container (100 cm^3) from all the plots in March, June, and October 2010. At each plot, four sampling points were randomly selected 1.5–2 m from the base of a tree trunk to minimize the direct physical and chemical influences of roots [4]. A total of 36 soil samples at a time were collected from the three stand types. The soils were oven-dried at 105 °C for 24 hours and the soil water content (%) was measured on a gravimetric basis. A particle-size analysis was conducted with the pipette method [24], and soil texture (standard of US Department of Agriculture) was determined. We measured soil organic matter contents by combusting it at 600 °C for 4 hours in a muffle furnace and weighing the residue [25]. For chemical analyses, the sampled soils were air-dried and sieved (<2 mm). We measured soil pH (soil:water = 1:5), total nitrogen content (TN) [25], cation exchange capacity (CEC) [26], and total phosphorus (TP; inductively coupled plasma ICPS-1000IV, Shimazu, Japan). Available soil phosphorous (P_2O_5) was determined by analyzing the Bray No. 1 extracts with an ICP (inductively coupled plasma) optical emission spectrometer (ICP-730 ES, Varian, USA). We used an elemental analyzer to measure the C:N ratio (Flash EA 1112, Thermo Electron Corporation, USA).

2.3. Leaf litter Decomposition Experiment

Leaf litter fall of Mongolian oak and Korean pine (5–6 kg per each species) was collected from forest floor of the Mongolian oak stand and Korean pine plantation, respectively, in the study sites in March 2009. Our study focused on in-situ condition and year-round changes in leaf litter on the forest floor. Leaf litter fall of Mongolian oak occurred in autumn and Korean pine was a conifer with leaf litter fall distributed over the whole year, so we collected forest floor litter in March before active decomposition began instead of autumn to avoid the combination of very fresh Mongolian oak litter fall and relatively less fresh Korean pine litter fall. We were careful to collect intact leaves without clear sign of decomposition from the top layer. The leaf litter sample was oven-dried at 80 °C for 48 hours. The oven-dried leaf litter was weighed into 10 g samples and put into litterbags, either pure or in a 1:1 ratio of the two leaf litters by mass [27,28]. Each litterbag was 20 cm × 10 cm and was made of nylon mesh (1 mm × 1.2 mm mesh size).

The litterbags were placed on topsoil parallel to the forest floor in each study plot in April 2009. A total of 972 litterbags were deployed at the study sites: 3 stand types × 3 plots at each stand type × 3 leaf litter types × 3 replicates × 12 retrieving dates from April 2009 to March 2011. Each litterbag was at least 15 cm away from other litterbags, and all litterbags at each site were tied together with a nylon string to minimize unnecessary movements. Every two months, 27 litterbags were retrieved from each study site. The retrieved leaf litter was oven-dried at 80 °C for 48 hours and brushed carefully to minimize soil contamination [29]. Oven-dried litter was weighed to the nearest 0.01 g to determine its mass loss, and milled for chemical analyses [30]. Total nitrogen and C concentrations (mg g^{-1}) were measured using a dry combustion method with a C-N analyzer (Yanaco MT-500, Kyoto, Japan) [28]. Total P in the leaf litter was determined with an automated ion analyzer (QuikChem AE, Lachat Inc., Loveland, Colorado, USA). The amount of nutrients in the leaf litter was calculated as follows:

$$N_t = \frac{C_t}{100} \times L_t \tag{1}$$

Where N_t is the amount of nutrient (g) at time t, C_t is the percent concentration of nutrient at time t, and L_t is the leaf litter remaining mass (g) at time t.

2.4. Data Analysis

The decomposition rate constant (k) was derived from the negative exponential decomposition equation proposed by Olson [31] as follows:

$$-\text{k} = \ln\left(\frac{\text{X}}{\text{X}_0}\right)t \tag{2}$$

Where X_0 is the initial mass, and X is the remaining mass at time t.

The predicted remaining mass for the mixed leaf litter was estimated using equation (3) based on the observed remaining mass of the single species to verify a mixed effect on leaf litter decomposition [29]. The predicted results were compared with the observed remaining mass of the mixed leaf litter.

$$E_{\mathrm{ML}} = \sum_{i=1}^{S} O_{\mathrm{ML}i} \times P_{\mathrm{IM}i} \quad (3)$$

Where E_{ML} is the predicted remaining mass of mixed litter (%); S is the species; $O_{\mathrm{ML}i}$ is the observed leaf litter mass of the single species (%); and $P_{\mathrm{IM}i}$ is the initial proportion of species i.

When the observed remaining mass of mixed leaf litter is not significantly different from the predicted results, mixing has an additive effect; otherwise, it has a non-additive effect. The non-additive effect is differentiated into a synergistic effect (decomposition rates are more than predicted values) or an antagonistic effect (decomposition rates are less than predicted values) [16]. The predicted and observed remaining mass were compared using a paired t-test to determine possible mixed leaf litter effects (*i.e.*, additive or non-additive).

One-way ANOVA was used to compare the decomposition rate constant (k) and the changes in leaf litter N, P, and C concentrations among the three stand types. A simple linear regression was used to assess the relationship between mass loss (%) and total C in decomposing leaf litters. The relationship between the total N, P, and C in each leaf litter type over 24 months was estimated using polynomial regression analysis.

A repeated-measures ANOVA was used to compare the soil physical and chemical characteristics in different seasons (March, June, and October 2010) and identify whether significant differences occurred in mass loss and changes in leaf litter N and P every two months by leaf litter type in a single stand or among the stand types. Tukey's test was applied for post-hoc comparisons, with significance set at $p < 0.05$. The SPSS statistical package program Version 18.0 was used for all statistical analyses (SPSS Software 2009, IBM SPSS Inc., Chicago, IL, USA).

3. Results

3.1. Soil Characteristics

All study sites had similar soil temperatures. The maximum soil temperature reached at 24 °C in August, and it stayed near 0 °C in January and February (Figure 1). The Korean pine plantation had a daily mean soil moisture content lower than the other two stand types in June, October, and November (Figure 1), and soil moisture content measurements lower than the other two stand types in March and October 2010 (Table 2; $p < 0.001$). All study sites showed similar daily soil moisture content in the summer.

The three stand types showed significant differences in soil TN concentrations, which were highest in the oak stand and lowest in the Korean pine plantation in spring and autumn ($p = 0.03$). However, soil TN did not significantly differ among stand types in the summer. The oak stand had the highest organic matter content (19.7%) in March 2010 ($p < 0.05$), but we found no differences in organic matter content in the summer and autumn (Table 2).

Figure 1. Daily mean soil temperatures (St, °C) and daily mean soil water contents (Sw, $m^3 m^{-3}$) of the Mongolian oak stand, the mixed stand, and the Korean pine plantation from 28 March 2010 to 31 March 2011.

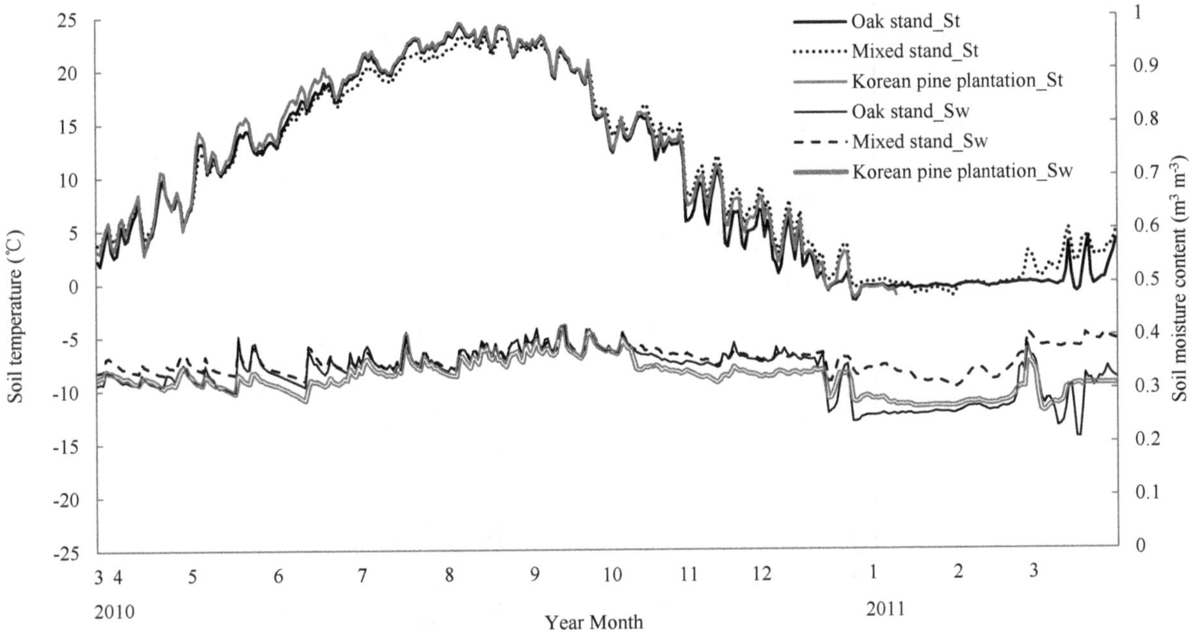

The soil P_2O_5 content was highest in the Korean pine plantation with 126.7 mg L^{-1}; was more than twice the 49 mg L^{-1} observed in the oak stand in March 2010. The largest soil N:P ratio was therefore in the oak stand (Figure 2), which had the lowest soil P_2O_5 and substantial seasonal differences in soil P_2O_5 and soil N:P ($p = 0.015$) between March and October, 2010, though few seasonal changes in soil TN were observed. Due to the high variability of soil P_2O_5 within the stands, soil P_2O_5 contents among the stands were not significantly different. Soil pH, TP, and CEC were not significantly different among the stand types.

Figure 2. Seasonal changes in the soil N:P ratio in soil (0–10 cm) at the study sites. Data are means ± SE ($n = 9$). Data in the same stand followed by the different letters are significantly different at $p < 0.05$ according to t-tests.

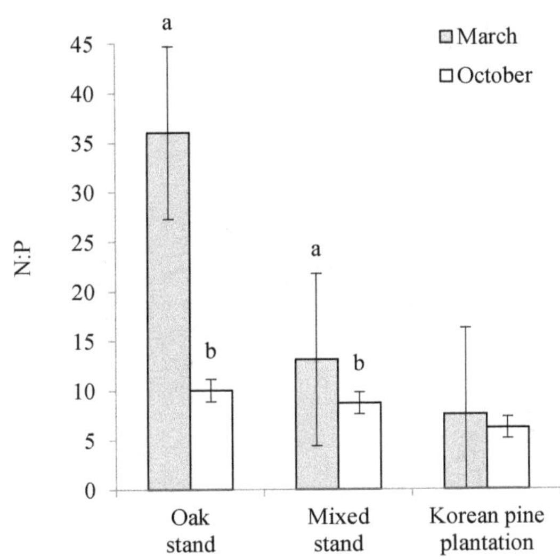

Table 2. Soil chemical characteristics of the three study sites in the Seoul National University Forest at Mt. Taehwa, Gyeonggi Province in March, June, and October 2010.

Month	Stand type	Soil water content (%)	Organic matter (%)	pH	Total N (%)	Total P (mg L^{-1})	P_2O_5 (mg L^{-1})	CEC (cmol kg^{-1})	C:N ratio
March	Oak stand	67.3 ± 11.5 ^{a}	19.7 ± 2.8 ^{a}	4.0 ± 0.2	0.60 ± 0.00 ^{a}	301.2 ± 12.8	49.0 ± 7.7	23.0 ± 0.8	14.6 ± 0.2 ^{ab}
	Mixed stand	53.4 ± 6.4 ^{b}	16.2 ± 2.1 ^{b}	4.0 ± 0.1	0.43 ± 0.02 ^{b}	308.5 ± 15.6	97.9 ± 20.1	21.4 ± 1.4	14.0 ± 0.0 ^{b}
	Korean pine plantation	42.7 ± 5.8 ^{c}	13.3 ± 2.7 ^{c}	4.3 ± 0.9	0.30 ± 0.03 ^{c}	278.1 ± 43.7	126.7 ± 34.2	18.6 ± 4.2	15.0 ± 0.3 ^{a}
June	Oak stand	-	-	-	0.44 ± 0.04	248.5 ± 16.8	60.2 ± 5.2	-	15.0 ± 0.6 ^{ab}
	Mixed stand	-	-	-	0.36 ± 0.02	269.1 ± 21.0	79.8 ± 12.1	-	13.8 ± 0.3 ^{b}
	Korean pine plantation	-	-	-	0.30 ± 0.05	238.3 ± 45.1	73.0 ± 20.0	-	16.1 ± 0.5 ^{a}
October	Oak stand	60.4 ± 7.1 ^{a}	16.6 ± 3.0	4.1 ± 0.3	0.50 ± 0.03 ^{a}	315.1 ± 19.3	129.9 ± 14.2	-	14.0 ± 0.2 ^{b}
	Mixed stand	49.3 ± 8.1 ^{b}	14.4 ± 2.0	4.0 ± 0.1	0.47 ± 0.02 ^{a}	337.3 ± 18.7	129.4 ± 7.1	-	15.4 ± 0.4 ^{a}
	Korean pine plantation	39.6 ± 7.0^{c}	13.8 ± 4.4	4.1 ± 0.1	0.30 ± 0.05 ^{b}	296.1 ± 53.0	143.4 ± 99.6	-	15.8 ± 0.3 ^{a}

Data in the same month followed by the same letter are not significantly different at $p < 0.05$ according to Tukey's multiple range tests; Values are means ± SE.

3.2. Leaf litter Decomposition and Mixed Leaf litter Effects

All leaf litter showed distinct seasonal changes in decomposition rates. About 20% of mass loss occurred in all study sites during the initial six months from April to October 2010 regardless of leaf litter type. Little change in the remaining mass was detected in the subsequent six months (during winter and spring). During the second summer, fourteen months after the experiment started, all types of leaf litter in each stand lost 14%–29% of the original mass (Figure 3).

The leaf litter decomposition rate in the oak stand was significantly slower than those in the Korean pine plantation and the mixed stand until the second summer ($p < 0.001$). The leaf litter decomposition rate was highest in the Korean pine plantation and lowest in the oak stand for 17 months. Although the leaf litter mass loss in the oak stand was less than that in the other stands prior to the second summer, the remaining leaf litter mass continuously decreased in the oak stand during the second autumn, whereas further decomposition did not progress in the other stands.

As a result, after 24 months the remaining leaf litter mass was similar among the stand types, ranging from approximately 41–47%. No significant difference was found in the decomposition rate among the leaf litter types in the same stand.

Figure 3. Percent remaining leaf litter mass for Mongolian oak leaf litter (black solid line), Korean pine leaf litter (gray solid line) and mixed leaf litter (dashed line) in (**A**) the oak stand, (**B**) the mixed stand, and (**C**) the Korean pine plantation. Data are means ± SE ($n = 9$).

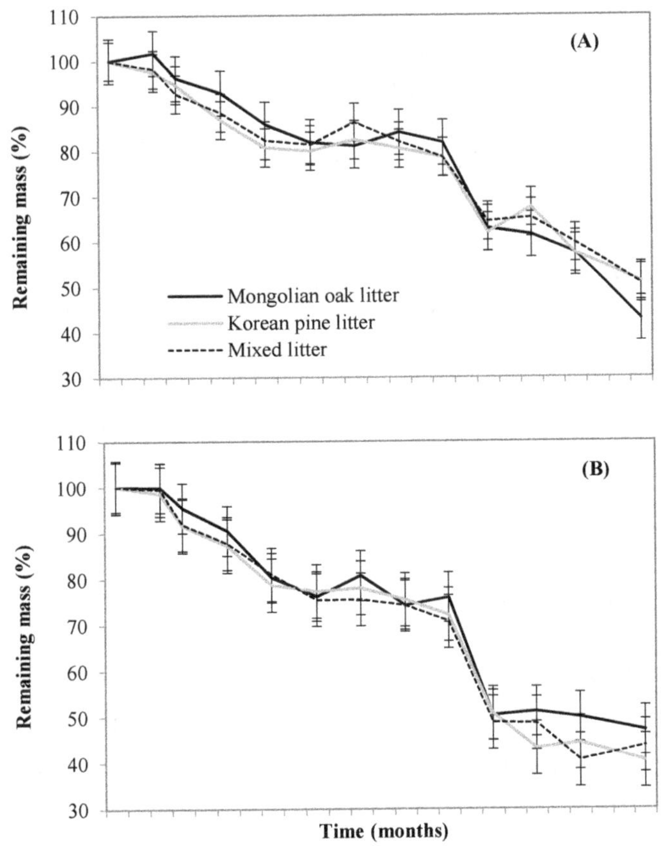

Figure 3. *Cont.*

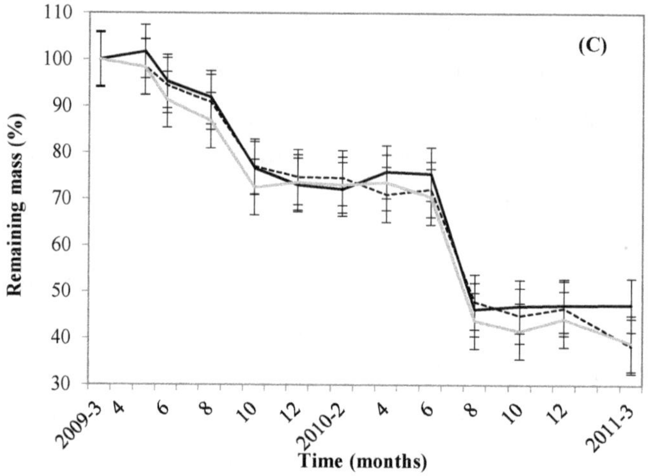

Observed and predicted decay constants in each stand were not significantly different, indicating that the leaf litter mixing of Korean pine and Mongolian oak had an additive effect on leaf litter decomposition. We found no significant synergistic or antagonistic interactions between those two species in leaf litter decomposition over 24 months (Figure 4).

Figure 4. Comparisons of predicted (white bars) and observed (grey bars) decay constant k at each site. Predicted values calculated from the corresponding monoculture litterbags (paired t- test). Data are means ± SE ($n = 9$).

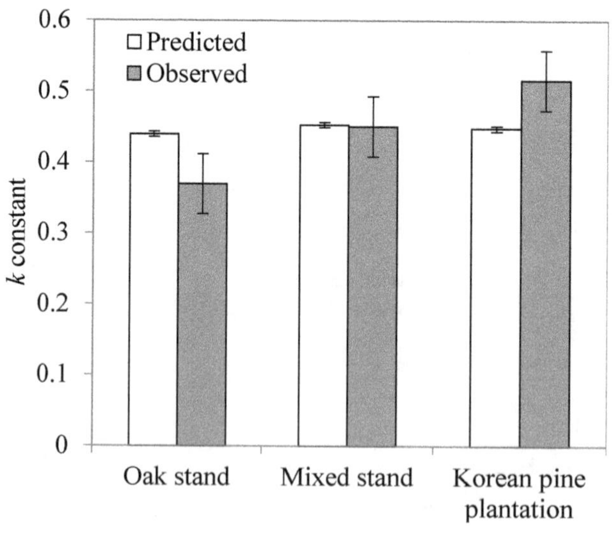

3.3. Leaf litter N, P, and C Changes and Relationships

Changes in the decomposition constant (k) for 24 months corresponded to changes in the amount of C in decomposing leaf litters, and those changes were highly correlated ($p < 0.001$; Table 3). The amount of leaf litter C decreased in a pattern similar to leaf litter mass loss (Figure 3). Consequently, the changes in leaf litter C mass did not vary among leaf litter types, but they were significantly different among stands: the oak stand retained significantly more leaf litter C than the other stands ($p < 0.001$).

Table 3. Leaf litter decomposition constant (k) and C percentage after 24 months relative to the initial leaf litter C ($n = 9$), and R^2-values of simple linear regression for the relationship between mass loss (%) and leaf litter C in the decomposing phase ($p < 0.05$).

Forest type	Leaf litter type	k	C (%)	Regression	R^2
Oak stand ***	Mongolian oak	0.4588	34.0	$C = -0.0935\,x + 7.8699$	0.86
	Mixed leaf litter	0.3694	32.5	$C = -0.0934\,x + 7.9997$	0.72
	Korean pine	0.3774	33.1	$C = -0.0927\,x + 8.1664$	0.87
Mixed stand	Mongolian oak	0.4035	29.7	$C = -0.0921\,x + 7.9344$	0.88
	Mixed leaf litter	0.4504	28.3	$C = -0.0902\,x + 8.1363$	0.98
	Korean pine	0.4886	25.2	$C = -0.1045\,x + 8.4264$	0.93
Korean pine plantation	Mongolian oak	0.4056	25.1	$C = -0.0894\,x + 7.9403$	0.97
	Mixed leaf litter	0.5159	23.1	$C = -0.0875\,x + 7.9612$	0.92
	Korean pine	0.5048	22.7	$C = -0.0947\,x + 8.3453$	0.93

The notation *** indicates that the leaf litter C percentage among the forest types after 24 months is significantly different at $p < 0.001$.

Percentage changes in leaf litter N and P showed that N and P were immobilized during the first year in all study sites. The immobilization was sustained until the beginning of the second summer, June 2010. Mineralization of N and P was observed after 14 months (Figure 5).

Figure 5. Percent of original N ((**A**), (**B**) and (**C**)) and P ((**D**), (**E**) and (**F**)) remaining for Mongolian oak leaf litter (black solid line), Korean pine leaf litter (gray solid line) and mixed leaf litter (dashed line) during the decomposition in (**A**), (**D**) the oak stand, (**B**), (**E**) the mixed stand, and (**C**), (**F**) the Korean pine plantation after 24 months. Data are means ± SE ($n = 9$).

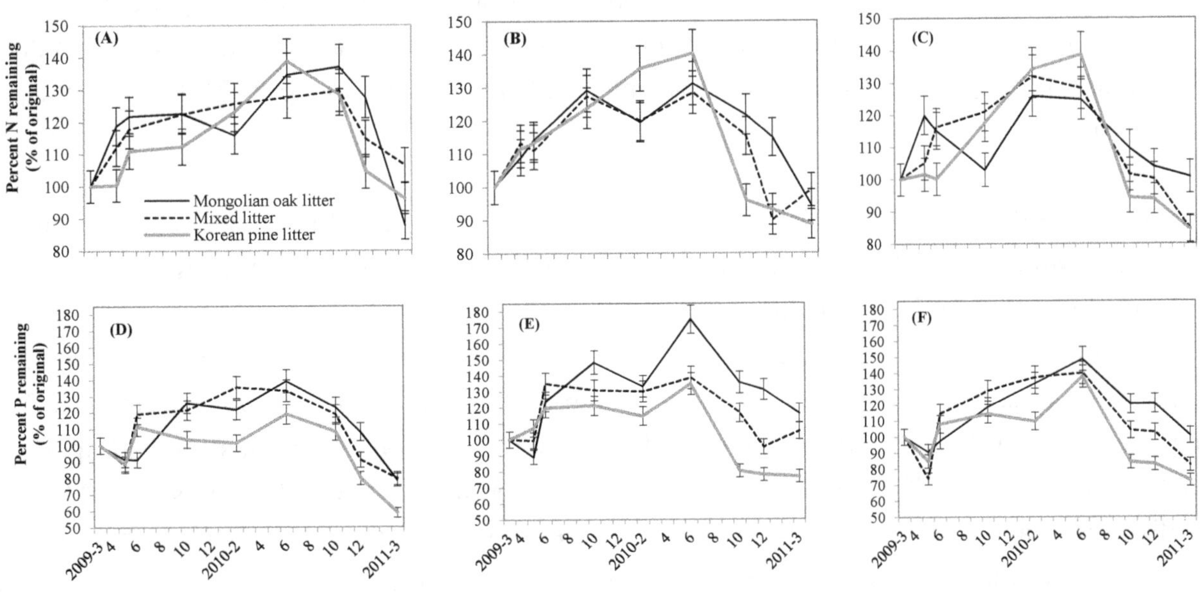

The shift from N immobilization to N mineralization in the mixed stand and the Korean pine plantation occurred in the 16th month, June 2010, whereas the process of N immobilization in the oak stand was sustained until October 2010. Although the oak stand had a longer N immobilization phase, it showed a higher rate of N mineralization in the second autumn, and the percent of N remaining in the leaf litter in the oak stand was similar to that in the other stands in the 24th month. Changes in the percent of N remaining in the oak leaf litter was significantly different from that of Korean pine leaf litter in the oak stand ($p < 0.05$), but the other stands did not show significant differences in the percent of N remaining by leaf litter type (Figure 5A–C).

Temporary leaching of P was shown at the beginning of the study, and then a process of P immobilization started within two months and continued until the 16th month in all study sites (Figure 5D–F). The highest P immobilization rate (175%) was observed in the oak leaf litter at the mixed stand (Figure 5E). The Korean pine leaf litter had the lowest P immobilization percentage in all stand types. Korean pine leaf litter showed the lowest percent of P remaining among leaf litter types in all study sites, indicating the greatest leaf litter P decomposition, and oak leaf litter showed the highest percent of P remaining ($p < 0.05$). Percent of P remaining in the leaf litter in the 24th month was 59–79% in the oak stand, 77–116% in the mixed stand, and 73–100% in the Korean pine plantation. The oak stand showed the greatest P mineralization among the stand types. The percent of P remaining in the Korean pine leaf litter was lower in the oak stand than in the other stands ($p = 0.018$).

The changes in leaf litter N and P had a polynomial correlation with leaf litter C (Figure 6, Table 4). The relationship among leaf litter N, P, and C differed among the leaf litter types. The leaf litter N amount relative to leaf litter C loss was significantly different in the order Korean pine leaf litter^{b} < mixed leaf litter^{ab} < Mongolian oak leaf litter^{a} (Figure 6A–C; $p < 0.001$). The inflection points between N immobilization and mineralization differed by stand type ($p < 0.05$). The maximum amount of leaf litter N peaked when the total amount of C ranged from 5 g to 6 g, which was interpreted to be 50–60% of the initial leaf litter C amount, in the mixed stand and the Korean pine plantation (Figure 5B,C) and 4 g of C in the oak stand (Figure 6A). After 24 months of the experiment, leaf litter C:N ranged from 25 to 46. Leaf litter C:N was not significantly different among stand types, but differed among the leaf litter types as Korean pine leaf litter had significantly higher C:N than the other litter types ($p < 0.001$).

Korean pine leaf litter had the greatest leaf litter P content among leaf litter types. The amount of leaf litter P in the Mongolian oak leaf litter was significantly smaller than that in the mixed leaf litter and Korean pine leaf litter until the remaining C reached 4 g in all the stand types. The amount of leaf litter P became similar among all leaf litter types in the mixed stand and the Korean pine plantation after remaining C was less than 4 g (Figure 6E,F). Similar amounts of C (4–5 g) were observed at the inflection point of P changes in all stand types (Figure 6D–F; $p < 0.05$). The Mongolian oak stand had significantly higher leaf litter C:P of 476 than the Korean pine plantation of 383 after 24 months ($p = 0.019$). The Mongolian oak stand had significantly higher leaf litter N:P of 14 than the other two sites of *ca.* 13 ($p < 0.05$). Three litter types showed significant difference in leaf litter N:P in 24 months in the order Korean pine leaf litter^{c} < mixed leaf litter^{b} < Mongolian oak leaf litter^{a} ($p < 0.05$).

Table 4. The relationship between amount of C, N and P in each decomposing leaf litter type as described by polynomial regression.

Forest type	Leaf litter type	Regression	R^2	*P*-value
Oak stand	Mongolian oak	$N = -0.0006C^2 + 0.0034C + 0.0924$	0.6046	0.156
		$P = -0.0001C^2 + 0.0006C + 0.0010$	0.7871	0.045
	Mixed leaf litter	$N = -0.0011C^2 + 0.009C + 0.0643$	0.9198	0.006
		$P = -0.0001C^2 + 0.0009C + 0.0009$	0.8509	0.022
	Korean pine	$N = -0.0012C^2 + 0.0104C + 0.0503$	0.7944	0.042
		$P = -0.00003C^2 + 0.0003C + 0.0024$	0.3239	0.457
Mixed stand	Mongolian oak	$N = -0.0016C^2 + 0.0145C + 0.0606$	0.9046	0.009
		$P = -0.0001C^2 + 0.001C + 0.0007$	0.8408	0.025
	Mixed leaf litter	$N = -0.0012C^2 + 0.012C + 0.0508$	0.8081	0.037
		$P = -0.0001C^2 + 0.0008C + 0.0012$	0.8119	0.035
	Korean pine	$N = -0.0019C^2 + 0.0203C + 0.017$	0.8809	0.014
		$P = -0.0001C^2 + 0.001C + 0.0004$	0.8722	0.016
Korean pine plantation	Mongolian oak	$N = -0.0011C^2 + 0.0108C + 0.0603$	0.5020	0.248
		$P = -0.0001C^2 + 0.0005C + 0.0014$	0.7598	0.058
	Mixed leaf litter	$N = -0.0017C^2 + 0.017C + 0.0368$	0.9258	0.006
		$P = -0.0001C^2 + 0.0009C + 0.001$	0.7812	0.048
	Korean pine	$N = -0.0017C^2 + 0.018C + 0.0233$	0.7957	0.042
		$P = -0.0001C^2 + 0.0009C + 0.0009$	0.6936	0.094

Figure 6. Changes in the total N ((A), (B), (C)) and P ((D), (E), (F)) relative to total C for Mongolian oak leaf litter, Korean pine leaf litter and mixed leaf litter in (A), (D) the oak stand, (B), (E) the mixed stand, and (C), (F) the Korean pine plantation.

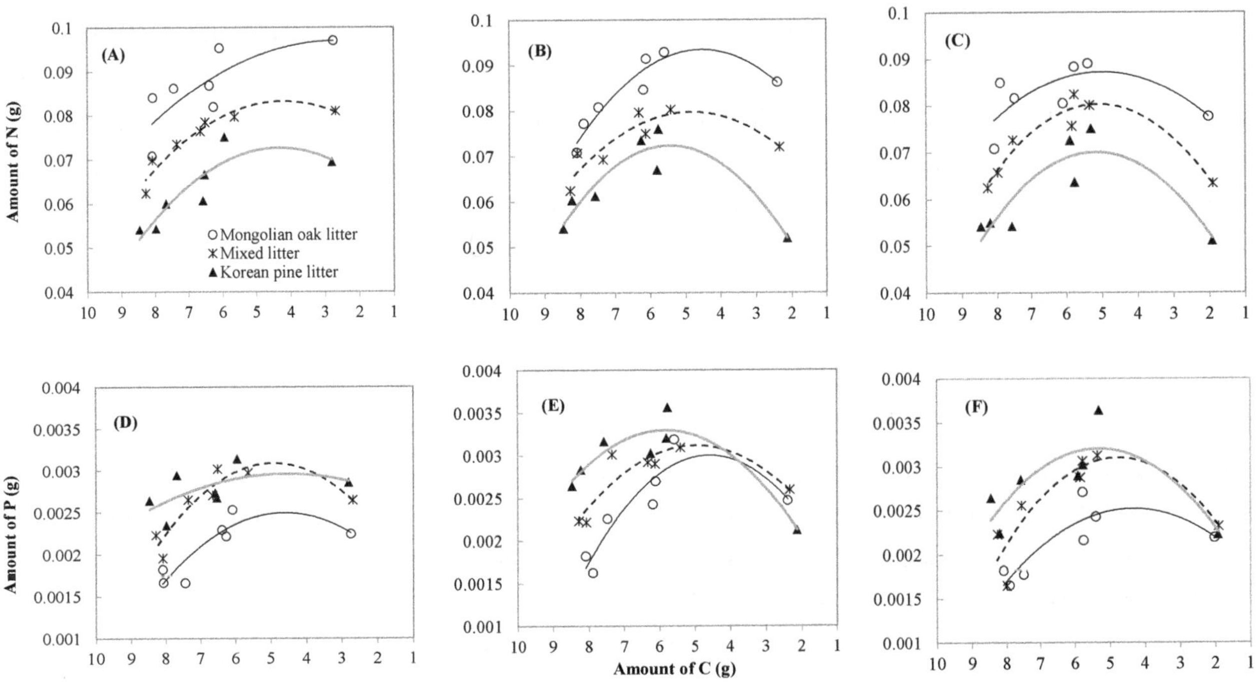

4. Discussion

4.1. Seasonal Leaf litter Decomposition Patterns

Seasonal weather changes have a large effect on leaf litter breakdown in temperate deciduous forests, especially when abundant rainfall and high summer temperatures contribute to a substantial leaf litter mass loss [32]. The majority of leaf litter decomposition in all study sites occurred from June to August, although the decomposition rates differed among the stand types. The seasonal effect on the decomposition rate of *Quercus* leaf litters in a temperate deciduous forest was also reported in Lee *et al.* [33]: the majority of leaf litter mass loss occurred in the summer, and little mass loss was detected from February to April.

In this study, the percent mass of the Korean pine leaf litter remaining after 24 months was 40–50%, which was similar to the 51.6% in 26 months reported in the Gwangneung Experimental Forest, in the same province as our study sites [30]. On Mt. Gyebang, which is north of Mt. Taehwa and has lower temperatures, the percent mass of leaf litter remaining after 24 months was *ca.* 52% [34]. The percent average leaf litter mass remaining after 24 months was 55–60% for Korean pine leaf litter in Jilin Province, northeast China, which has a higher latitude than Mt. Taehwa [19]. About 50% of the leaf litter was decomposed during the initial two years, and the decomposition rate lowered slightly as the latitude increased, showing that temperature was a determining factor for leaf litter decomposition rate: a hotter humid summer corresponds to faster leaf litter decomposition in temperate deciduous forests in NE Asia [35].

All study sites had the same soil texture (loam) and similar soil temperatures. Soil moisture contents were similar in the summer and different in spring and autumn. The oak stand and mixed stand were on a north slope, and the Korean pine plantation was mostly on a south slope, which explained why the lowest soil moisture content was found in the Korean pine plantation in the dry seasons of spring and autumn [36,37]. Leaf litter decomposition rate generally increases with soil moisture content [38]. However, our study sites had similar decomposition rates after 24 months, indicating that the soil moisture content in summer was the key determinant of leaf litter decomposition and confirming the importance of summer weather in leaf litter decomposition [39].

4.2. Mixed Leaf Litter Effect

The leaf litter decomposition rate of broadleaved species is generally higher than that of conifers [40,41]. However, the decomposition rates of Mongolian oak leaves and Korean pine needles in the same stand were not different in this study. Mongolian oak leaves have thick hard blades and low nutrient content compared with other oak species, retarding the leaf litter decomposition rate [34] and contributing to the similar decomposition rates of the two species.

The mixed leaf litter of Korean pine and Mongolian oak showed no mixed litter effect on leaf litter decomposition, indicating that the interaction strength between the leaf litters of the two species was weak. The leaf litter decomposition rate of the mixed leaf litter could therefore be predicted from the decomposition rate of each individual species [15]. Mixed leaf litter could have a higher decomposition rate than a single-species leaf litter because nutrient exchange among leaf litters of different species through microbes could enhance the nutrient condition of low-quality leaf litters, accelerating the decomposition rate [27,42]. At the beginning, Mongolian oak leaves contained more N, and Korean pine needles had more P (Figure 6). We could observe more physical breakdown of the Mongolian oak leaf litter in the mixed leaf litter. However, contrary to our expectations that transport of nutrients between the Mongolian oak leaf litter and the Korean pine leaf litter would induce complementary interactions and a synergistic effect, an actual mixed effect on the remaining mass was not detected within 24 months, suggesting that synergistic effects of mixed leaf litters of contrasting qualities is questionable [29,43]. Moreover, the seasonal climate was a primary control on the leaf litter decomposition pattern, making species difference less powerful in leaf litter decomposition in a temperate deciduous forest [44]. However, the possibility that an additive effect is temporary and switches to a non-additive effect as decomposition progresses still exists because the non-additive effect of mixed leaf litters for mass loss could be time dependent [45].

4.3. Leaf Litter Nutrient Dynamics

After two years of experiments, only 10–20% of net N mineralization was detected due to the immobilization process in the initial decomposition period, which lasted longer than one year. Nutrients in decaying foliar leaf litter temporarily increase due to nutrient immobilization by microbes in the initial mass loss process. The leaf litter N and P immobilization phase prolonged more than 14 months in this study, which was similar to the N and P immobilization phase of more than 15 months in the foliar leaf litter decomposition of oak and Korean pine in a broadleaf-Korean pine mixed forest in NE China [19].

Different leaf litter types had a similar period for the immobilization and mineralization of N and P to diminishing C as well as similar leaf litter decomposition rates at the same stand. Net N mineralization rates of leaf litter are greatly influenced by the site conditions rather than leaf litter species, causing the same species to show different mineralization rates depending on the site [46].

Although soil N is an important determinant in leaf litter decay, soil P becomes a stronger control for leaf litter decay when it is limited under abundant soil N [47]. Seasonal P dynamics are more variable than N in leaf litter decomposition [48]. High precipitation in the summer promotes the movement of phosphate in the soil, inducing seasonal changes in the distribution of soil phosphate [33]. Thus, the proportions of P with other nutrients, such as C:P and N:P, as well as the concentration of P itself, were major regulators that could inhibit or further decay rates by providing a balanced nutritional pool for decomposers [28,49]. Seasonal changes in soil N:P in the oak stand were related to changes in available P content. The lowest soil P_2O_5 and spatial and seasonal variances of soil P_2O_5 (Table 2) and N:P (Figure 2) might contribute to a fluctuation of nutrient supply that could explain the lowest leaf litter decomposition rate prior to the second summer and the different patterns of immobilization and mineralization of N and P in the oak stand compared with the other stands. Therefore, leaf litter nutrient dynamics were more affected by site conditions than by leaf litter types, suggesting that leaf litter decomposition and leaf litter nutrient dynamics patterns between a secondary forest and an adjacent plantation would not be much different if the site condition and stand age were similar, due to the overwhelming seasonal climatic effect on leaf litter decomposition in temperate deciduous forests.

Oak leaf litter showed the smallest P content among leaf litter types, which might be related to the occurrence of least soil P_2O_5 in the oak stand. The Korean pine plantation had the lowest soil TP, but it had the highest soil P_2O_5 among the stand types, and the Korean pine leaf litter had the greatest leaf litter P content and mineralization rate, suggesting that after 40 years of occupation of a Korean pine plantation might increase soil P_2O_5 [50]. Lee *et al.* [33] reported that rapid P cycling in black locust communities resulted in increased membrane phosphate in soil compared with adjacent Mongolian oak communities. Although the effects of species composition might be less powerful than climate on leaf litter decomposition in temperate deciduous forests, effects of species change on detailed leaf litter decomposition process, such as leaf litter nutrient dynamics and soil nutrients, can be expected [10]. We could suggest different leaf litter decomposition patterns relevant to leaf litter N and P dynamics and soil properties between the Korean pine plantation and the oak stand. We expect that further studies on decomposer communities would clarify the decomposition mechanism in temperate deciduous forests.

5. Conclusions

This study compared the leaf litter decomposition patterns and nutrient dynamics between a Korean pine plantation and an adjacent native Mongolian oak–dominant broadleaved temperate forest. Seasonal climate, especially hot and humid summer weather, was a more powerful determinant in leaf litter decomposition patterns than leaf litter types, resulting in no mixed leaf litter effect on leaf litter decomposition. However, we did find effects of the changes in dominant species on leaf litter nutrient dynamics and soil nutrients. The conversion of natural forests into plantations may have long-term effects on nutrients available in the soil, eventually modifying soil nutrient budgets.

Acknowledgments

This study was supported by the Korea Forest Service (Project No. S111214L020130 and S211314L020100). We thank all the participants in the field survey and data processing. We thank the Seoul National University Forest for providing access to the study sites and weather data. We acknowledge the Research Institute of Agriculture and Life Sciences, Seoul National University for grammar assistance and anonymous reviewers for their valuable comments.

Author Contributions

Jaeeun Sohng and Pil Sun Park conceived and designed the experiments. Jaeeun Sohng carried out field work, data analysis and prepared the manuscript. Ah Reum Han, Mi-Ae Jeong, and Yunmi Park contributed to field work and data analysis. Byung Bae Park contributed to the chemical analysis of soil and leaf litter. Pil Sun Park supervised the study, did additional data analysis and reviewed and edited the work.

References

1. Gosz, J. Organic matter and nutrient dynamics of the forest and forest floor in the Hubbard Brook Forest. *Oecologia* **1976**, *22*, 305–320.

2. Meentemeyer, V. An approach to the biometereology of decomposer organisms. *Int. J. Biometeorol.* **1978**, *22*, 94–111.
3. Mudrick, D.A.; Hoosein, M.; Ray, R.H.J.; Townsend, E.C. Decomposition of leaf litter in an Appalachian forest: Effects of leaf species, aspect, slope position and time. *For. Ecol. Manage.* **1994**, 68, 231–250.
4. Sariyildiz, T.; Anderson, J.M.; Kucuk, M. Effects of tree species and topography on soil chemistry, leaf litter quality, and decomposition in Northeast Turkey. *Soil Biol. Biochem.* **2005**, *37*, 1695–1706.
5. Fogel, R.; Cromack, K.J. Effect of habitat and substrate quality on Douglas-fir leaf litter decomposition in western Oregon. *Can. J. Bot.* **1977**, *55*, 1632–1640.
6. Valachovic, Y.S.; Caldwell, B.A.; Cromack, K. Jr.; Griffiths, R.P. Leaf litter chemistry controls on decomposition of Pacific Northwest trees and woody shrubs. *Can. J. For. Res.* **2004**, *34,* 2131–2147.
7. Prescott, C.E. Influence of forest floor type on rates of leaf litter decomposition in microcosms. *Soil Biol. Biochem.* **1996**, *28*, 1319–1325.
8. Zhou, G.; Guan, L.; Wei, X.; Tang, X.; Liu, S.; Liu, J.; Zhang, D.; Yan, J. Factors influencing leaf litter decomposition: An intersite decomposition experiment across China. *Plant Soil* **2008**, *311*, 61–72.
9. Rouifed, S.; Handa, I.T.; David, J.-F.; Hättenschwiler, S. The importance of biotic factors in predicting global change effects on decomposition of temperate forest leaf litter. *Oecologia* **2010**, *163*, 247–256.
10. Aponte, C.; García, L.V.; Marañón, T. Tree species effect on leaf litter decomposition and nutrient release in Mediterranean oak forests changes over time. *Ecosystems* **2012**, *15*, 1204–1218.
11. Madritch, M.D.; Cardinale, B.J. Impacts of tree species diversity on leaf litter decomposition in northern temperate forests of Wisconsin, USA: A multi-site experiment along a latitudinal gradient *Plant Soil* **2007**, *292*, 147–159.
12. Wang, Q.; Wang, S.; Xu, G.; Fan, B. Conversion of secondary broadleaved forest into Chinese fir plantation alters leaf litter production and potential nutrient returns. *Plant Ecol.* **2010**, *209*, 269–278.
13. Lian, Y.; Zhang, Q. Conversion of a natural broad-leafed evergreen forest into pure and mixed plantation forests in a subtropical area: Effects on nutrient cycling. *Can. J. For. Res.* **1998**, *28*, 1518–1529.
14. Rothe, A.; Binkley, D. Nutritional interactions in mixed species forests: A synthesis. *Can. J. For. Res.* **2001**, *31*, 1855–1870.
15. Hoorens, B.; Aerts, R.; Stroetenga, M. Does initial leaf litter chemistry explain leaf litter mixture effects on decomposition? *Oecologia* **2003**, *137*, 578–586.
16. Gartner, T.B.; Cardon, Z.G. Decomposition dynamics in mixed-species leaf litter. *Oikos* **2004**, *104*, 230–246.
17. Gholz, H.; Wedin, D.; Smitherman, S.; Harmon, M.; Parton, W. Long-term dynamics of pine and hardwood leaf litter in contrasting environments: Toward a global model of decomposition. *Glob. Change Biol.* **2000**, *6*, 751–765.

18. Laganiere, J.; Pare, D.; Bradley, R.L. How does a tree species influence leaf litter decomposition? Separating the relative contribution of leaf litter quality, leaf litter mixing, and forest floor conditions. *Can. J. For. Res.* **2010**, 40, 465–475.
19. Li, X.; Han, S.; Zhang, Y. Foliar decomposition in a broadleaf-mixed Korean pine, *Pinus koraiensis* Sieb. et Zucc. plantation forest, the impact of initial leaf litter quality and the decomposition of three kinds of organic matter fraction on mass loss and nutrient release rates. *Plant Soil* **2007**, *295*, 151–167.
20. Korea Forest Service. *Statistical Yearbook of Forestry*; Korea Forest Service: Daejeon, Korea, 2011.
21. Zhang, P.; Shao, G.; Zhao, G.; Le Master, D.C.; Parker, G.R.; Dunning, J.B. Jr.; Li, Q. China's forest policy for the 21st century. *Science* **2000**, *288*, 2135–2136.
22. Weather information. Available online: http://www.kma.go.kr/weather/climate/average_30years.jsp (accessed on 3 October 2013).
23. Buol, S.W.; Southar, R.J.; Graham, R.C.; McDaniel, P.A. *Soil Genesis and Classification*; 5th ed.; Iowa State Press: Iowa, IA, USA, 2003.
24. Miller, W.P.; Miller, D.M. A micro - pipette method for soil mechanical analysis. *Commun. Soil Sci. Plan.* **1987**, *18*, 1–15.
25. Konen, M.E.; Jacobs, P.M.; Burras, C.L.; Talaga, B.J.; Mason, J.A. Equations for predicting soil organic carbon using Loss-on-Ignition for North Central U.S. soils. *Soil Sci. Soc. Am. J.* **2002**, *66*, 1878–1881.
26. Sumner, M.E.; Miller, W.P. Cation exchange capacity, and exchange coefficients. In *Methods of Soil Analysis, Part 3. Chemical methods. Soil Science of Society of America Book Series no. 5*; 3rd ed.; Sparks, D.L., Ed.; Soil Science Society of America: Madison, WI, USA, 1996; pp. 1201–1229.
27. Wardle, D.A.; Bonner, K.I.; Nicholson, K.S. Biodiversity and plant leaf litter: Experimental evidence which does not support the view that enhanced species richness improves ecosystem function. *Oikos* **1997**, *79,* 247–258.
28. Xu, X.; Hirata, E. Decomposition patterns of leaf litter of seven common canopy species in a subtropical forest: N and P dynamics. *Plant Soil* **2005**, *273*, 279–289.
29. Bonanomi, G.; Incerti, G.; Antignani, V.; Capodilupo, M.; Mazzoleni, S. Decomposition and nutrient dynamics in mixed leaf litter of Mediterranean species. *Plant Soil* **2010**, *331*, 481–496.
30. You, Y.H.; Namgung, J.; Lee, Y.Y.; Kim, J.H.; Lee, J.Y.; Mun, H.T. Mass loss and nutrients dynamics during the leaf litter decomposition in Kwangnung experimental forest. *Jour. Korean For. Soc.* **2000**, *89*, 41–48.
31. Olson, J.S. Energy stores and the balance of producers and decomposers in ecological systems. *Ecology* **1963**, *44*, 322–331.
32. Anaya, C.A.; Jaramillo, V.J.; Martínez-yrízar, A.; García-oliva, F. Large rainfall pulses control leaf litter decomposition in a tropical dry forest: Evidence from an 8-year study. *Ecosystems* **2012**, *15*; 652–663.
33. Lee, Y.C.; Nam, J.M.; Kim, J.G. The influence of black locust, *Robinia pseudoacacia*. flower and leaf fall on soil phosphate. *Plant Soil* **2011**, *341*, 269–277.

34. Lee, I.K.; Lim, J.H.; Kim, C.S.; Kim, Y.K. Nutrient dynamics in decomposing leaf litter and leaf litter production at the long-term ecological research site in Mt. Gyebangsan. *J. Ecol. Field Biol.* **2006**, *29*, 585–591.
35. Zhang, D.; Hui, D.; Luo, Y.; Zhou, G. Rates of leaf litter decomposition in terrestrial ecosystems: Global patterns and controlling factors. *J. Plant Ecol.* **2008**, *1*, 85–93.
36. Han, A.R.; Lee, S.K.; Suh, G.U.; Park, Y.; Park, P.S. Wind and topography influence the crown growth of *Picea jezoensis* in a subalpine forest on Mt. Deogyu, Korea. Agric. *For. Meteorol.* **2012**, *166–167*, 207–214.
37. Hinckley, E-L. S.; Ebel, B.A.; Barnes, R.T.; Anderson, R.S.; Williams, M.W.; Anderson, S.P. Aspect control of water movement on hillslopes near the rain-snow transition of the Colorado Front Range. *Hydrolo. Proc.* **2014**, *28*, 74–85.
38. Wang, S.; Ruan, H.; Han, Y. Effects of microclimate, leaf litter type, and mesh size on leaf litter decomposition along an elevation gradient in the Wuyi Mountains, China. *Ecol. Res.* **2010**, *25*, 1113–1120.
39. Austin, A.T.; Vitousek, P.M. Precipitation, decomposition and leaf litter decomposability of *Metrosideros polymorpha* in native forests on Hawaii. *J. Ecol.* **2000**, *88*, 129–138.
40. Mun, H.T.; Joo, H.T. Leaf litter production and decomposition in the *Quercus acutissima* and *Pinus rigida* forests. *Korean J. Ecol.* **1994**, *17*, 345–353.
41. Hisabae, M.; Sone, S.; Inoue, M. Breakdown and macrointervebrate colonization of needle and leaf litter in conifer plantation streams in Shikoku, southwestern Japan. *J. For. Res.* **2011**, *16*, 108–115.
42. Bardgett, R.D.; Shine, A. Linkages between plant leaf litter diversity, soil microbial biomass and ecosystem function in temperate grasslands. *Soil Biol Biochem* **1999**, *31*, 317–321.
43. De Marco, A.; Meola, A.; Maisto, G.; Giordano, M.; De Santo, A.V. Non-additive effects of leaf litter mixtures on decomposition of leaf litters in a Mediterranean maquis. *Plant Soil* **2011**, *344*, 305–317.
44. García-Palacios, P.; Maestre, F.T.; Kattge, J.; Wall, D.H. Climate and leaf litter quality differently modulate the effects of soil fauna on leaf litter decomposition across biomes. *Ecol. Lett.* **2013**, *16*, 1045–1053.
45. Chapman, S.K.; Newman, G.S.; Hart, S.C.; Schweitzer, J.A.; Koch, G.W. Leaf litter mixtures alter microbial community development: mechanisms for non-additive effects in leaf litter decomposition. *PLoS One* **2013**, *8*, doi:10.1371/journal.pone.0062671.
46. Prescott, C.E.; Vesterdal, L.; Pratt, J.; Venner, K.H.; Montigny, L.M.D. Trofymow, J.A. Nutrient concentrations and nitrogen mineralization in forest floors of single species conifer plantations in coastal British Columbia. *Can. J. For. Res.* **2000**, *30*, 1341–1352.
47. Hobbie, S.E.; Vitousek, P.M. Nutrient limitation of decomposition in Hawaiian forests. *Ecology* **2000**, *81*, 1867–1877.
48. Aerts, R.; Callaghan, T.V.; Dorrepaal, E.; van Logtestijn, R.S.P.; Cornelissen, J.H.C. Seasonal climate manipulations have only minor effects on leaf litter decomposition rates and N dynamics but strong effects on leaf litter P dynamics of sub-arctic bog species. *Oecologia* **2012**, *170*, 809–819.

49. Moore, T.R.; Trofymow, J.A.; Prescott, C.E.; Titus, B.D.; Group, C.W. Nature and nurture in the dynamics of C, N and P during leaf litter decomposition in Canadian forests. *Plant Soil* **2011**, *339*, 163–175.

50. Celi, L.; Cerli, C.; Turner, B.L.; Santoni, S.; Bonifacio, E. Biogeochemical cycling of soil phosphorus during natural revegetation of *Pinus sylvestris* on disused sand quarries in Northwestern Russia. *Plant Soil* **2013**, *367*, 121–134.

7

Native and Alien Plant Species Richness Response to Soil Nitrogen and Phosphorus in Temperate Floodplain and Swamp Forests

Richard Hrivnák [1,*], Michal Slezák [1,2,3], Benjamín Jarčuška [4], Ivan Jarolímek [1] and Judita Kochjarová [1,5]

[1] Institute of Botany, Slovak Academy of Sciences, Dúbravská cesta 9, SK-845 23 Bratislava, Slovakia; E-Mails: slezak.miso@gmail.com (M.S.); ivan.jarolimek@savba.sk (I.J.); kochjarova@rec.uniba.sk (J.K.)

[2] Department of Biology and Ecology, Faculty of Education, Catholic University, Hrabovská cesta 1, SK-034 01 Ružomberok, Slovakia

[3] Department of Phytology, Faculty of Forestry, Technical University in Zvolen, T. G. Masaryka 24, SK-960 53 Zvolen, Slovakia

[4] Institute of Forest Ecology, Štúrova 2, SK-960 53 Zvolen, Slovakia; E-Mail: jarcuska@savzv.sk

[5] Botanical Garden, Comenius University, SK-038 15 Blatnica, Slovakia

* Author to whom correspondence should be addressed; E-Mail: richard.hrivnak@savba.sk;

Academic Editor: Heinz Rennenberg

Abstract: Soil nitrogen and phosphorus are commonly limiting elements affecting plant species richness in temperate zones. Our species richness-ecological study was performed in alder-dominated forests representing temperate floodplains (streamside alder forests of *Alnion incanae* alliance) and swamp forests (alder carrs of *Alnion glutinosae* alliance) in the Western Carpathians. Species richness (*i.e.*, the number of vascular plants in a vegetation plot) was analyzed separately for native and alien vascular plants in 240 vegetation plots across the study area covering Slovakia, northern Hungary and southern Poland. The relationship between the species richness of each plant group and total soil nitrogen content, plant-available phosphorus and carbon to nitrogen (C/N) ratio was analyzed by generalized linear mixed models (GLMM) with Poisson error distribution and log-link function. The number of recorded native and alien species was 17–84 (average 45.4) and 0–9 (average 1.5) species per plot, respectively. The GLMMs were statistically significant

($p < 0.001$) for both plant groups, but the total explained variation was higher for native (14%) than alien plants (9%). The richness of native species was negatively affected by the total soil nitrogen content and plant-available phosphorus, whereas the C/N ratio showed a positive impact. The alien richness was predicted only by the total soil nitrogen content showing a negative effect.

Keywords: alder-dominated forests; diversity; native and alien species; soil nutrients; vascular plants; vegetation

---

1. Introduction

Variation of plant species richness in plant communities can be explained by more than one hundred plausible ecological hypotheses and theories [1,2] with little consensus regarding the nature of causal processes [3]. Experimental and observational biodiversity research recognized habitat productivity as one of the major determinants controlling local species richness of vascular plants (e.g., [4,5]). This relationship originally showed a typical hump-shaped pattern, *i.e.*, the highest species richness is at the intermediate productivity level and gradually declined towards both marginal parts of the productivity gradient [6]. However, this response appeared to be vegetation- and scale-dependent [2,7] and richness has been recently seen, not simply as a function of productivity, but as feedback to influence productivity [8].

While the relationship between habitat productivity and species richness attracted much of the research effort aimed to understand species richness patterns [3], the relationship between species richness and other measured environmental factors is far less well studied [4]. Soil nitrogen and phosphorus play an important role in the regulation of plant growth and the functioning of several ecosystem processes [9]. These nutrients are commonly absorbed from soluble inorganic sources in soil solutions. However, plant uptake strategies and mechanisms related to nutrient acquisition promote resource partitioning and may lead to local species coexistence [9,10]. Changes in soil nitrogen and phosphorus availability are reflected either in their concentrations in plant tissues, ratios with other nutrients [11] or in species richness patterns. They are considered to be a limiting soil element affecting species richness of vegetation in temperate zone [10,12,13]. The increasing soil nitrogen and phosphorus content usually reduces the richness of herb layer vascular plants (e.g., [14,15]), but unimodal, positive or non-significant effects were also found in empirical studies (e.g., [4,16,17]). These responses are most likely a consequence of diverse nitrogen and phosphorus pools among study sites. While the total soil nitrogen content corresponds especially to its supply in mineral soil and/or thickness of organic layers, the level of phosphorus is driven mainly by abiotic resources and soil age [10]. The effect of soil phosphorus on species richness is much more ambiguous because of interactions with other covariant factors modifying the supply, chemical form and availability of nutrients (e.g., soil reaction, chemistry of geological substrates). The carbon to nitrogen (C/N) ratio of soil organic matter is a useful indicator of site quality with effects on both forest species composition and richness [18,19]. However, plant species richness in floodplain forests is also driven by several other environmental factors such as soil moisture and reaction, landscape configuration and dynamic of water regime [20–22]. These factors can account for a major part of explained variation in the species richness-environmental relationship [20].

Black alder (*Alnus glutinosa* (L.) Gaertn.) and grey alder (*Alnus incana* (L.) Moench) represent indigenous tree species to Central Europe, which dominate in floodplains (streamside alder forests, phytosociological alliance *Alnion incanae* Pawłowski *et al*. 1928) and swamp forests (alder carrs, *Alnion glutinosae* Malcuit 1929) [23] (hereinafter alder forests). While *Alnus glutinosa* can dominate in both vegetation types, *Alnus incana* almost exclusively prefers streamside alder forests [23]. They grow in habitats with relatively broad ecological gradients including mineral and organic soils [24–26]. Both alder species have constant effect on understory vegetation, as they improve the soil nitrogen pool and consequently modify the C/N ratio of the soil organic matter by symbiotic N_2-fixation [27]. The alder forests are, thus, commonly referred to as non-nitrogen-limited ecosystems, although a mineralization of organic matter is locally disabled by permanently flooded soils [28]. Alder species also positively affect diversity and activity of soil microbial communities leading to a larger increase in soil phosphorus availability [29]. Therefore, the alder forests represent a proper base for the study of relationships among species richness and soil characteristics.

Invasions of alien plant species, especially neophytes, are considered one of the major threats to the diversity of natural ecosystems including floodplain forests [30–32]. Among natural and semi-natural vegetation types, floodplain forests belong to the most invaded habitats [33]. They usually contain a higher number of alien species compared with all other forest vegetation types [32,34–36]. Plant invasions may, thus, enrich the overall species richness and modify the structure of species composition. The streamside alder forests are particularly highly prone to invasions of alien species, as rivers frequently act as propagule dispersal corridors [37,38]. Moreover, there is a strong human impact on these ecosystems (e.g., forestry, agricultural, both human utilization and activities in forests and surrounding landscape), which enhances the expansion rate of alien species into disturbed habitats.

Alder forests commonly create fragmented and small-scale forest stands due to the changes and transformation of the hydrological regime and fragmentation. These fragments are obviously influenced by adjacent landscape-scale factors [39]. Although these medium to species-rich forests [28,40] belong to threatened habitats at European or national levels (European Habitat Directive 92/43/EEC) [41,42], they represent a relatively frequent habitat in the Western Carpathians [23].

The aim of present study is to assess the relationship between soil nitrogen, plant-available phosphorus, C/N ratio and plant species richness in alder forests. Our main questions were: (i) what is the role of soil nitrogen, plant-available phosphorus and C/N ratio in the variation of native and alien plant species richness? and (ii) are there differences in responses of both plant groups to these factors? As native and invasive species show differences in their competitive abilities for nutrients [43], we expect different responses of the native and invasive species richness to the explanatory variables.

2. Material and Methods

2.1. Study Area

The species richness-edaphic relationship in floodplains (streamside alder forests; *Alnion incanae*) and swamp forests (alder carrs; *Alnion glutinosae*) was studied across the central part of the Western

Carpathians, including Slovakia, northern Hungary and southern Poland (Figure 1). Vegetation plots of this territory cover a wide latitudinal (47.829°–49.685°) and altitudinal (131–917 m) range, and include two ecologically and floristically distinctive European biogeographical areas—the Pannonian and Western Carpathian regions. The Pannonian climate shows drier and warmer vegetation seasons and affects the southern part of the study area, whereas the Carpathian climate is more humid and affects the central and northern part. The streamside alder forests prefer riparian zones and spring-fed areas from lowland to mountain regions, with the understory formed mainly by meso to hygrophilous species. In contrast, the crucial plant components of the understory of alder carr forests, frequent especially within lowland catchments, are mainly sedges and perennial grasses [23].

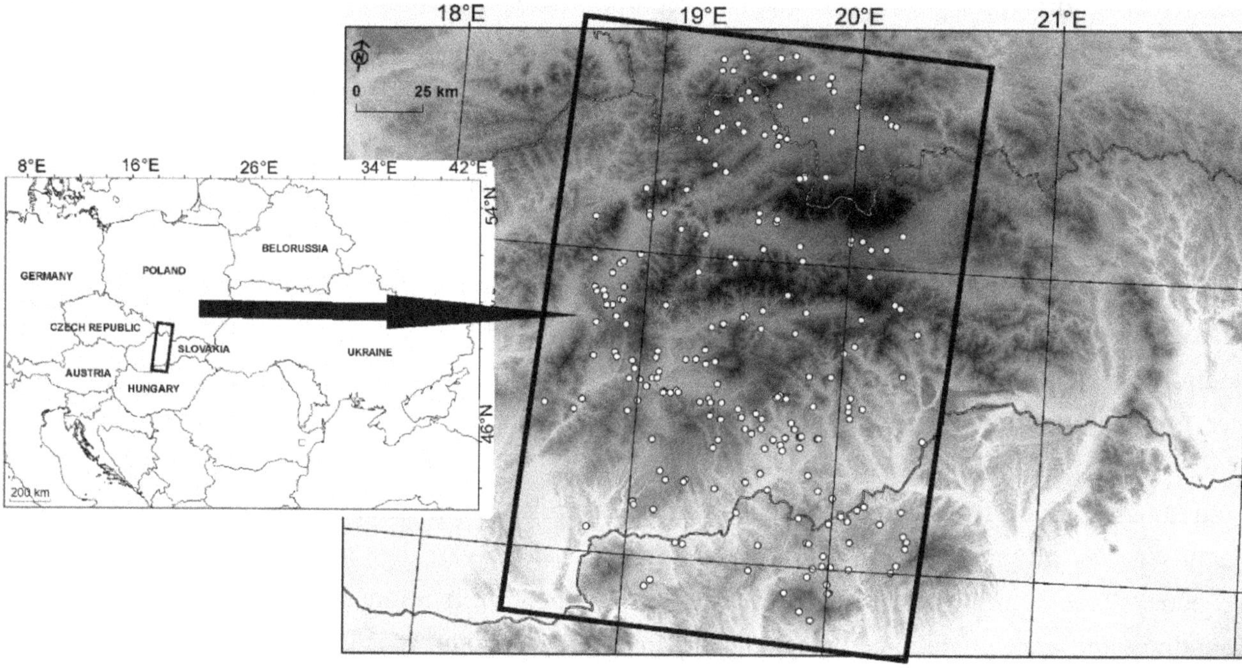

Figure 1. Study area and distribution of vegetation plots.

2.2. Field Sampling and Laboratory Analyses

Vegetation and soil variables were sampled in the centre of physiognomically homogeneous forest stands with a dominance of species *Alnus glutinosa* or *A. incana* in the tree layer (*i.e.*, canopy cover more than 50%). Sampling was carried out in plots with a uniform size of 400 m^2. The intensity of sampling in each alder-dominated site corresponded to the environmental heterogeneity and presence of different vegetation types [23]. In more detail, the distance between the two nearest plots of the same vegetation type was usually at least 4 km. In case the habitat conditions of the alder-dominated site were considerably variable, which was obvious in the species composition pattern, we decided to sample more than one plot. Two hundred and forty plots were sampled between June and August from 2010–2014. They were originally collected for the purpose of phytosociological classification using the standard European phytosociological approach [44]; results were published recently in Slovakia [23]. Each plot contained a list of all present vascular plant species. Species nomenclature and taxonomy were unified according to the checklist of Marhold and Hindák [45]. The plant species were divided into two groups, native and alien species based on a checklist of non-native plants in Slovakia [46]. Species richness (*i.e.*, the number of species per sampling plot) was determined for both plant groups. This dataset was stored in the Institute of Botany SAS as an xls-file.

Soil samples were randomly taken in three places in various parts of each plot from the uppermost mineral horizon (0–10 cm depth, litter removed) and mixed to form a single sample (weight approximately 1 kg) per plot in order to reduce the soil heterogeneity. They were dried at a laboratory temperature, crushed and passed through a 2 mm sieve. Soil analyses were performed on air-dried samples following standard protocols. Total carbon (C) and total nitrogen (N) contents were determined using an NCS-FLASH 1112 analyzer (CE Instruments, UK) and subsequently, the C/N-ratio as an appropriate surrogate for the mineralization rate [47] was calculated. Plant-available phosphorus (P) was extracted using the Mehlich II solution [48] and measured by spectrophotometry (AES-ICP).

2.3. Data Analysis

The variation in species richness of native and alien vascular plants explained by the soil variables was analyzed using the generalized linear models (GLM) and generalized linear mixed models (GLMM). Poisson error distribution and log-link function were initially applied in model specification. Since over dispersion and presence of some spatial structure in the GLMs' residuals were observed (for more details see [49,50]), spatial GLMM with penalized quasi-likelihood (glmmPQL; library "MASS") [51] was used finally. This modeling procedure allows the use of spatially structured random effects. Gaussian correlation structure was used in the GLMMs. However, differences between these models—accounting for spatial autocorrelation—and GLM-models were slight.

Quadratic terms for all explanatory variables (after centering) were also meant to be included in the models. However, only the squared C/N-ratio entered the models due to collinearity ($R > 0.60$) of quadratic terms. Two models (i) containing only main effects and (ii) with second order interactions between explanatory variables were created for both native and alien species richness.

The graphic distribution of model's residuals was assessed by plotting of the standardized residuals against the values predicted by the model, as well as against each explanatory variable. The spatial autocorrelation of model's residuals was checked by the global Moran's I test (library "ape") [52]. The spatial autocorrelation in response to distance was assessed by a Moran's I correlogram with an increment of 2 (library "ncf") [53]. The proportion of explained variation in glmmPQL (pseudo-R2 value; hereinafter, R2) was calculated as the squared Pearson correlation coefficient between observed and predicted values [54]. All statistical analyses were performed in the R version 3.0.0 [55] through the integrated development environment RStudio [56].

3. Results

Altogether, 490 vascular plant species were recorded in 240 vegetation plots of alder forests. The total number of species ranged from 17 to 92 per plot. Native species created a substantial part of the total species richness, whereas the number of alien species was considerably lower, but had obvious variance among particular plots. The values of studied soil characteristics were relatively variable in plots (Table 1).

Table 1. Descriptive statistics of species richness characteristics and ecological variables ($n = 240$).

	Average ± S.D.	Min.	Max.	C.V.
Vegetation Characteristics				
Number of native species	45.4 ± 11.8	17	84	25.9
Number of alien species	1.5 ± 1.8	0	9	121.8
Soil Variables				
Total N (%)	0.599 ± 0.468	0.080	2.549	79.6
P (mg/kg)	17.702 ± 9.546	0.500	64.290	53.9
Total C (%)	7.651 ± 6.153	1.000	31.727	82.0
C/N ratio	12.882 ± 3.170	4.929	25.189	24.6

S.D.: standard deviation, CV: coefficient of variation.

The main effects models and the models with second order interactions between soil variables accounted for a comparable proportion of data variability (14% and 16%, respectively) (Tables 2 and 3). Total explained variability in species richness of vascular plants was higher for native (14%) than alien (9%) species. While the species richness of native plants was significantly related to all analyzed variables, the richness pattern of aliens was affected by only one predictor (Table 2). Higher concentrations of plant-available phosphorus and total nitrogen reduced the species richness of native plants, whereas the C/N ratio showed a positive effect. The alien species richness was negatively related to the total nitrogen content (Figure 2). Second order interactions between the soil variables did not explain the variability of species richness either in native or alien plant species (Table 3).

Table 2. Main effects models with the number of native and alien plants as a response to ecological variables, respectively.

	Native Species				Alien Species			
	Estimate	**S.E.**	***t***	***P***	**Estimate**	**S.E.**	***t***	***P***
(Intercept)	3.8243	0.0182	210.02	<0.001	0.4036	0.0850	4.75	<0.001
Total N	−0.0790	0.0340	−2.32	**0.021**	−0.6019	0.1506	−4.00	**<0.001**
P	−0.0068	0.0018	−3.87	**<0.001**	0.0088	0.0071	1.25	0.213
C/N	0.0217	0.0061	3.53	**<0.001**	−0.0188	0.0232	−0.81	0.419
C/N^2	−0.0016	0.0009	−1.77	0.08	−0.0043	0.0039	−1.11	0.269
R^2	0.14	$p < 0.001$			0.09	$p < 0.001$		

S.E.: standard error; df = 235; for R^2 calculation see Data analysis.

Table 3. Second order interactions models with the number of native and alien plants as response variables, respectively.

	Native Species				Alien Species			
	Estimate	**S.E.**	***t***	***P***	**Estimate**	**S.E.**	***t***	***P***
(Intercept)	3.8247	0.0189	202.81	<0.001	0.4176	0.0872	4.79	<0.001
Total N (1)	−0.1057	0.0354	−2.99	**0.003**	−0.4653	0.1494	−3.11	**0.002**
P (2)	−0.0060	0.0020	−2.97	**0.003**	0.0029	0.0082	0.36	0.722
C/N (3)	0.0074	0.0086	0.86	0.392	−0.0620	0.0376	−1.65	0.100
C/N^2 (4)	−0.0021	0.0015	−1.40	0.164	−0.0125	0.0068	−1.85	0.066
1 × 2	−0.0051	0.0041	−1.26	0.207	−0.0087	0.0176	−0.50	0.620

Table 3. *Cont.*

	Native Species				Alien Species			
	Estimate	**S.E.**	***t***	***P***	**Estimate**	**S.E.**	***t***	***P***
1 × 3	−0.0297	0.0205	−1.44	0.149	0.0170	0.0930	0.18	0.855
1 × 4	0.0019	0.0041	0.49	0.626	−0.0107	0.0192	−0.56	0.577
2 × 3	−0.00002	0.0008	−0.02	0.985	−0.0040	0.0030	−1.34	0.183
2 × 4	−0.0002	0.0001	−1.20	0.230	0.00004	0.0005	0.09	0.932
3 × 4	−0.0002	0.0002	0.96	0.336	0.0012	0.0007	1.56	0.120
R^2	0.16	$p < 0.001$			0.08	$p < 0.001$		

S.E.: standard error; df = 229; for R^2 calculation see Data analysis.

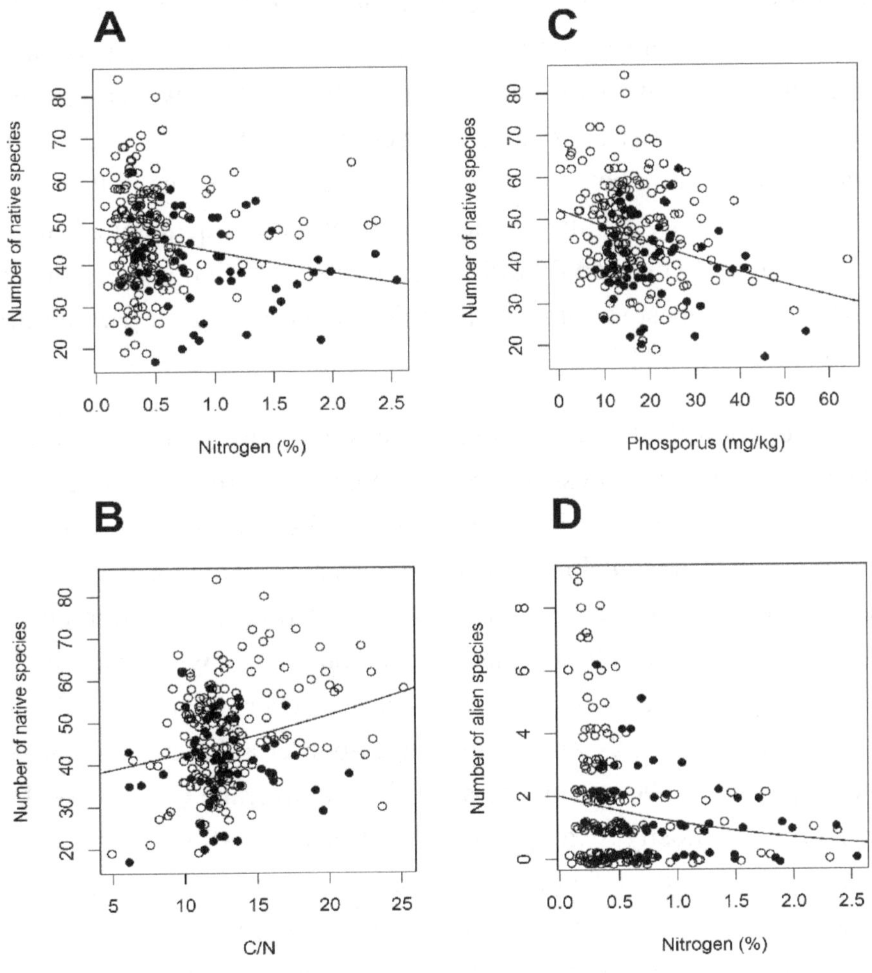

Figure 2. Relationship between species richness of native (**A–C**) and alien (**D**) vascular plants and ecological variables, which were statistically significant in the used model (empty circles—streamside alder forests, full circles—alder carrs).

4. Discussion

The present study showed a negative effect of total nitrogen on species richness for both native and alien plant species. Although some previous studies focusing on diversity research in forest understory vegetation found the unimodal response (e.g., [6]), this negative trend (*i.e.*, a higher amount of soil nitrogen led to a lower number of species) was most often identified in temperate deciduous forests [14,16,17]. The differences in the species richness trends along the nitrogen gradient can be partially addressed to habitat-related biases (e.g., restricted length of analyzed nitrogen gradient). Thus, the final response of plant species richness along the productivity gradient generally appears to be dependent on the inclusion of extreme habitats on the gradient [6,14]. We do not suppose this pattern to be relevant in our study, as the analyzed data covered the whole vegetation and nitrogen gradient of the alder forests in the Western Carpathians.

The negative effect of soil nitrogen content on species richness can be explained by a potential positive effect of nitrogen on certain highly competitive plants [16,21]. These species may have suppressed competitively weaker species. In our data, the nitrogen-rich sites are usually covered by dominant species such as *Caltha palustris*, *Carex acutiformis*, *Petasites hybridus* or *Phragmites australis*. We expected differences in total nitrogen-species richness pattern for the native and alien species, as they differ in their strategies for growth and survival in habitats [43]. However, the negative response of species richness to soil nitrogen was consistent between groups. Similarly, Jones and Chapman [17] found a negative decreasing trend in both cases of native and alien species in Pennsylvania oak forests. In addition, the highest variability of alien species richness was documented at lower parts of the nitrogen gradient in our study, which could affect the number of alien species-nitrogen relationships. Similarly to alder forests, the highest variability of species richness at lower part of nitrogen gradient has also been observed in other forests [17].

The C/N ratio had a positive effect on the species richness of native plants. Our findings differed from those observed in German floodplain forests [19,57], where no relationship between C/N ratio and plant species richness was found. Similarly, positive trends and/or slightly convex curves were also identified in deciduous forests by Schuster and Diekmann [4]. The C/N ratio of soil organic matter was related to a nitrogen immobilization and mineralization during decomposition by microorganisms [18,58]. Alder forests of European temperate zone create two main vegetation groups, which vary in their environmental conditions, including their affinity for soil substrates. Species-richer streamside alder forests show better soil mineralization [28] and higher C/N ratio than species-poorer alder carrs. In our data, the ten most species-rich sites corresponded to the streamside alder forests (average C/N ratio = 14.15), whereas ten most species-poor sites were mostly swamp forests with a C/N ratio of 11.12 (see also Figure 2). Increasing species richness along the C/N ratio gradient in the total data set of alder forests is therefore logical and expected, although it does not agree with findings reported from mesic deciduous forests. In addition, a positive correlation between C/N ratio and species richness most likely corresponds to a negative effect of nitrogen on species richness.

Plant-available phosphorus was the most variable soil characteristic in our study (Table 1). Similar soil phosphorus concentrations were also documented in temperate floodplain and swamp alder forests of the south-western Poland [59]. Diverse forms of soil phosphorus support niche differentiation and facilitate plant species coexistence [60], but plant species richness commonly declines along an increasing soil phosphorus gradient [15,61]. The negative species richness-phosphorus relationship was also found in the present study for native vascular plants. High soil phosphorus content promotes vigorous growth, especially in fast-colonizing, competitive species, and reduces the species richness of vascular plants in broadleaved deciduous forests [16,62]. Phosphorus rich sites in studied alder forests were dominated mainly by highly productive clonal species, most often by *Urtica dioica*. The high availability of soil phosphorus favors species *Urtica dioica* during successful development and strong competition [63]. Indeed, clonal species with lateral spreading are able to reallocate resources among individual ramets growing in patches with diverse nutrient-availability [22]. Thus, their competitive advantage at high phosphorus levels can suppress the colonization and successful establishment of other plant species and, consequently, can lead to the species richness decline. Although these species occurred at phosphorus-poorer sites as well, they reached only an additional cover value within the herb layer vegetation. Changes in dominant species at phosphorus-rich grassland sites are usually accompanied by changes in overall species composition and vegetation [61], but we did not observe this pattern in alder forests.

It is important to note that the soil variables (total nitrogen, plant-available phosphorus and C/N ratio) used as explanatory variables in our study explained together only between 8 and 16% of the variability in alien and native species richness in alder forests' understories, respectively. For a complex interpretation of variation in species richness or composition of vascular plants in streamside and alder carr forests, other habitat quality and landscape configuration-related factors have to be taken into account as well [20,21,39]. In more detail, habitat quality parameters, such as hydrological characteristics (e.g., water table, flooding), soil nutrient/acidity complex, temperature and light availability represent significant variables controlling vegetation structure of alder forests [20,22]. Moreover, the portion of surrounding country, distance to settlements or distance from the stream source plays the important role, as well [20,22,39]. These drivers could explain the major part of the variation in species richness-environmental relationships within temperate floodplain and swamp forests.

Acknowledgments

We would like to thank Dušan Senko for the preparation of the map. The research was supported by the Science Grant Agency of the Ministry of Education of the Slovak Republic and the Slovak Academy of Sciences (VEGA 2/0019/14 and 2/0051/15) and COST project "European Information System for Alien Species".

Author Contributions

We declare the following concerning the authors' contributions to the research: fieldwork: Richard Hrivnák, Michal Slezák, Judita Kochjarová, Ivan Jarolímek; manuscript preparation: Richard Hrivnák, Michal Slezák, Benjamín Jarčuška, Ivan Jarolímek; data analyses: Benjamín Jarčuška.

References

1. Palmer, M.W. Variation in species richness: Towards a unification of hypotheses. *Folia Geobot. Phytotaxon.* **1994**, *29*, 511–530.
2. Grace, J.B. The factors controlling species density in herbaceous plant communities: An assessment. *Perspect. Plant Ecol. Evol. Syst.* **1999**, *2*, 1–28.

3. Gillman, L.N.; Wright, S.D. The influence of productivity on the species richness of plants: A critical assessment. *Ecology* **2006**, *87*, 1234–1243.
4. Schuster, B.; Diekmann, M. Species richness and environmental correlates in deciduous forests of Northwest Germany. *For. Ecol. Manag.* **2005**, *206*, 197–205.
5. Axmanová, I.; Chytrý, M.; Zelený, D.; Li, C.-F.; Vymazalová, M.; Danihelka, J.; Horsák, M.; Kočí, M.; Kubešová, S.; Lososová, Z.; *et al.* The species richness-productivity relationship in the herb layer of European deciduous forests. *Glob. Ecol. Biogeogr.* **2012**, *21*, 657–667.
6. Pausas, J.G.; Austin, M.P. Patterns of plant species richness in relation to different environments: An appraisal. *J. Veg. Sci.* **2001**, *12*, 153–166.
7. Waide, R.B.; Willing, M.R.; Steiner, C.F.; Mittelbach, G.; Gough, L.; Dodson, S.I.; Juday, G.P.; Parmenter, R. The relationship between productivity and species richness. *Annu. Rev. Ecol. Syst.* **1999**, *30*, 257–300.
8. Adler, P.B.; Seabloom, E.W.; Borer, E.T.; Hillebrand, H.; Hautier, Y.; Hector, A.; Harpole, W.S.; O'Halloran, L.R.; Grace, J.B.; Anderson, T.M.; *et al.* Productivity is a poor predictor of plant species richness. *Science* **2011**, *333*, 1750–1753.
9. Lambers, H.; Brundrett, M.C.; Raven, J.A.; Hopper, S.D. Plant mineral nutrition in ancient landscape: High plant species diversity on infertile soils is linked to functional diversity for nutritional strategies. *Plant Soil* **2010**, *334*, 11–31.
10. Lambers, H.; Raven, J.A.; Shaver, G.R.; Smith, S.E. Plant nutrient-acquisition strategies change with soil age. *Trends Ecol. Evol.* **2008**, *23*, 95–103.
11. Güsewell, S. N:P ratios in terrestrial plants: variation and functional significance. *New Phytol.* **2004**, *164*, 243–266.
12. Attiwill, P.M.; Adams, M.A. Nutrient cycling in forests. *New Phytol.* **1993**, *124*, 561–582.
13. Vitousek, P.M.; Porder, S.; Houlton, B.Z.; Chadwick, O.A. Terrestrial phosphorus limitation: Mechanisms, implications, and nitrogen-phosphorus interactions. *Ecol. Appl.* **2010**, *20*, 5–15.
14. Dupré, C.; Wessberg, C.; Diekmann, M. Species richness in deciduous forests: Effects of species pools and environmental variables. *J. Veg. Sci.* **2002**, *13*, 505–516.
15. Dumortier, M.; Butaye, J.; Jacquemyn, H.; van Camp, N.; Lust, N.; Hermy, M. Predicting vascular plant species richness of fragmented forests in agricultural landscapes in central Belgium. *For. Ecol. Manag.* **2002**, *158*, 85–102.
16. Hofmeister, J.; Hošek, J.; Modrý, M.; Roleček, J. The influence of light and nutrient availability on herb layer species richness in oak-dominated forests in central Bohemia. *Plant Ecol.* **2009**, *205*, 57–75.
17. Jones, R.O.; Chapman, S.K. The roles of biotic resistance and nitrogen deposition in regulating non-native understory plant diversity. *Plant Soil* **2011**, *345*, 257–269.
18. Yakamura, T.; Sahunalu, P. Soil carbon/nitrogen ratio as a site quality index for some South-east Asian forests. *J. Trop. Ecol.* **1990**, *6*, 371–378.
19. Härdtle, W.; von Oheimb, G.; Westphal, C. The effects of light and soil conditions on the species richness of the ground vegetation of deciduous forests in northern Germany (Schleswig-Holstein). *For. Ecol. Manag.* **2003**, *182*, 327–338.
20. Douda, J. The role of landscape configuration in plant composition of floodplain forests across different physiographic areas. *J. Veg. Sci.* **2010**, *21*, 1110–1124.

21. Mölder, A.; Schneider, E. On the beautiful diverse Danube? Danubian floodplain forest vegetation and flora under the influence of river eutrophication. *River Res. Appl.* **2011**, *27*, 881–894.
22. Douda, J.; Doudová-Kochánková, J.; Boublík, K.; Drašnarová, A. Plant species coexistence at local scale in temperate swamp forest: Test of habitat heterogeneity hypothesis. *Oecologia* **2012**, *169*, 523–534.
23. Slezák, M.; Hrivnák, R.; Petrášová, A. Numerical classification of alder carr and riparian alder forests in Slovakia. *Phytocoenologia* **2014**, *44*, 283–308.
24. McVean, D.N. Biological flora of the British Isles. *Alnus glutinosa* (L.) Gaertn. *J. Ecol.* **1953**, *41*, 447–466.
25. Schwabe, A. Monographie *Alnus incana*-reicher Waldgesellschaften in Europa Variabilität und Ähnlichkeit einer azonal verbreiteten Gesellschaftsgruppe. *Phytocoenologia* **1985**, *13*, 197–302.
26. Claessens, H.; Oosterbaan, A.; Savill, P.; Rondeux, J. A review of the characteristics of black alder (*Alnus glutinosa* (L.) Gaertn.) and their implications for silvicultural practises. *Forestry* **2010**, *83*, 163–175.
27. Eickenscheidt, T.; Heinichen, J.; Augustin, J.; Freibauer, A.; Drösler, M. Nitrogen mineralization and gaseous nitrogen losses from waterlogged and drained organic soils in a black alder (*Alnus glutinosa* (L.) Gaertn.) forest. *Biogeosciences* **2014**, *11*, 2961–2976.
28. Ellenberg, H. *Vegetation Ecology of Central Europe*; Cambridge University Press: New York, NY, USA, 2009; pp. 1–731.
29. Lõhmus, K.; Truu, M.; Truu, J.; Ostonen, I.; Kaar, E.; Vares, A.; Uri, V.; Alama, S.; Kanal, A. Functional diversity of culturable bacterial communities in the rhizosphere in relation to fine-root and soil parameters in alder stands on forests, abandoned agricultural, and oil-shale mining areas. *Plant Soil* **2006**, *283*, 1–10.
30. Kowarik, I. *Biologische Invasionen: Neophyten und Neozoen in Mitteleuropa*; Ulmer: Stuttgart, Germany, 2003; pp. 1–492.
31. Weber, E. *Invasive Plant Species of the World: A Reference Guide to Environmental Weeds*; CAB International Publ.: Wallingford, UK, 2003; pp. 1–548.
32. Pyšek, P.; Bacher, S.; Chytrý, M.; Jarošík, V.; Wild, J.; Celesti-Grapow, L.; Gassó, N.; Kenis, M.; Lambdon, P.W.; Nentwig, W.; *et al.* Contrasting patterns in the invasions of European terrestrial and freshwater habitats by alien plants, insects and vertebrates. *Glob. Ecol. Biogeogr.* **2010**, *19*, 317–331.
33. Richardson, D.M.; Holmes, P.M.; Esler, K.J.; Galatowitsch, S.M.; Stromberg, J.G.; Kirkman, S.P.; Pyšek, P.; Hobbs, R.J. Riparian vegetation: Degradation, alien plant invasions, and restoration prospects. *Divers. Distrib.* **2007**, *13*, 126–139.
34. Medvecká, J.; Jarolímek, I.; Senko, D.; Svitok, M. Fifty years of plant invasion dynamics in Slovakia along a 2500 m altitudinal gradient. *Biol. Invasions* **2014**, *16*, 1627–1638.
35. Walter, J.; Essl, F.; Englisch, T.; Kiehn, M. Neophytes in Austria: Habitat preferences and ecological effects. *Neobiota* **2005**, *6*, 13–25.
36. Vilá, M.; Pino, J.; Font, X. Regional assesment of plant invasions across different habitat types. *J. Veg. Sci.* **2007**, *18*, 35–42.
37. Burkart, M. River corridor plants (Stromtalpflanzen) in Central European lowland: A review of a poorly understood plant distribution pattern. *Glob. Ecol. Biogeogr.* **2001**, *10*, 449–468.

38. Zając, A.; Tokarska-Guzik, B.; Zając, M. The role of rivers and streams in the migration of alien plants into the Polish Carpathians. *Biodivers. Res. Conserv.* **2011**, *23*, 43–56.
39. Pielech, R.; Anioł-Kwiatkowska, J.; Szczęśniak, E. Landscape-scale factors driving plant species composition in mountain streamside and spring riparian forests. *For. Ecol. Manag.* **2015**, *347*, 217–227.
40. Chytrý, M.; Dražil, T.; Hájek, M.; Kalníková, V.; Preislerová, Z.; Šibík, J.; Ujházy, K.; Axmanová, I.; Bernátová, D.; Blanár, D.; *et al.* The most species-rich plant communities of the Czech Republic and Slovakia (with new world records). *Preslia* **2015**, *87*, 217–278.
41. Stanová, V.; Valachovič, M. *Katalóg Biotopov Slovenska*; DAPHNE—Inštitút aplikovanej ekológie: Bratislava, Slovakia, 2002; pp. 1–225.
42. Bölöni, J.; Molnár, Z.; Kun, A. *Magyarország Élőhelyei Vegetációtípusok Leírása és Határozója. Ánér 2011*; MTA Ökológiai és Botanikai Kutatóintézete: Vácrátót, Hungary, 2011; pp. 1–441.
43. Davis, M.A.; Grime, J.P.; Thomson, K. Fluctuating resources in plant communities: A general theory of invisibility. *J. Ecol.* **2000**, *88*, 528–534.
44. Dengler, J.; Chytrý, M.; Ewald, J. Phytosociology. In *Encyclopedia of Ecology*; Jørgensen, S.E., Fath, B.D., Eds.; Elsevier: Oxford, UK, 2008; Volume 4, pp. 2767–2779.
45. Marhold, K.; Hindák, F. *Zoznam nižších a vyšších rastlín Slovenska*; Veda: Bratislava, Slovakia, 1998; pp. 1–688.
46. Medvecká, J.; Kliment, J.; Májeková, J.; Halada, Ľ.; Zaliberová, M.; Gojdičová, E.; Feráková, V.; Jarolímek, I. Inventory of alien species of Slovakia. *Preslia* **2012**, *84*, 257–309.
47. Finzi, A.C.; van Breemen, N.; Canham, C.D. Canopy tree-soil interactions within temperate forests: species effects on soil carbon and nitrogen. *Ecol. Appl.* **1998**, *8*, 440–446.
48. Mehlich, A. New extractant for soil test evaluation of phosphorus, potassium, magnesium, calcium, sodium, manganese, and zinc. *Commun. Soil Sci. Plant Anal.* **1978**, *9*, 477–492.
49. Dormann, C.F.; McPherson, J.M.; Araújo, M.B.; Bivand, R.; Bolliger, J.; Carl, G.; Davies, R.G.; Hirzel, A.; Jetz, W.; Kissling, W.D.; *et al.* Methods to account for spatial autocorrelation in the analysis of species distributional data: a review. *Ecography* **2007**, *30*, 609–628.
50. Beale, C.M.; Lennon, J.J.; Yearsley, J.M.; Brewer, M.J.; Elston, D.A. Regression analysis of spatial data. *Ecol. Lett.* **2010**, *13*, 246–264.
51. Venables, W.N.; Ripley, B.D. *Modern Applied Statistics with S*; Springer: New York, NY, USA, 2002; pp. 1–495.
52. Paradis, E.; Claude, J.; Strimmer, K. APE: Analyses of phylogenetics and evolution in R language. *Bioinformatics* **2004**, *20*, 289–290.
53. Bjornstad, O.N. *Ncf: Spatial Nonparametric Covariance Functions*; R package version 1.1-5, 2013. Available online: http://onb.ent.psu.edu/onb1/Rs (accessed on 2 April 2015).
54. Kissling, W.D.; Carl, G. Spatial autocorrelation and the selection of simultaneous autoregressive models. *Glob. Ecol. Biogeogr.* **2008**, *17*, 59–71.
55. R Core Team. *R: A Language and Environment for Statistical Computing*; R Foundation for Statistical Computing: Vienna, Austria, 2013. Available online: http://www.R-project.org/ (accessed on 13 June 2014).

56. RStudio. *RStudio: Integrated Development Environment for R (version 0.98.1091)*; Computer Software: Boston, MA, USA, 2014. Available online: http://www.rstudio.org/ (accessed on 31 March 2015).
57. Härdtle, W.; von Oheimb, G.; Meyer, H.; Westphal, C. Patterns of species composition and species richness in moist (ash-alder) forests of northern Germany (Schleswig-Holstein). *Feddes Repert.* **2003**, *114*, 574–586.
58. Swift, M.J.; Heal, O.W.; Anderson, J.M. *Decomposition in Terrestrial Ecosystems*; Blackwell Scientific Publications: Oxford, UK, 1979; pp. 1–372.
59. Orczewska, A. The impact of former agriculture on habitat conditions and distribution patterns of ancient woodland plant species in recent black alder (*Alnus glutinosa* (L.) Gaertn.) woods in south-western Poland. *For. Ecol. Manag.* **2009**, *258*, 794–803.
60. Turner, B.L. Resource partitioning for soil phosphorus: A hypothesis. *J. Ecol.* **2008**, *96*, 698–702.
61. Merunková, K.; Chytrý, M. Environmental control of species richness and composition in upland grasslands of the southern Czech Republic. *Plant Ecol.* **2012**, *213*, 591–602.
62. Honnay, O.; Hermy, M.; Coppin, P. Impact of habitat quality on forest plant species colonization. *For. Ecol. Manag.* **1999**, *115*, 157–170.
63. Pigott, C.D. Analysis of the response of *Urtica dioica* to phosphate. *New Phytol.* **1971**, *70*, 953–966.

8

Biomass, Carbon and Nutrient Storage in a 30-Year-Old Chinese Cork Oak (*Quercus Variabilis*) Forest on the South Slope of the Qinling Mountains, China

Yang Cao [1,2] and Yunming Chen [1,2,*]

[1] State Key Laboratory of Soil Erosion and Dryland Farming on the Loess Plateau, Northwest A & F University, Yangling 712100, China; E-Mail: yang.cao@nwsuaf.edu.cn

[2] Institute of Soil and Water Conservation, Chinese Academy of Sciences and Ministry of Water Resources, Yangling 712100, China

* Author to whom correspondence should be addressed; E-Mail: ymchen@ms.iswc.ac.cn;

Academic Editors: Heinz Rennenberg and Mark A. Adams

Abstract: Chinese cork oak (*Quercus variabilis*) forests are protected on a large-scale under the Natural Forest Protection (NFP) program in China to improve the ecological environment. However, information about carbon (C) storage to increase C sequestration and sustainable management is lacking. Biomass, C, nitrogen (N) and phosphorus (P) storage of trees, shrubs, herb, litter and soil (0–100 cm) were determined from destructive tree sampling and plot level investigation in approximately 30-year old Chinese cork oak forests on the south slope of the Qinling Mountains. There was no significant difference in tree components' biomass estimation, with the exception of roots, among the available allometric equations developed from this study site and other previous study sites. Leaves had the highest C, N and P concentrations among tree components and stems were the major compartments for tree biomass, C, N and P storage. In contrast to finding no difference in N concentrations along the whole soil profile, higher C and P concentrations were observed in the upper 0–10 cm of soil than in the deeper soil layers. The ecosystem C, N, and P storage was 163.76, 18.54 and 2.50 t ha^{-1}, respectively. Soil (0–100 cm) contained the largest amount of C, N and P storage, accounting for 61.76%, 92.78% and 99.72% of the total ecosystem, followed by 36.14%, 6.03% and 0.23% for trees, and 2.10%, 1.19% and 0.03% for shrubs, herbs and litter, respectively. The equations accurately estimate ecosystem biomass, and the knowledge of the distribution of C, N and P storage will contribute to increased C sequestration and sustainable management of Chinese cork oak forests under the NFP program.

Keywords: biomass allocation; carbon concentration; nutrient element; Natural Forest Protection (NFP) program

---

1. Introduction

Protecting existing forests and planting new forests through reforestation and afforestation are important measures to enhance carbon (C) sequestration capacity and potential in terrestrial ecosystems [1–4]. A growing number of studies have addressed C storage and sequestration in key national ecological restoration programs in China, such as the Sloping Land Conversion program, Three-North Sheltbelt Forest program, Changjiang (Yangtze) River Basin Forest Protection program, Beijing-Tianjin Sandstorm Source Control program, Natural Forest Protection program (NFP), and other programs [2,5–9].

The NFP program was implemented in 1998 with the goal of promoting natural forest resource protection and cultivation to improve the ecological environment, has been implemented since 1998 [6,10]. It covers about 6.0×10^7 ha of natural forest land, which accounts for 50% of the total natural forest area within 17 provincial-level administrative units in China [10]. Hu and Liu [11] used the volume-biomass method and National Forestry Statistics to calculate C storage of the NFP program from 1998–2002 and found that in total it sequestered 44.07 Tg, including 21.32 Tg from reforestation and afforestation, and 22.75 Tg from reduced the timber production. The results of Wei *et al.* [10] showed that the tree C pool under the NFP program in northeastern China increased from 1998 to 2008, by 6.3 Tg C, which was mainly sequestrated by natural forests (5.1 Tg C $year^{-1}$). It has been projected that biomass C storage from afforestation under the NFP program could potentially increase from 33.67 Tg in 2011 to 96.03 Tg in 2020 [6]. However, most of these previous studies were mainly conducted on forest biomass C storage at the national and regional scales with different estimation methods and different forest resource data. Moreover, there are few precise studies concerning direct plot investigations for various forest types, C storage estimates that include understory, forest floor, and soil, and the relationship between climatic factors and forest types on regional scales [12–16]. An age-related study on C storage in a black locust forest ecosystem on the Loess Plateau showed that tree C storage increased from 5 to 38 years, but significantly decreased from 38 to 56 years owing to high tree mortality. Moreover, storage in the shrub layer increased with stand age, but it was age-independent in the herb layer and litter. Storage in the topsoil (0–20 cm) increased at a constant rate with stand age, while it was age-independent in sub-top soil [7]. With an increased numbers of local studies, we can gain a more comprehensive understanding of the complex nature of ecosystem C storage in order to scale up to regional and global levels.

Quercus species are a keystone species in a wide range of habitats from Mediterranean semi-desert woodlands to subtropical rainforest in Europe, North America, and Southeast Asia [17]. For example, the cork oak is the second most important Portuguese forest species both in terms of the country forest area and in terms of forest industry product exports [18]. Many studies have explored the temporal and spatial distribution of biomass and nutrient accumulation in *Quercus* species for suitable management and conservation, especially in Spain and Portugal [17–26]. Guyette, Dey and Stambaugh [17] documented the temporal distribution in C storage of oak wood at floodplains in northern Missouri, USA. In this study, we chose Chinese cork oak (*Quercus variabilis*) as a model system to carefully evaluate biomass and nutrient pools in different ecosystem components. Chinese cork oak is one of the major *Quercus* species in warm-temperate and subtropical forests, ranging from 22°–42° N to 99°–122° E. There were only a few studies that developed allometric equations (Table 1; Equations (1)–(3)) and addressed tree biomass allocation of Chinese cork oak in different sites in China [27–29]. The Qinling Mountains in the Shanxi Province is one of the major distribution areas of Chinese cork oak forests. Chinese cork oak forests in the Qinling Mountains make up the largest forest vegetation carbon sink based on the Shanxi Province forest resource inventory data. However, these forests have been under serious threat owing to excessive overexploitation and inappropriate management for timber and charcoal production, cultivation of edible wild mushrooms, and dye products. Therefore, many studies have been conducted to provide some recommendations toward sustainable management of the Chinese cork oak forests in the Qinling Mountains under the NFP program [30–33]. However, there is still a lack of information about C stocks, especially below-ground, for Chinese cork oak forests in the Qinling Mountain, China.

The focus of this study was to develop suitable allometric equations to calculate various tree biomass components, and to quantify the distribution patterns and quantities of C, N, and P among the major tree components, shrubs, herbs, litter, and soil (0–100 cm) in Chinese cork oak forest along the south slope of the Qinling Mountains. Results of this study may provide a crucial complement to previous studies on the understanding of C storage and forest management under the NFP program.

2. Materials and Methods

2.1. Study Site and Sampling

This study was conducted in the Shan Yang Country, Southern Shanxi Province, China (33°09′–33°42′ N, 109°32′–110°29′ E; Figure 1). The Shan Yang Country is located on the south slope of the Qinling Mountains, China. The study area is situated in the transitional area between the subtropical zone and the warm temperate zone, with a mean annual temperature 13.1 °C, a mean annual rainfall of 709 mm, and a mean frost-free period of 207 days. Forest coverage is 62.6%, and more than 80% of the forest area is dominated by pure Chinese cork oak forest, which originated from seedlings, with a few *Sabina chinensis*, and *Pinus armandii* species. Shrub species at the site include *Pyrus betulifolia*, *Lespedeza bicolor*, and *Platycarya strobilacea*, and the main herb species are *Carextristachya*, *Imperata cylindrical* and *Ophiopogon japonicas*.

Table 1. Collection of the available allometric equations used to calculate biomass of the various tree components for Chinese cork oak forests in China.

NO.	Site	Stand [a]	Components	Allometric equation	Correlation coefficient (*R*)	Sources
1.	Xiaolong Mountains, Gansu Province (33°30′–34°49′ N, 104°22′–106°43′ E)	20–84 6.0–18.6 4.3–13.1	Stem wood	$LnW_S = -3.7447 + 0.9679\ Ln\ (DBH^2H)$	0.9981	[28]
			Stem bark	$LnW_{BA} = -3.2565 + 0.7156\ Ln\ (DBH^2H)$	0.9852	
			Branches	$LnW_{BR} = -4.8449 + 1.0013\ Ln\ (DBH^2H)$	0.9917	
			Leaves	$LnW_L = -3.3569 + 0.6050\ Ln\ (DBH^2H)$	0.9611	
			Roots	$LnW_R = -2.9066 + 0.8144\ Ln\ (DBH^2H)$	0.9941	
2.	Baotianman Natural Reserve, Henan Province (33°25′–33°33′ N, 111°53′–112° E)	45 6.0–30.5 8.3–18.5	Stem wood	$LgW_S = -0.5440 + 0.6796\ Lg\ (DBH^2H)$	0.9969	[27]
			Stem bark	$LgW_{BA} = -0.8246 + 0.5896\ Lg\ (DBH^2H)$	0.9983	
			Branches	$LgW_{BR} = -2.5609 + 1.1092\ Lg\ (DBH^2H)$	0.9750	
			Leaves	$LgW_L = -2.0038 + 0.7460\ Lg\ (DBH^2H)$	0.9851	
			Roots	$LgW_R = -0.2645 + 0.5173\ Lg\ (DBH^2H)$	0.9986	
3.	Xishan Mountains, Beijing (39°34′ N, 116°28′ E)	26 3.1–10.3 6.0–8.0	Stems	$W_{S+BA} = 0.0508(DBH^2H)^{0.92}$	0.9889	[29]
			Branches	$W_{BR} = 0.0197\ (DBH^2H)^{0.8944}$	0.9412	
			Leaves	$W_L = 0.0029\ (DBH^2H)^{0.9125}$	0.9557	
			Roots	$W_R = 0.0458\ (DBH^2H)^{0.7484}$	0.9485	
4.	Shanyang Country, Shanxi Province (33°9′–34°42′ N, 109°32′–110°29′ E)	30 5.0–34.2 4.0–15.0	Stem wood	$W_S = 0.0335\ (DBH^2H)^{0.9579}$	0.9949	This study
			Stem bark	$W_{BA} = 0.0458\ DBH^{2.1044}$	0.9836	
			Branches	$W_{BR} = 0.0047\ DBH^{2.9836}$	0.9830	
			Leaves	$W_L = 0.0128\ DBH^{2.1013}$	0.9916	
			Roots	$W_R = 0.0831\ DBH^{2.1980}$	0.9634	

[a] referring to stand age (year), DBH (cm) and H (m), respectively.

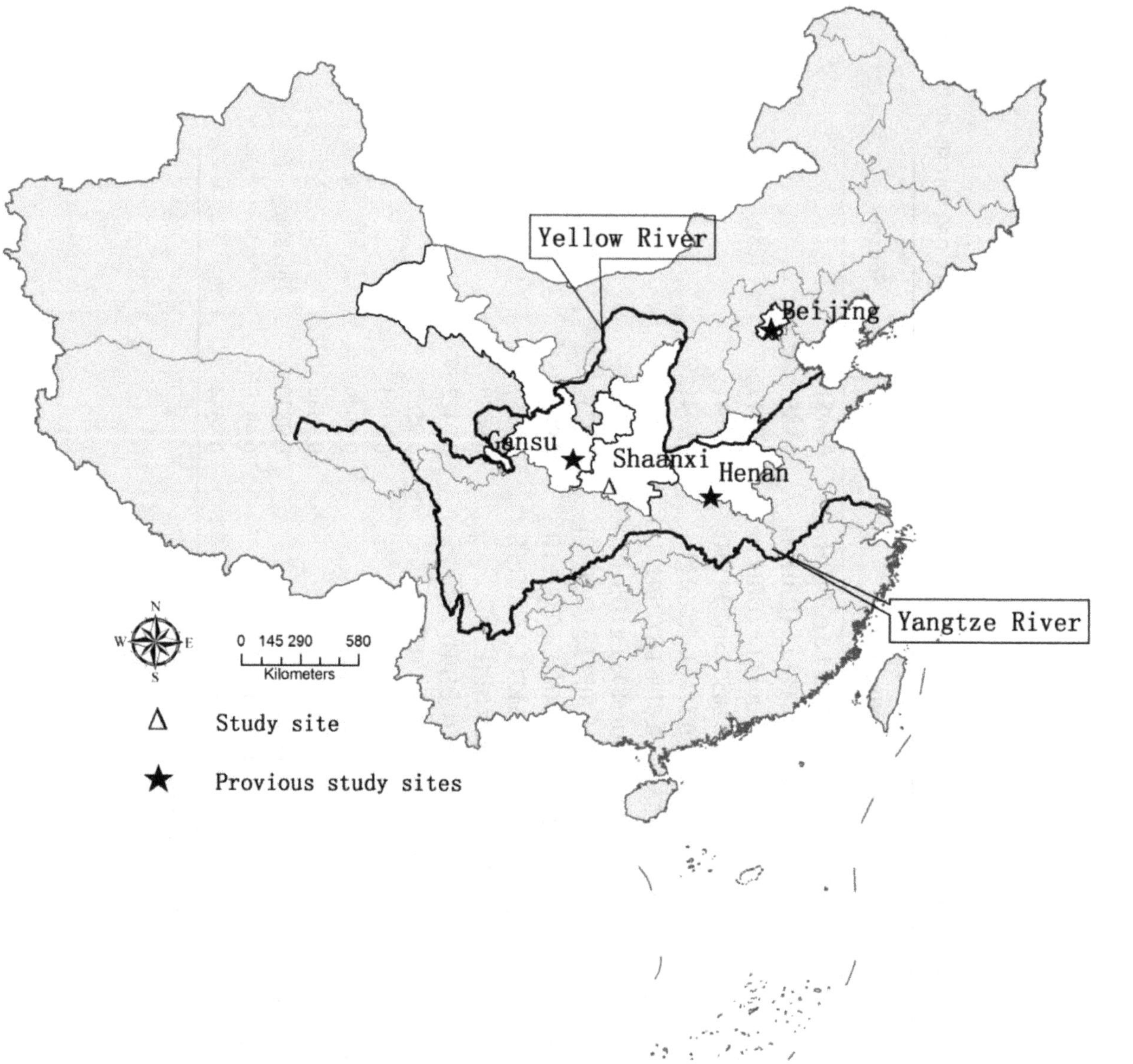

Figure 1. Site distribution of studies on Chinese cork oak allometric equations include Xiaolong Mountains, Gansu Province; the Xishan Mountains, Beijing; the Baotianman Natural Reserve, Henan Province; and the present study at Shanyang Country, Shanxi Province.

With the guidance of local forestry bureau staff, three widely distributed Chinese cork oak stands with similar site conditions, little or no human disturbance, and approximately 30 years old were selected under from the Shan Yang Country. The sites were all located near the middle of slopes and there was little difference among the sites with regard to aspect (North West), gradient (30–35°), and elevation (845–1068 m). The distance between each stand was about 5–8 km. Historically, all study stands naturally regenerated after the natural Chinese cork oaks were harvested. A 20 m × 20 m plot was constructed in the central area of each stand for sampling. Diameter at breast height (DBH) and height (H) were measured for all trees (DBH ≥ 5.0 cm) in each plot. In early August 2013, five Chinese cork oak trees within representative stand-specific DBH range were selected and harvested destructively in each stand following a previously published harvest method [19,27–29,34–36]. Trees were cut near the ground surface, after measurement of the total H of each tree, the tree stem was first cut open at 1.3 m, and then the top part of stem (from 1.3 m to the tips) was divided into 1-m-long sections. All branches and leaves of each stem section were clipped from the tree stems and branches, respectively and weighed. Stem discs (5 cm thickness) were collected from of each section, and the stem bark was separated from each disc to measure the weight of the fresh stem bark and the stem wood without bark. The whole root system was manually excavated, washed lightly to remove soil particles, and weighed.

Shrub and herb biomass was determined using total harvesting destructive sampling techniques [34]. Sampling of the shrub layer and herb layer was conducted in five 2 m × 2 m subplots and 1 m × 1 m subplots, respectively. These subplots were randomly selected within each plot. Shrub plants were separated into leaves, stems and roots, and herbs were separated into aboveground and belowground components. Litter was sampled by collecting the entire organic material within five 1 m × 1 m subplots randomly chosen in each plot [34].

Subsamples of tree components (stem wood, stem bark, branches, foliage, and roots), shrub (leaves, stems and roots), herb (above- and belowground) and litter were sealed in plastic bags, and then oven dried at 80 °C in the laboratory to constant weight to obtain wet-to-dry mass conversion factors. The dried samples were ground and used to determine plant C and N concentrations by an elemental analyzer (Carlo Erba 1106, Milan, Italy). Total P was determined by the $HClO_4$-H_2SO_4 colorimetric method [37].

In each plot, five soil cores (5 cm in diameter) were randomly collected at 0–10 cm, 10–20 cm, 20–30 cm, 30–50 cm and 50–100 cm layers. After removing the plant roots, fauna and debris by hand, the soil was air dried at room temperature around 20 °C, and then ground and passed through a 0.25 mm sieve for determination of soil chemical properties. C concentration was determined by the $K_2Cr_2O_7$-H_2SO_4 method [38], total N concentration by the Kjeldahl method [39] and total P by the $HClO_4$-H_2SO_4 colorimetric method [37].A soil profile (1 m × 1 m × 1 m) in each plot was dug for measuring soil bulk density (g cm^{-3}). After excluding recognizable soil surface litter, stainless cutting rings (5 cm in diameter) were used to sample five replicated 100 cm^3 of soil at each layer at same depth intervals of 0–10 cm, 10–20 cm, 20–30 cm, 30–50 cm and 50–100 cm layers. The ring soil samples were scraped out and roots manually removed. The soils were dried at 105 °C to constant weight to calculate bulk density (oven-dried soil sample/volume of the metal ring). All of the samples were analyzed at the central laboratory of the Institute of Soil and Water Conservation, Chinese Academy of Sciences and Ministry of Water Resources (Yangling, China).

2.2. Data Calculation and Analysis

Before establishing the allometric equation, scatter plots were used to visualize whether the relationship between the independent (dry biomass (kg) of each tree component) and dependent variables (DBH and/or H of each tree) was linear. A power function was selected as an appropriate model in this study, with the following equations:

$$W = aDBH^b \quad (1)$$

$$W = a(DBH^2 \times H)^b \quad (2)$$

where W is the dry biomass (kg) of each tree component; a and b are allometric parameters; least squares linear regression was used to estimate the value of a and b. Ordinary least squares regression was performed to determine the coefficient of determination (*R*).

The developed allometric equations from this study (Table 1; Equations (4)) and the available allometric equations (Table 1; Equations (1)–(3)) developed previously in other study sites (Figure 1) were used to compare the differences in tree biomass (stem wood, stem bark, branches, leaves, and roots) estimations. Then, the total ecosystem C, N and P storage in the present study site were calculated based on the combination of tree (stem wood, stem bark, branches, leaves, and roots), shrub (leaves, stem and roots), herbs (above- and belowground), litter layers, and soil (0–100 cm) pools. C, N, and P concentrations of leaves, branches, stem bark, stem wood, and roots were multiplied by each tree component biomass from the developed species-specific allometric equations from the present site (Table 1, Equation (4)) to partition C, N, and P stocks among the tree components, and then summed the stocks for each tree and site to calculate the stand level stocks. For shrubs, herbs and litter, the C, N, and P concentrations were multiplied by their component mass at the plot level to calculate the stand level C, N and P stocks. The stock of soil C, N, and P in the different layers of soil were calculated by multiplying the soil bulk density with soil depth and C, N, and P concentrations in each layer. The carbon storage results in each layer of soil (0–10 cm, 10–20 cm, 20–30 cm, 30–50 cm and 50–100 cm) were then summed to compute the total C, N, and P storage to a depth of 100 cm.

All statistical analyses were performed with SPSS (version 20.0, SPSS Inc., Chicago, IL, USA) and the accepted significance level was $\alpha = 0.05$. Allometric equations were developed with linear regression. Goodness of fit was based on the coefficient of determination (R) and the level of probability (p). All comparisons of biomass among the different allometric equations were performed using ANOVA, followed by multiple comparisons (LSD tests).

3. Results

Allometric Equations and Tree Biomass

The allometric equations developed in this study were all power functions (Table 1). Only significant parameters were included in the equations; therefore, DBH was the only significant parameter for bark, leaves, and roots components. DBH squared multiplied H was the significant parameter for stem wood. All equations had significant linear relationships between dependent and independent variables, and their correlation coefficients ranged from 0.9634 to 0.9949 (Table 1).

Although allometric equations of different forms are available from the present and previous studies, there was no significant difference in the biomass estimation for the various tree components in this study, with the exception of roots (Table 2). The root biomass calculated by the equation of Bao *et al.* [19] was significantly lower (17.59 t ha^{-1}) than the values generated by the other equations. Therefore, biomass from other equations was distributed similarly in the following manner: stem wood > roots > branches > stem bark > leaves. According to the tree biomass distribution pattern in the present sites, stem wood made the largest contribution to the total tree biomass, accounting for around 50%, while the proportion of leaves to total tree biomass was only around 3%, and the ratio of below- to aboveground was 0.33 (Table 2).

Table 2. Comparison of the biomass estimates of the various tree components by different allometric equations for a Chinese cork oak forest in the south slope of the Qinling Mountains, China.

Tree component	Allometric Equations (1) [28]		Allometric Equations (2) [27]		Allometric Equations (3) [29]		Allometric Equations (4) (This Study)	
	Biomass ($t\ ha^{-1}$)	Proportion (%)	Biomass ($t\ ha^{-1}$)	Proportion (%)	Biomass ($t\ ha^{-1}$)	Proportion (%)	Biomass ($t\ ha^{-1}$)	Proportion (%)
Stem wood	50.56 ± 4.97a	40.96	64.53 ± 5.16a	46.42			66.20 ± 6.46a	49.13
Stem bark	11.48 ± 0.94a	9.30	16.99 ± 1.28a	12.22			14.48 ± 1.12a	10.75
Stem	62.05 ± 5.89a	50.26	81.54 ± 6.44a	58.65	74.46 ± 7.06a	62.22	80.68 ± 7.54a	59.87
Branches	21.92 ± 2.21a	17.76	18.07 ± 1.98a	13.00	23.61 ± 2.19a	19.73	16.30 ± 1.64a	12.10
Leaves	4.45 ± 0.34a	3.60	3.74 ± 0.31a	2.70	4.01 ± 0.38a	3.35	4.01 ± 0.31a	2.98
Aboveground	88.41 ± 8.42a	71.62	103.35 ± 8.61a	74.34	102.08 ± 9.63a	85.30	100.99 ± 9.47a	74.95
Root	35.05 ± 3.07a	28.39	35.67 ± 2.59a	25.66	17.59 ± 1.47b	14.70	33.76 ± 2.68a	25.05
Total tree	123.45 ± 11.48a		139.02 ± 11.14a		119.67 ± 11.09a		134.75 ± 12.11a	

Data is reported as mean ± standard error ($n = 3$); within a line, values followed by the same lowercase letter indicate that the estimated biomass of the same tree component did not differ significantly among different allometric equations ($p < 0.05$).

Higher C, N, and P concentrations were observed in the shrub leaves than in shrub roots (Table 3). Similarly, the C and N concentrations in aboveground herbs were significantly higher than in belowground herbs (Table 3). The C:N and C:P mass ratios in shrub leaves (19.63 and 369.08, respectively) were significant lower than that in branches (60.91 and 742.61) and roots (63.64 and 717.74) of shrubs, whereas the N:P mass ratio in shrub leaves (18.92) was significantly higher than in branches (12.51) and roots (13.52) of shrubs. However, there was no significant difference in the C:N, C:P and N:P mass ratios between aboveground and belowground herbs. In contrast, no significant differences were found in the N concentrations in the different soil layer. The highest C and P concentrations were in the surface 0–10 cm soil layer, and the values significant decreased with soil depth (Table 3). The C:N, C:P and N:P ratios did not change significantly with soil depths.

C, N and P storage in the Chinese cork oak ecosystem were 163.76, 18.54 and 2.50 t ha^{-1}, respectively. The C storage distribution among different tree components was in the following order: stem wood > roots > branches > stem bark > leaves. Whereas, the order was stem wood > roots > leaves > stem bark > branches for the N storage distribution, and stem wood > roots > branches > leaves > stem bark for the P storage distribution (Table 3). C, N and P storage in tree accounted 36.14%, 6.03%, and 0.23% of the total ecosystem C, N, and P pools, respectively. The C, N, and P storage in the shrub, herb and litter pools accounted 2.10%, 1.19% and 0.03% of the total ecosystem C, N, and P pools, respectively. Soil (0–100 cm) contained the largest amount of C, N, and P storage, accounting for 61.76%, 92.78% and 99.72% of total ecosystem C, N, and P pools, respectively. C and P storage in the 0–50 cm soil depthswere significantly greater than that in deeper soil, accounting for 61.35% and 60.02% of the entire soil profile (0–100 cm); whereas, N storage between 0 and 50 cm was almost equal to that in deeper soil, accounting for 47.40% of the entire soil profile (0–100 cm).

Table 3. C, N and P concentrations and storage distribution of a Chinese cork oak forest in the south slope of the Qinling Mountains, China.

Layer	Component	C Concentration (g kg^{-1})	C Storage (t ha^{-1})	N Concentration (g kg^{-1})	N Storage (kg ha^{-1})	P Concentration (g kg^{-1})	P Storage (kg ha^{-1})
Tree	Stem wood	433.28 ± 9.43c	28.68 ± 2.80a	2.63 ± 0.24c	174.11 ± 16.98a	0.35 ± 0.06b	23.17 ± 2.26a
	Stem bark	454.60 ± 2.20b	6.58 ± 0.51c	5.05 ± 0.10b	73.11 ± 5.66b	0.29 ± 0.01b	4.20 ± 0.32c
	Branches	433.81 ± 1.36c	7.07 ± 0.72c	4.50 ± 0.02bc	73.36 ± 7.42b	0.62 ± 0.08b	10.11 ± 1.02b
	Leaves	483.07 ± 2.59a	1.94 ± 0.15c	19.31 ± 1.19a	77.48 ± 5.99b	1.26 ± 0.16a	5.06 ± 0.39c
	Root	441.96 ± 4.03bc	14.92 ± 1.18b	4.54 ± 0.06bc	153.25 ± 12.17a	0.39 ± 0.02b	13.17 ± 1.04b
	Subtotal		59.19 ± 5.31		551.31 ± 47.76		55.70 ± 4.99
Shrub	Leaves	459.12 ± 4.96a	0.09 ± 0.02b	24.08 ± 0.81a	4.60 ± 0.90a	1.30 ± 0.06a	0.25 ± 0.05a
	Branch	449.30 ± 5.89a	0.37 ± 0.07ab	8.20 ± 0.60b	6.80 ± 1.34a	0.69 ± 0.06b	0.57 ± 0.11ab
	Root	428.63 ± 5.70b	0.59 ± 0.17a	7.85 ± 0.93b	10.84 ± 3.06a	0.85 ± 0.11b	1.18 ± 0.33a
	Subtotal		1.06 ± 0.22		22.24 ± 4.36		2.00 ± 0.42
Herb	Aboveground	428.73 ± 6.44a	0.08 ± 0.01a	14.83 ± 0.85a	2.81 ± 0.45a	1.34 ± 0.08a	0.25 ± 0.04a
	Belowground	262.03 ± 13.08b	0.11 ± 0.03a	7.61 ± 0.40b	3.14 ± 0.91a	1.02 ± 0.15a	0.42 ± 0.12a
	Subtotal		0.19 ± 0.04		5.95 ± 1.14		0.67 ± 0.14
Litter		364.17 ± 10.69	2.18 ± 0.44	12.49 ± 0.21	74.88 ± 15.20	0.97 ± 0.07	5.82 ± 1.18
Soil	0–10 cm	16.63 ± 1.19a	18.05 ± 1.86b	1.69 ± 0.21a	1.83 ± 0.25t ha^{-1}b	0.43 ± 0.11a	0.46 ± 0.11 t ha^{-1}b
	10–20cm	10.12 ± 1.41bc	11.57 ± 2.05b	1.47 ± 0.27a	1.68 ± 0.34 t ha^{-1}b	0.29 ± 0.06ab	0.33 ± 0.07 t ha^{-1}b
	20–30cm	10.31 ± 0.64b	11.54 ± 0.81b	1.42 ± 0.29a	1.64 ± 0.38 t ha^{-1}b	0.27 ± 0.02ab	0.30 ± 0.02 t ha^{-1}b
	30–50 cm	8.63 ± 1.34bc	20.89 ± 3.77b	1.35 ± 0.33a	3.33 ± 0.87 t ha^{-1}b	0.16 ± 0.03b	0.38 ± 0.06 t ha^{-1}b
	50–100cm	5.76 ± 0.53c	39.10 ± 6.44a	1.26 ± 0.37a	9.41 ± 3.19 t ha^{-1}a	0.16 ± 0.03b	0.96 ± 0.13 t ha^{-1}a
	Subtotal		101.14 ± 11.03		17.89 ± 4.92 t ha^{-1}		2.43 ± 0.20 t ha^{-1}
Total ecosystem		163.76 ± 11.25		18.54 ± 4.96 t ha^{-1}		2.50 ± 0.21t ha^{-1}	

Data is reported as mean ±standard error ($n = 3$); within a column, values followed by the same lowercase letter indicate that they did not differ significantly within the same layer ($p < 0.05$).

4. Discussion

Allometric equations are crucial in order to accurately estimate forest biomass for C accounting. A number of previous studies demonstrated that power function allometric equations based on DBH or squared DBH multiplied by H can be used to estimate tree biomass [29,35]. Allometric equations based on DBH are recommended because the measurement of H is time-consuming and less accurate than DBH [19,34,40,41]. In this study, we observed that DBH was the only significant parameter for bark, leaf, and root components; whereas, the squared DBH multiplied by H as the significant parameter for stem wood. Different models (Table 1) were used to estimate biomass distribution for Chinese cork oak in China due to various biotic and abiotic environmental factors. However, there was no significant difference in tree components biomass estimation, with the exception of roots, among all the available allometric equations. In some cases the power function failed, and then transformed models were needed to develop significant allometric equations for different tree species, locations, and specific-components [7,25,27,28,42–47].

The biomass distribution among tree components was as follows: stem wood > roots > branches > stem bark > leaves. Our findings were similar to the previous reports for Chinese cork oak in the temperate region of China [27–29]. However, the different biomass distribution also observed for Chinese cork oak. For example, the biomass distribution of a 20-year-old Chinese cork oak forest was distributed as follows: stem wood > branches > roots > stem bark > leaves at; while that in 30- and 40-year-old in hilly region of Taihang Mountain, the distribution was stem wood > stem bark > branches > roots > leaves at [48]. The total Chinese cork oak tree biomass was 158.84 t ha^{-1} in the Baotianman Natural Reserve, Henan Province [27], 134.75 t ha^{-1} in the south slope of the Qinling Mountains of the present study, 79.80 t ha^{-1} in the Xiaolong Mountains, Gansu Province [28] and 53.64 t ha^{-1} in the Xishan Mountains, Beijing [29]. These obvious differences were mainly caused by age, tree density, and climate factors. Zhao *et al.* [48] reported that in a hilly region of Taihang Mountain, tree biomass significantly increased from 131.65 t ha^{-1} in a 20 year old stand to 202.96 and 291.15 t ha^{-1} in 30 and 40 years old stands, respectively.

The nutrient concentrations among plant components were significantly different. The concentrations of C, N and P in tree leaves were the highest, but the lowest values of C and N were observed in tree stems. In addition to C, N and P, the highest mobile nutrient concentrations, such as Mg, K, and Mn, were also detected in the leaves of oak forests in Spain [19]. Moreover, the N, P, and K concentrations were the highest in spring and then decreased throughout the vegetative period [19]. Similarly, the highest concentrations were in shrub leaves, but the lowest values were in shrub roots. The C, N and P concentrations in aboveground portion of herb were higher than in the belowground portion. The leaves N:P ratio, which is relatively easy to determine, has been widely used to indicate limitations in soil N (N:P < 14) and soil P (N:P > 16) [49,50]. The leaf N (19.31 g kg^{-1}), P (1.26 g kg^{-1}) concentrations, and N:P ratio (16.10) of Chinese cork oak forests in the present study were similar with the average values of leaves N (18.33 g kg^{-1}) and P (1.18 g kg^{-1}) concentrations and the N:P ratio (16.56) in China [51]. Moreover, the N:P ratio of shrub leaves was also higher than 16. Therefore, soil P is a limiting nutrient in this study area for the growth demands of Chinese cork oak forests. In contrast, the leaves N:P ratios of sharptooth oak, Chinese pine, and Armand pine indicated that there is soil N limitation in the Qinling Mountains [52]. However, the nutrient concentrations in live leaves decreased and the N:P ratio increased during the growing seasons were observed in a Mediterranean cork oak forest in southwestern Spain [26]. Moreover, the leaf N and P concentrations of Chinese cork oak forests also decreased during stand development. It was shown that there is an disconnect bewteen soil P supply and plant growth demand indicated by an increased N:P ratio from 13.84 in 20-year-old stand to 16.05 and 19.75 in 30- and 40-year-old stands, respectively, in a hilly region of the Taihang Mountains [48]. Therefore, quantifying soil nutrient limitation by only using the leaf N:P ratio presents challenges because of the limited amount of data and only a partial understanding of the processes involved [50,53]. Although significant differences in leaves N and P concentrations were detected across all *Quercus* species in China, leaf N:P ratio (13.96) was well constrained to a relatively stable range for *Quercus* species and was less influenced by environmental variables across China [51].

Abundant precipitation, full sunshine and nutrient-rich soil were considered to be suitable for supporting plant growth in the southern slope of the Qinling Mountains [52]. C, N, and P were enriched at 0–10 cm soil depth, while C and P decreased with an increase in soil depth. However, N showed a stable trend. C, N, and P concentrations of 0–10 cm soil depth in this study area (16.63, 1.69 and 0.43 g kg^{-1}, respectively) were higher than the mean values (12.28, 0.94 and 0.38 g kg^{-1}, respectively) of soils in China. As a consequence, the C:N, C:P and N:P (10.55, 51.51 and 4.59 in molar, respectively) ratios of 0–10 cm soil depth in this study area were smaller than those mean values (14.4, 136 and 9.3, respectively) of soils in China [54]. C, N, and P concentrations increased during stand development of Chinese cork oak forests in the hilly region of the Taihang Mountains, possibly due to a larger accumulation of organic matter in older stands [48]. Under Mediterranean climate conditions in Mainland Spain, the soil C storage of cork oak forests was favored by large organic matter inputs, high soil clay contents, a calcium-saturated soil matrix and reduced summer aridity [23].

The proportion of stem to total tree biomass was around 60% in this study. The highest percentage was 73% for four Mediterranean oak forests in Spain [19]. The proportion of stem to total tree biomass has been used to infer the light conditions, soil nutrient and age stage at study sites [35,36]. For example, the ratios in Chinese pine were 46.9%, 72.2%, 70.6% and 70.7% for young, middle-aged, immature, and mature stands, respectively [34]. Although the low percentage of nutrients accumulated in the leaves because leaf biomass represents only around 3% of the total, the amount of nutrients accumulated in leaves was of great importance because the nutrients were subject to internal annual cycles within the tree, and some of them return to the soil in the form of leaf litter [19]. At the same time, root biomass accounted for a large proportion of the total tree biomass. The ratio of below- to aboveground biomass was 0.33 for Chinese cork oak forests in this study.The ratio value was 0.28, 0.29 and 0.23 for the Baotianman Natural Reserve of Henan Province [27], the Xiaolong Mountainsof Gansu Province [28], and the Xishan Mountains of Beijing [29], respectively. In contrast to the result [55] that there is a relatively constant ratio of below- to aboveground biomass during stand development globally, Zhao *et al.* [48] found that the ratio decreased during stand development, from 0.23 in a 20-year-old stand to 0.16 and 0.13 in 30- and 40-year-old stands, respectively, in a hilly region of the Taihang Mountains. The decreased below- to aboveground biomass ratio may demonstrate different strategies in nutrient cycling and water uptake potential [1]. Our study agrees with previous studies that soil is the largest C and nutrient element storage component, followed by tree, understory, and litter [21]. Soil C, N, and P storage were 1.62-, 27.52- and 34.71-fold higher than vegetation C, N, and P storage. In addition, soil C storage was higher in the topsoil than in deeper soil owing to soil organic matter is the main source of C stored in topsoil. Moreover, soil organic matter is essential to ecosystem productivity and regeneration. Recently, many studies have focused on biomass, C, and nutrient storage during stand development [1,7,34–36,48,55]. These studies provided a comprehensive understanding of the importance of considering the succession development of forest ecosystem C pools and the interaction with other nutrient elements, especially when estimating C sink potential over a life cycle. Because human disturbances can have huge impacts on certain age stage forests, we only selected 30-year-old Chinese cork oak stands with little or no human disturbance to describe the basic characteristics of biomass, C and nutrient storage distribution. Therefore, future studies should focus on improving our understand of the effects of age and disturbance in biomass, C concentrations and stocks for devising optimum forest management strategies aimed at mitigating climate change.

5. Conclusions

In this study, we presented the basic C, N and P storage distribution patterns among trees, shrubs, herbs, litter, and soil (0–100 cm) in a 30-year-old Chinese cork oak forest in the south slope of the Qinling Mountains, China. The biomass of tree components can be better predicted from allometric equations using DBH as the independent variable. There was no significant difference in various tree components biomass estimations, with the exception of roots, between different Chinese cork oak allometric equations developed from different sites. Stems and roots were the main proportion of total tree biomass. The proportion of stem to total tree biomass was around 60% and the ratio of below- to aboveground biomass was 0.33. There were significant differences in C, N and P concentrations between plant components. Plant leaves had the highest C, N, and P concentrations than other plant components. The N:P ratio was similar to the national level and indicated soil P limitation. However, C, N and P concentrations of the 0–10 cm soil depth were higher than those mean values of soils in China. C, N, and P storage of this Chinese cork oak ecosystem were 163.76, 18.54 and 2.50 t ha^{-1}, respectively. Soil was the largest C and nutrient element storage component, followed by trees. C, N, and P stocks in shrubs, herbs, and litter contributed only little to ecosystem C, N and P stocks. This study demonstrated that large-scale Chinese cork oak forests in the Qinling Mountains play an important role in C sequestration under NFP program. Furthermore, we suggest that further research on human-caused, natural disturbance and other influence factors on the continuous accumulation of C in both the plants and soils are especially critical for developing recommendations on appropriate forest management practices under the NFP program.

Acknowledgments

This research was supported by the National Nature Science Foundation of China (No. 41201088 and 41371506), the Strategic Priority Research Program and West Light Foundation of The Chinese Academy of Sciences (XDA05050203 and K318021304), the Doctoral Fund of Ministry of Education of China (20120204120014), Specialized Research Fund for the Doctoral Program of Higher Education (2014YB056) and Institute of Soil and Water Conservation Funding (A315021380).

Author Contributions

The paper was written by Yang Cao with a contribution by Yunming Chen. Yunming Chen conceptualized the research design and site selection, and submitted the article. Yang Cao conducted field data collection and data processing.

References

1. Laclau, P. Biomass and carbon sequestration of ponderosa pine plantations and native cypress forests in northwest Patagonia. *For. Ecol. Manag.* **2003**, *180*, 317–333.
2. Chen, X.G.; Zhang, X.Q.; Zhang, Y.P.; Wan, C.B. Carbon sequestration potential of the stands under the grain for green program in Yunnan province, China. *For. Ecol. Manag.* **2009**, *258*, 199–206.
3. Bonner, M.T.L.; Schmidt, S.; Shoo, L.P. A meta-analytical global comparison of aboveground biomass accumulation between tropical secondary forests and monoculture plantations. *For. Ecol. Manag.* **2013**, *291*, 73–86.
4. He, Y.J.; Qin, L.; Li, Z.Y.; Liang, X.Y.; Shao, M.X.; Tan, L. Carbon storage capacity of monoculture and mixed-species plantations in subtropical China. *For. Ecol. Manag.* **2013**, *295*, 193–198.
5. Zeng, X.H.; Zhang, W.J.; Liu, X.P.; Cao, J.S.; Shen, H.T.; Zhao, X.; Zhang, N.N.; Bai, Y.R.; Yi, M. Change of soil organic carbon after cropland afforestation in "Beijing-Tianjin sandstorm source control" program area in China. *Chin. Geogra. Sci.* **2014**, *24*, 461–470.
6. Zhou, W.M.; Lewis, B.J.; Wu, S.N.; Yu, D.P.; Zhou, L.; Wei, Y.W.; Dai, L.M. Biomass carbon storage and its sequestration potential of afforestation under natural forest protection program in China. *Chin. Geogra. Sci.* **2014**, *24*, 406–413.
7. Li, T.; Liu, G. Age-related changes in carbon accumulation and allocation in plants and soil of a black locust forest on the loess plateau. *Chin. Geogra. Sci.* **2014**, *24*, 414–422.
8. Deng, L.; Liu, G.B.; Shangguan, Z.P. Land-use conversion and changing soil carbon stocks in China's "grain-for-green" program: A synthesis. *Glob. Change Biol.* **2014**, *20*, 3544–3556.
9. Chang, R.; Fu, B.; Liu, G.; Liu, S. Soil carbon sequestration potential for "grain for green" project in loess plateau, China. *Environ. Manag.* **2011**, *48*, 1158–1172.
10. Wei, Y.W.; Yu, D.P.; Lewis, B.J.; Zhou, L.; Zhou, W.M.; Fang, X.M.; Zhao, W.; Wu, S.N.; Dai, L.M. Forest carbon storage and tree carbon pool dynamics under natural forest protection program in northeastern China. *Chin. Geogra. Sci.* **2014**, *24*, 397–405.
11. Hu, H.; Liu, G. Carbon sequestration of China's national natural forest protection project. *Acta Ecol. Sin.* **2006**, *26*, 291–296.
12. Fang, J.; Brown, S.; Tang, Y.; Nabuurs, G.-J.; Wang, X.; Shen, H. Overestimated biomass carbon pools of the northern mid- and high latitude forests. *Clim. Chang.* **2006**, *74*, 355–368.
13. Liu, G.; Fu, B.; Fang, J. Carbon dynamics of Chinese forests and its contribution to global carbon balance. *Acta Ecol. Sin.* **2000**, *20*, 733–740.

14. Zhao, M.; Zhou, G. Forest inventory data (FID)-based biomass models and their prospects. *Chin. J. Appl. Ecol.* **2004**, *15*, 1468–1472.
15. Wei, Y.W.; Li, M.H.; Chen, H.; Lewis, B.J.; Yu, D.P.; Zhou, L.; Zhou, W.M.; Fang, X.M.; Zhao, W.; Dai, L.M. Variation in carbon storage and its distribution by stand age and forest type in boreal and temperate forests in northeastern China. *PLoS ONE* **2013**, *8*, doi:10.1371/journal.pone.0072201.
16. Du, L.; Zhou, T.; Zou, Z.H.; Zhao, X.; Huang, K.C.; Wu, H. Mapping forest biomass using remote sensing and national forest inventory in China. *Forests* **2014**, *5*, 1267–1283.
17. Guyette, R.P.; Dey, D.C.; Stambaugh, M.C. The temporal distribution and carbon storage of large oak wood in streams and floodplain deposits. *Ecosystems* **2008**, *11*, 643–653.
18. Borges, J.G.; Oliveira, A.C.; Costa, M.A. A quantitative approach to cork oak forest management. *For. Ecol. Manag.* **1997**, *97*, 223–229.
19. Santa-Regina, I. Biomass estimation and nutrient pools in four *Quercus pyrenaica* in sierra de gata mountains, Salamanca, Spain. *For. Ecol. Manag.* **2000**, *132*, 127–141.
20. Gardiner, E.S.; Hodges, J.D. Growth and biomass distribution of cherrybark oak (*Quercus pagoda raf.*) seedlings as influenced by light availability. *For. Ecol. Manag.* **1998**, *108*, 127–134.
21. Vallet, P.; Meredieu, C.; Seynave, I.; Belouard, T.; Dhote, J.F. Species substitution for carbon storage: Sessile oak *versus* corsican pine in France as a case study. *For. Ecol. Manag.* **2009**, *257*, 1314–1323.
22. Lopez, B.C.; Sabate, S.; Gracia, C.A. Thinning effects on carbon allocation to fine roots in a *Quercus ilex* forest. *Tree Physiol.* **2003**, *23*, 1217–1224.
23. Gonzalez, I.G.; Corbi, J.M.G.; Cancio, A.F.; Ballesta, R.J.; Cascon, M.R.G. Soil carbon stocks and soil solution chemistry in quercus ilex stands in mainland Spain. *Eur. J. For. Res.* **2012**, *131*, 1653–1667.
24. Cerasoll, S.; Scartazza, A.; Brugnoli, E.; Chaves, M.M.; Pereira, J.S. Effects of partial defoliation on carbon and nitrogen partitioning and photosynthetic carbon uptake by two-year-old cork oak (*Quercus suber*) saplings. *Tree Physiol.* **2004**, *24*, 83–90.
25. Fonseca, T.J.F.; Parresol, B.R. A new model for cork weight estimation in northern Portugal with methodology for construction of confidence intervals. *For. Ecol. Manag.* **2001**, *152*, 131–139.
26. Andivia, E.; Fernandez, M.; Vazquez-Pique, J.; Gonzalez-Perez, A.; Tapias, R. Nutrients return from leaves and litterfall in a mediterranean cork oak (*Quercus suber* L.) forest in southwestern Spain. *Eur. J. For. Res.* **2010**, *129*, 5–12.
27. Liu, Y.; Wu, M.; Guo, Z.M.; Jiang, Y.; Liu, S. Biomass and net productivity of *Quercus variabilis* forest in Baotianman natural reserve. *Chin. J. Appl. Ecol.* **1998**, *9*, 569–574.
28. Cheng, T.; Ma, Q.; Feng, Z.; Luo, X. Research on forest biomass in Xiaolong Mountains, Gansu province. *J. Beijing For. Univ.* **2007**, *29*, 31–36.
29. Bao, X.; Chen, L.; Chen, Q.; Ren, J.; Hu, Y.; Li, Y. The biomass of planted oriental oak (*Quercus variabilis*) forest. *Acta Phytoecol. et Geobot. Sin.* **1984**, *8*, 313–320.
30. Zhang, W.; Lu, Z. A study on the biological and ecological property and geographical distribution of *Quercus variabilis* population. *Acta Bot. Boreali-Occidential Sin.* **2001**, *22*, 1093–1101.
31. Ran, R.; Zhang, W.; Zhou, J.; He, J. Effects of thinning intensities on population regeneration of natural *Quercus variabilis* forest on the south slope of Qinling Mountains. *Chin. J. Appl. Ecol.* **2014**, *25*, 695–701.

32. Ran, R.; Zhang, W.; Zhou, J.; He, J. Effects of thinning intensity on the seed bank and seedling growth of *Quercus variabilis* on the south slope of Qinling mountains, northwest China. *Chin. J. Appl. Ecol.* **2013**, *24*, 1494–1500.
33. Yi, Q.Z.; Zhang, W.; Tang, D. Effects of reserved sprout number per stump on sprout development and biomass accumulation of *Quercus variabilis*. *Sci. Silvae Sin.* **2013**, *7*, 34–39.
34. Zhao, J.L.; Kang, F.F.; Wang, L.X.; Yu, X.W.; Zhao, W.H.; Song, X.S.; Zhang, Y.L.; Chen, F.; Sun, Y.; He, T.F.; *et al.* Patterns of biomass and carbon distribution across a chronosequence of Chinese pine (*Pinus tabulaeformis*) forests. *PLoS ONE* **2014**, *9*, doi:10.1371/journal.pone.0094966.
35. Li, H.; Li, C.; Zha, T.; Liu, J.; Jia, X.; Wang, X.; Chen, W.; He, G. Patterns of biomass allocation in an age-sequence of secondary *Pinus bungeana* forests in China. *For. Chron.* **2014**, *90*, 169–176.
36. Noh, N.J.; Son, Y.; Kim, R.H.; Seo, K.W.; Koo, J.W.; Park, I.H.; Lee, Y.J.; Lee, K.H.; Son, Y.M. Biomass accumulations and the distribution of nitrogen and phosphorus within three *Quercus acutissima* stands in central Korea. *J. Plant Biol.* **2007**, *50*, 461–466.
37. Parkinson, J.; Allen, S. A wet oxidation procedure suitable for the determination of nitrogen and mineral nutrients in biological material. *Commun. Soil Sci. Plant Anal.* **1975**, *6*, 1–11.
38. Nelson, D.W.; Sommers, L.E.; Sparks, D.; Page, A.; Helmke, P.; Loeppert, R.; Soltanpour, P.; Tabatabai, M.; Johnston, C.; Sumner, M. Total carbon, organic carbon, and organic matter. *Methods of Soil Analysis. Part 3—Chemical Methods,* SSSA Book Series 5.3; Soil Science Society of America, American Society of Agronomy: Madison, WI, UAS, 1996; pp. 961–1010.
39. Bremner, J.M. Total Nitrogen. In *Methods of Soil Analysis. Part 2. Chemical and Microbiological Properties, Agronomy Monograph 9.2*; SSSA: Madison, WI, USA, 1965; pp. 1149–1178.
40. Peichl, M.; Arain, M.A. Allometry and partitioning of above- and belowground tree biomass in an age-sequence of white pine forests. *For. Ecol. Manag.* **2007**, *253*, 68–80.
41. Gower, S.T.; Kucharik, C.J.; Norman, J.M. Direct and indirect estimation of leaf area index, fAPAR, and net primary production of terrestrial ecosystems. *Remote Sens. Environ.* **1999**, *70*, 29–51.
42. Basuki, T.M.; van Laake, P.E.; Skidmore, A.K.; Hussin, Y.A. Allometric equations for estimating the above-ground biomass in tropical lowland dipterocarp forests. *For. Ecol. Manag.* **2009**, *257*, 1684–1694.
43. Makungwa, S.D.; Chittock, A.; Skole, D.L.; Kanyama-Phiri, G.Y.; Woodhouse, I.H. Allometry for biomass estimation in *Jatropha* trees planted as boundary hedge in farmers' fields. *Forests* **2013**, *4*, 218–233.
44. Sitoe, A.A.; Mandlate, L.J.C.; Guedes, B.S. Biomass and carbon stocks of Sofala bay mangrove forests. *Forests* **2014**, *5*, 1967–1981.
45. Ritchie, M.W.; Zhang, J.W.; Hamilton, T.A. Aboveground tree biomass for *Pinus ponderosa* in northeastern California. *Forests* **2013**, *4*, 179–196.
46. Beets, P.N.; Kimberley, M.O.; Oliver, G.R.; Pearce, S.H.; Graham, J.D.; Brandon, A. Allometric equations for estimating carbon stocks in natural forest in New Zealand. *Forests* **2012**, *3*, 818–839.
47. Mate, R.; Johansson, T.; Sitoe, A. Biomass equations for tropical forest tree species in Mozambique. *Forests* **2014**, *5*, 535–556.
48. Zhao, Y.; Wang, P.; Fan, W.; Zhu, Y. Nutrient cycling in *Quercus varlabilis* plantations of different ages classes in hilly region of Taihang mountain. *Sci. Soil Water Conserv.* **2009**, *7*, 66–71.

49. Gusewell, S. N:P ratios in terrestrial plants: Variation and functional significance. *New Phytol.* **2004**, *164*, 243–266.
50. Townsend, A.R.; Cleveland, C.C.; Asner, G.P.; Bustamante, M.M.C. Controls over foliar N:P ratios in tropical rain forests. *Ecol.* **2007**, *88*, 107–118.
51. Wu, T.G.; Dong, Y.; Yu, M.K.; Wang, G.G.; Zeng, D.H. Leaf nitrogen and phosphorus stoichiometry of *Quercus* species across China. *For. Ecol. Manag.* **2012**, *284*, 116–123.
52. Liu, G.; Zhao, S.; Tu, X. Distributional characteristics on biomass and nutrient elements of pine-oak forest belt in Mt. Qinling. *Sci. Silvae Sin.* **2001**, *37*, 28–36.
53. Goll, D.S.; Brovkin, V.; Parida, B.R.; Reick, C.H.; Kattge, J.; Reich, P.B.; van Bodegom, P.M.; Niinemets, U. Nutrient limitation reduces land carbon uptake in simulations with a model of combined carbon, nitrogen and phosphorus cycling. *Biogeosciences* **2012**, *9*, 3547–3569.
54. Tian, H.Q.; Chen, G.S.; Zhang, C.; Melillo, J.M.; Hall, C.A.S. Pattern and variation of C:N:P ratios in China's soils: A synthesis of observational data. *Biogeochemistry* **2010**, *98*, 139–151.
55. Yang, Y.; Luo, Y. Carbon:Nitrogen stoichiometry in forest ecosystems during stand development. *Glob. Ecol. Biogeogr.* **2011**, *20*, 354–361.

9

Increased Biomass of Nursery-Grown Douglas-Fir Seedlings upon Inoculation with Diazotrophic Endophytic Consortia

Zareen Khan, Shyam L. Kandel, Daniela N. Ramos, Gregory J. Ettl, Soo-Hyung Kim and Sharon L. Doty *

School of Environmental and Forest Sciences, College of the Environment, University of Washington, Seattle, WA 98195, USA; E-Mails: zareen@uw.edu (Z.K.); sk66@uw.edu (S.L.K.); dnavil9@uw.edu (D.R.); ettl@uw.edu (G.J.E.); soohkim@uw.edu (S.-H.K.)

* Author to whom correspondence should be addressed; E-Mail: sldoty@uw.edu;

Academic Editor: Heinz Rennenberg

Abstract: Douglas-fir (*Pseudotsuga menziesii*) seedlings are periodically challenged by biotic and abiotic stresses. The ability of endophytes to colonize the interior of plants could confer benefits to host plants that may play an important role in plant adaptation to environmental changes. In this greenhouse study, nursery-grown Douglas-fir seedlings were inoculated with diazotrophic endophytes previously isolated from poplar and willow trees and grown for fifteen months in nutrient-poor conditions. Inoculated seedlings had significant increases in biomass (48%), root length (13%) and shoot height (16%) compared to the control seedlings. Characterization of these endophytes for symbiotic traits in addition to nitrogen fixation revealed that they can also solubilize phosphate and produce siderophores. Colonization was observed through fluorescent microscopy in seedlings inoculated with *gfp*- and *mkate*-tagged strains. Inoculation with beneficial endophytes could prove to be valuable for increasing the production of planting stocks in forest nurseries.

Keywords: Douglas-fir; conifers; nurseries; endophytes; diazotrophic; phosphate solubilization; siderophore; biomass; sustainable forestry

---

1. Introduction

Millions of Douglas-fir (*Pseudotsuga menziesii*) seedlings are grown in nurseries for reforestation purposes in North America, Europe and elsewhere [1]. The majority of seedlingsare grown for more than a year as bare root seedlings, although some are also grown in containers. Environmental disturbances and stresses contribute to seedling failure when transplanted in the field, and in order to maximize production, nurseries apply intensive cultural practices, fertilization being one of the most common [1,2]. This can result in additional costs to nursery operators and could also have negative environmental repercussions if nitrogen leaches from nurseries situated in areas near lakes and rivers and/or where ground water reservoirs are found. An inexpensive and environmentally-benign alternative for enhancing the productivity of newly-established forest plantations involves nursery inoculation of seedlings with plant growth-promoting microorganisms. In forestry, this has traditionally been restricted to inoculation with mycorrhizal fungi [3,4]; however, it is reported that in many cases, mycorrhizal inoculation resulted in low to no measurable benefit in outplanting success. Recently, root-associated bacteria have been shown to stimulate tree seedling growth in addition to improving mycorrhizal colonization and may be valuable in current reforestation efforts [5,6].

Some species of conifers are able to inhabit harsh and nutrient-poor subalpine sites where few other plants grow. A key component of their ecological success may be attributed to the presence of endophytes. Endophytes are microbes (bacteria and fungi) that reside inside plants and do not cause disease [7–10], and there is increasing evidence of their profound impact on plant development, physiology, evolution and adaptation [11]. An important trait of some endophytes is the ability to supply nitrogen to their host plant through biological nitrogen fixation. Nitrogen-fixing endophytes have been isolated from a variety of species, such as sugarcane [12], wild rice [13], corn [14,15], African sweet potato [16], kallar grass [17], coffee [18], cactus [19] and woody plants, including poplar, willow and coniferous trees [20–23]. It has been demonstrated that N fixed by diazotrophic bacteria can be utilized by plants [24]. Endophytic bacteria have been found in different parts of tree tissues, such as roots [25], stems [26], shoots [27], leaves [28] and buds [29]. More recently, diazotrophic endophytes were isolated from conifers growing in nitrogen-poor soils. Inoculation of lodgepole pine and western red cedar seedlings with the isolate *Paenibacillus polymyxa* P2b2R resulted in significant foliar ^{15}N dilution, indicative of biological nitrogen fixation [30,31]. The importance of endophytes to the establishment and persistence of long-lived western conifers is unknown; however, endophytes may allow establishment and growth on nutrient-limited and environmentally-stressed sites; for example, similar endophytes were found in subalpine conifers from distant sites in California and Colorado [32].

Phosphorous is another essential macronutrient promoting plant growth and development. Many woody plants are dependent on ectomycorrhizal fungi for their growth and survival. Mycorrhizae inoculation of Douglas-fir seedlings is a common practice in commercial nurseries [33,34]. However, commercial nursery treatments, including frequent addition of fertilizer and water, may not be favorable for mycorrhizae colonization of container-grown seedlings [35]. The importance of phosphate-solubilizing endophytes in increasing the phosphorous availability by solubilization of inorganic phosphate and mineralization of organic phosphate has recently been demonstrated as one of the important mechanisms of increased plant growth [36]. Siderophores are low molecular weight ferric iron-specific chelating agents produced by bacteria and fungi [37]. Microbial siderophores have been reported to have a positive correlation with plant growth promotion, and the production of siderophores is being considered as one of the key traits for primary screening of beneficial endophytes [38].

In this preliminary study, we chose a consortia of diazotrophic endophytes, originally isolated from poplar and willow trees [20–22], because they had shown robust growth-promoting activity with agricultural crops, grasses and poplar plants grown under abiotic stress [22,39–41]. Our objective was to determine if these endophytes enhance the growth of Douglas-fir seedlings under nutrient-poor conditions.

2. Experimental Section

2.1. Plant Material and Endophytes

Douglas-fir container (1 + 0) (4 cu. in ; Ray Leach "cone-tainers") seedlings from a local western WA provenance obtained from Silvaseed Company Foresters (Roy, WA, USA) were individually planted into plastic pots containing Sunshine Mix #2 (Steubers, WA, USA). The propagation protocol for seedling production is described elsewhere [42]. Seven bacterial strains and one yeast strain (Table 1) were chosen based on strong growth in nitrogen-free media, the presence of the nitrogen fixation gene (*nifH*) [20,21], production of phytohormones [43] and/or their robust plant growth-promoting abilities.

Table 1. Poplar and willow endophytes used in this study and their phosphate solubilization index (PSI) in the plate assay [20,21].

Endophyte	Closest 16SrDNA Match	PSI
WP1	*Rhodotorula graminis*	1.05
WP5	*Rahnella* sp.	1.64
WP9	*Burkholderia* sp.	1.50
WP19	*Acinetobacter calcoaceticus*	3.10
PTD1	*Rhizobium tropici* bv *populus*	1.40
WW5	*Sphingomonas yanoikuyae*	1.10
WW6	*Pseudomonas putida*	1.64
WW7	*Sphingomonas* sp.	1.82

2.2. Methods

Inoculation suspensions were prepared by first growing the individual endophyte strains on nutrient-rich media from −80 °C glycerol stocks. Bacteria were grown on mannitol glutamate/Luria–Bertani (MG/L) medium [44]. The yeast strain was grown on yeast extract peptone dextrose (YPD) medium. Single colonies were inoculated into liquid broth and grown overnight on a rotary shaker (150 rpm) and agitated for 24 h at 30 °C. Bacteria were harvested by centrifugation at 8000 rpm for 10 min, resuspended in nitrogen-free Murashige and Skoog (NFMS) (Caisson Labs, USA) broth and washed three times in NFMS. The endophyte consortia treatment was prepared by mixing equal concentrations of each inoculation suspension as determined by measuring the optical density at 600 nm (OD_{600}) of the individual strains and adjusting to a final OD_{600} of 0.1. Using a sterile conical tube, a 50-mL sample of the endophyte consortia was then delivered 5–10 cm below the surface and in close proximity to the roots. Mock-inoculated control seedlings received the same volume of sterile liquid broth. A total of 24 pots, 12 replications of inoculated or mock-inoculated control plants, were arranged in a randomized fashion. Seedlings were kept in an environment with an 18h photoperiod, a 20/14 °C day/night temperature cycle and a relative humidity of 70%–85%. Drip trays were placed under each individual pot to prevent cross-contamination through run-off. Each pot was individually irrigated with tap water. Irrigation frequency and duration were adjusted, such that the pots were not allowed to flood nor dry out between watering. Additionally, each pot was fertigated with received 100 mL of 1/2 strength, 1/8 nitrogen modified Hoagland's medium containing (g L^{-1}): $CaCl_2 \cdot 2H_2O$, 0.22; K_2SO_4, 0.17; $MgSO_4 \cdot 7H_2O$, 0.26; KH_2PO_4, 0.136; NaFeEDTA(10% Fe), 0.015; with 1 $mL \cdot L^{-1}$ micronutrient solution containing ($g \cdot L^{-1}$): H_3BO_3, 0.773; $MnSO_4$, 0.169; $ZnSO_4 \cdot 7H_2O$, 0.288; $CuSO_4 \cdot 5H_2O$, 0.062; H_2MoO_4 (83% MoO_3), 0.04; weekly for the first two months post-inoculation, then 250 mL twice monthly until the end of the experiment.

Height measurements were recorded after inoculation and every month thereafter. After 15 months, the plants were removed from the soil, and roots and shoots were gently washed to remove dirt and debris. Roots and stems were then blotted dry, cut into root and shoot sections, and the lengths and wet weights were determined. Foliar samples were taken from each plant and oven dried at 65 °C, ground to a fine powder and analyzed for nitrogen content using a PE 2400 series II CHN elemental analyzer (Perkin-Elmer, MA, USA) at the University of Washington's School of Environmental and Forest Sciences Analytical Soils laboratory.

Potential plant growth-promoting traits, such as phosphate solubilization and siderophore production, were determined for the selected endophytes. The ability of the isolates to solubilize tricalcium orthophosphate (TCP) was tested on the National Botanical Research Institute's growth medium [45] containing $g \cdot L^{-1}$: glucose, 10; $Ca_3(PO_4)_2$, 5; $MgCl_2 \cdot 6H_2O$, 5; $MgSO_4 \cdot 7H_2O$, 0.25; KCl, 0.2; $(NH_4)_2SO_4$, 0.1. The inoculum was added in quadruplicate on the medium in Petri dishes. The halo and colony diameters were measured after 14 days of incubation of the plates at 25 °C. The ability of the microbes to solubilize the insoluble phosphate is described by the solubilization index (the ratio of the total diameter (halo + colony diameter (mm) to the colony diameter (mm)). Analysis of siderophore production by the endophyte isolates was performed following the chrome azurol S (CAS) method of Alexander and Zuberer [46]. For each isolate, 10 μL of inoculum were placed in quadruplicate on the medium. After incubation at 25 °C for 3 days, the discoloration of the medium (blue to orange) indicated siderophore-producing endophytes.

To evaluate endophytic colonization by some of the isolates, we used strains that were labeled with fluorescent markers with a broad host range plasmid. The *mkate* and *gfp* plasmids were introduced into *Acinetobacter calcoaceticus* strain WP19 and *Rahnella* sp. strain WP5, respectively, via triparental mating using *E. coli* DH5α as the donor and *E. coli* HB101 (pRK 2073) as the helper strain [47]. The inoculations were done as described above, and the colonization of root tissue and needles was verified through fluorescent microscopy using a Zeiss Imager M2 equipped with an AxioCam MRM and recorded with Zeiss AxioVision software (Karl Zeiss, LLC, Thornwood, NY, USA).

2.3. Statistical Analysis

One-way ANOVA (analysis of variance) and Tukey's honestly significant difference (HSD) test were used to identify significant differences between inoculated and mock-inoculated control seedlings.

3. Results and Discussion

Differences in plant heights and the health of inoculated and mock-inoculated control plants were apparent after fifteen months of growth in the greenhouse under N-limited conditions. As shown in Figure 1, the mock-inoculated control plants were stunted, with some being chlorotic and showing signs of nutrient deficiency, whereas the inoculated plants appeared taller, more robust and greener than the controls. The inoculated plants were 16% (p = 0.008) taller, had 13% longer roots (p = 0.024) (Figure 2) and a significant 48% (p = 0.019) increase in biomass (Figures 2 and 3) when compared to their mock-inoculated control plants. The concentration of N in plant tissues was higher in the inoculated plants compared to the controls, although it was not statistically significant (p = 0.156) (Figure 4).

Figure 1. Representative photograph of mock-inoculated (left) and inoculated (right) Douglas-fir seedling after 15 months of growth in nutrient-poor soil.

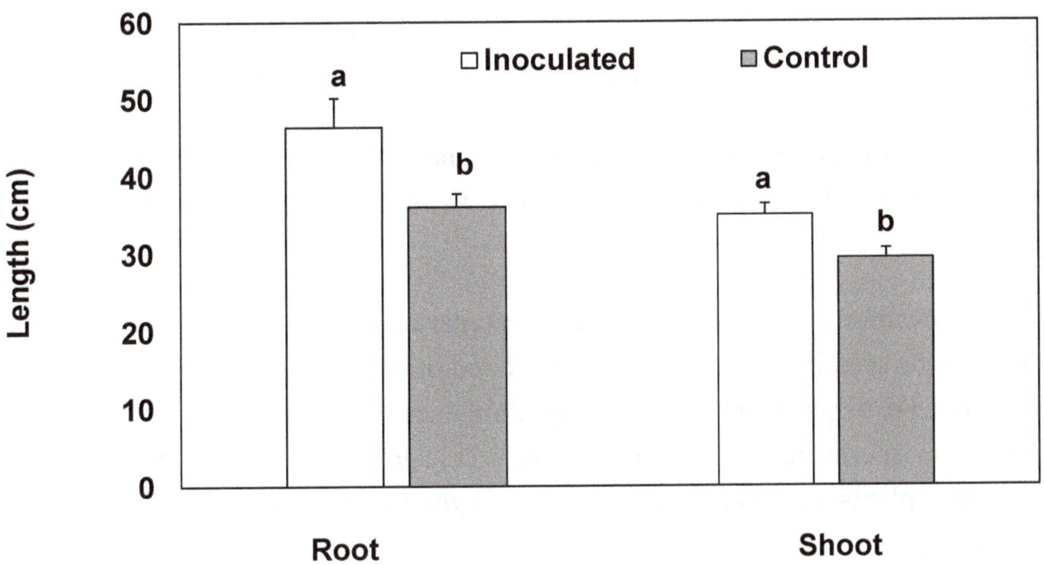

Figure 2. Root and shoot lengths of inoculated and control Douglas-fir seedlings at harvest. Data are the mean of 12 replicates. Error bars show the standard deviation. Means designated with different letters are significantly different.

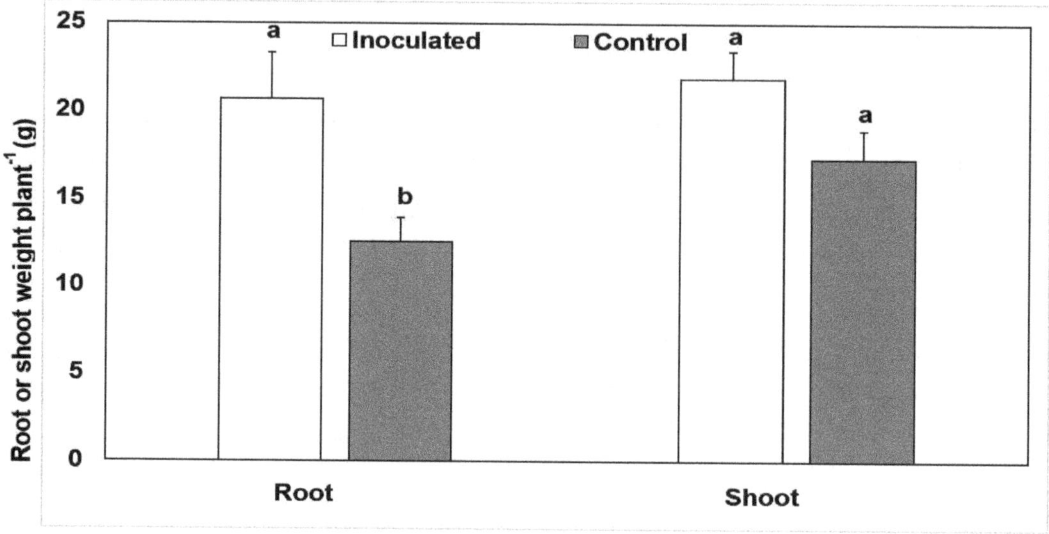

Figure 3. Root and shoot weights (g) of inoculated and control Douglas-fir seedlings at harvest. Data are the mean of 12 replicates. Error bars show the standard deviation. Means designated with different letters are significantly different.

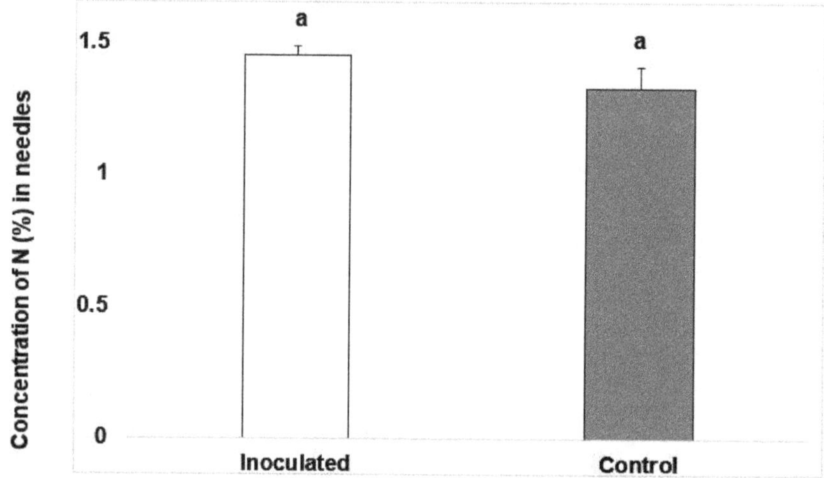

Figure 4. Total N concentration in the inoculated and control Douglas-fir seedlings needles at harvest. Data are the mean of 12 replicates. Error bars show the standard deviation. ($p = 0.156$).

Our results demonstrate that inoculation with the poplar and willow endophyte consortia stimulated growth of Douglas-fir seedlings under low nutrient conditions. It was previously concluded that these endophytes are not host-specific colonizers [39]. Poplar and willow plants are angiosperms, which diverged from gymnosperms more than 100 million years ago [48]. Remarkably, the endophytes from poplar and willow plants colonized Douglas-fir, a gymnosperm. It is possible that the required plant-microbe communication may be ancient, pre-dating the divergence of angiosperms and gymnosperms. Interestingly, the response of Douglas-fir to inoculation with the endophytes took more than a year to develop, which suggests that this plant species may require a period of growth under controlled conditions to facilitate the establishment of the symbiotic relationship with these endophytes. Parker and Dangerfield [49] also reported a delayed response to microbial inoculation in Douglas-fir. In studies done by Chanway *et al.* [50], rhizospheric bacteria isolated from perennial rye grass and white clover stimulated the growth of container Douglas-fir seedlings after 12 weeks of inoculation. More recently, Anand *et al.* [31] showed that seedling growth enhancement in other conifers, lodgepole pine and western red cedar occurred only after 13 months.

To assess other potential mechanisms in addition to N fixation for the improved growth of the inoculated plants, we investigated the potential of phosphate solubilization by the endophytes. The majority of the isolates were able to produce a clear zone in minimal medium containing insoluble phosphorous. As shown in Table 1, the PSI of isolate WP19 was the highest, indicating the high solubilization capacity of this endophyte compared to the rest of the endophytes. There have been a number of reports on plant growth promotion by bacteria that have the ability to convert the insoluble inorganic forms of P into soluble forms through acidification, secretion of organic acids or protons and exchange reactions [36]. Many symbiotic phosphate-solubilizing microorganisms belonging to species of the genera *Bacillus, Pseudomonas*, *Rhizobium*, *Enterobacter* and *Burkholderia* and certain fungi produce phytase to mineralize organic phosphate [51]. In our study, phosphate solubilization may be one of the important mechanism through which these endophytes promoted growth of Douglas-fir seedlings; however, additional studies including phosphorous quantification and mutant analysis will be necessary to determine what symbiotic factors contribute the most to growth enhancement. The role of mycorrhizal symbiosis in increasing phosphorous uptake has been studied in a number of conifer species, with recent studies showing how mycorrhizal effects on seedling establishment change with soil conditions [52]. In dry Douglas-fir forests, reduced mycorrhizal diversity and abundance have been associated with reduced survival and growth of newly-planted conifer seedlings [53,54]. Inoculating nursery seedlings with beneficial endophytes may help offset some of these disadvantages. Furthermore, endophytes may increase mycorrhizal development by affecting root colonization, as well as by enhancing N and P uptake.

Rungin *et al.* [55] recently demonstrated plant growth-enhancing effects by a siderophore-producing endophytic streptomycete isolated from Thai jasmine rice plants. Significant increases in total biomass of rice and mung bean plants were observed in the inoculated controls compared to the untreated controls and siderophore-deficient mutant treatment. Most of the endophytes in our study produced siderophores, which may have been beneficial in plant growth promotion. However, this is largely speculative and warrants further research. Siderophore-producing microbes may also have significant activity against plant pathogens [56,57]. Interestingly, we found that Douglas-fir seeds inoculated with these endophytes were significantly less affected by Fusarium, a common Douglas-fir pathogen [58].Therefore, these endophytes merit further investigation, as reducing mortality by inoculations could prove valuable to tree nurseries.

To evaluate endophytic colonization, Douglas-fir seedlings were inoculated with fluorescent-labelled strains of WP5 and WP19. As seen from the fluorescent microscope images, *mkate*-tagged cells of *Acinetobacter* strain WP19 were localized in the intercellular spaces of root tissue of the inoculated seedlings (Figure 5A). In addition, GFP-tagged cells of *Rahnella* sp. strain WP5 colonized the needles

(Figure 5B), suggesting that this endophyte is able to move up the plant into aerial parts after inoculation. Similar colonization zones have been reported for other endophytes [30].

The enhanced growth in inoculated seedlings seen in this preliminary study merits further research in larger greenhouse trials and field trials to assess the long-term benefits of endophyte inoculations.

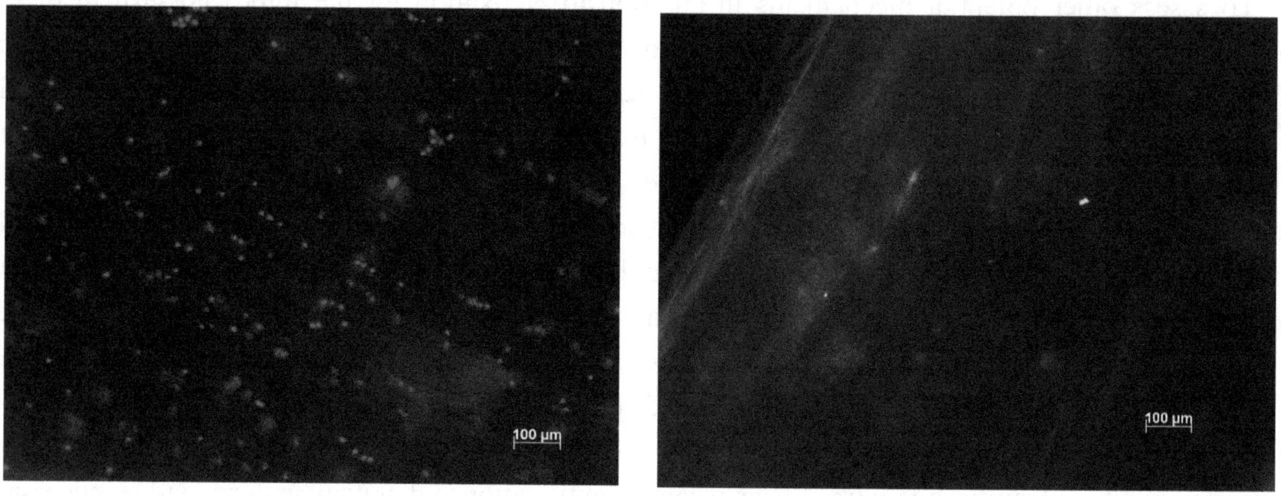

(A) (B)

Figure 5. Fluorescence micrographs showing intercellular colonization of *mkate*-tagged strain WP19 in the root tissue of the inoculated seedling of Douglas-fir (**A**) and GFP-tagged strain WP5 in the needle of the inoculated Douglas-fir (**B**). Samples were surface-sterilized and visualized (1000×) three weeks after inoculation.

4. Conclusions

The results of the present study indicate that inoculation with a diazotrophic endophytic consortia significantly increased Douglas-fir seedling growth under nutrient-limited conditions. Most commercial nursery treatments add high rates of fertilizers to maximize seedling shoot and total growth to increase establishment and success in the field [59] which is not only expensive, but also suppresses mycorrhiza development of container seedlings [60]. Therefore, endophyte inoculation of seedlings may be considered as a cost-effective technique to increase growth in the early stage for the nursery stock, which can lead to later stage successes in transplanting and establishment after transplanting.

Acknowledgments

This work was funded by the Mc Intire-Stennis and USDA National Institute of Food and Agriculture grants (# 2012-00931). The authors would like to thank Mike Gerdes and David Gerdes from Silvaseed Company Foresters for providing the seedlings and Sam Miller (University of Washington, Microbiology) for providing the plasmids.

Author Contributions

Zareen Khan carried out the experiments and wrote the manuscript. Shyam L Kandel and Daniela N Ramos contributed to the experiments and data analysis. Doty served as the Principal

Investigator (Mc Intire-Stennis and NIFA grants) and directed the project. Ettl and Kim served as co-PI's for the project. Doty, Ettl and Kim edited the manuscript.

References

1. Van den Driessche, R. Growth, survival and physiology of Douglas-fir seedlings following root wrenching and fertilization. *Can. J. For. Res.* **1983**, *13*, 270–278.
2. Aldhous, I. Nursery Practice. *For. Comm. Bull.* **1972**, *43*, 184.
3. Cram, M.M.; Dumroese, K.R. Mycorhizae in Forest Tree nurseries. In *Forest Nursery Pests. Agriculture Handbook*; Cram, M.M.F., Mallams, K.M., Eds.; Forest Service: Washington, DC, USA, 2012; pp. 20–25.
4. Kropp, B.R.; Langlois, C.G. Ectomycorrhizae in reforestation. *Can. J. For. Res.* **1990**, *20*, 438–451.
5. Tien, T.M.; Gaskins, M.H.; Hubbell, D.H. Plant growth substances produced by *Azospirillum brasilense* and their effect on the growth of Pearl Millet. *Appl. Environ. Microbiol.* **1979**, *37*, 1016–1024.
6. Chanway, C.P.; Holl, F.B.; Rurkington, R. Genotypic coadaptation in growth promotion of forage species by *Bacillus polymyxa*. *Plant Soil* **1988**, *106*, 281–284.
7. Ryan, R.P.; Germaine, K.; Franks, A.; Ryan, D.J.; Dowling, D.N. Bacterial endophytes: Recent developments and applications. *FEMS. Microbiol. Lett.* **2008**, *278*, 1–9.
8. Hardoim, P.R.; van Overbeek, L.S.; van Elsas, J.D. Properties of bacterial endophytes and their proposed role in plant growth. *Trends Microbiol.* **2008**, *16*, 463–471.
9. Gaiero, J.R.; McCall, C.A.; Thompson, K.; Day, N.J.; Best, A.S.; Dunfield, K.E. Inside the root microbiome: Bacterial root endophytes and plant growth promotion. *Am. J. Bot.* **2013**, *100*, 1738–1750.
10. *Symbiotic Endophytes*; Springer-Verlag Berlin Heidelberg: New York, NY, USA, 2014.
11. Compant, S.; van der Heijden, M.G.A.; Sessitch, A. Climate change on beneficial plant microorganism interactions. *FEMS Microbiol. Ecol.* **2010**, *73*, 197–214.
12. Welbaum, G.E.; Sturz, A.V.; Dong, Z.; Nowak, J. Endophytic nitrogen fixation in sugarcane: Present knowledge and future applications. *Plant Soil* **2003**, *252*, 139–149.
13. Peng, G.; Zhang, W.; Luo, H.; Xie, H.; Lai, W.; Tan, Z. Enterobacter oryzae sp. nov., a nitrogen-fixing bacterium isolated from the wild rice species *Oryza latifolia*. *Int. J. Syst. Evol. Microbiol.* **2009**, *59*, 1650–1655.
14. Caballero-Mellado, J.; Martinez-Aguilar, L.; Paredes-Valdez, G.; Estrada-de Los Santos, P. *Burkholderia unamae* sp. nov., an N_2-fixing rhizospheric and endophytic species. *Int. J. Syst. Evol. Microbiol.* **2004**, *54*, 1165–1172.
15. Montanez, A.; Abreu, C.; Gill, P.R.; Hardarson, G.; Sicardi, M. Biological nitrogen fixation in maize (*Zea mays* L.) by N-15 isotope-dilution and identification of associated culturable diazotrophs. *Biol. Fert. Soils* **2009**, *45*, 253–263.
16. Reiter, B.; Burgmann, H.; Burg, K.; Sessitsch, A. Endophytic nifH gene diversity in African sweet potato. *Can. J. Microbiol.* **2003**, *49*, 549–555.
17. Reinhold-Hurek, B.; Hurek, T. Life in grasses: Diazotrophic endophytes. *Trends Microbiol.* **1998**, *6*, 139–144.

18. Jimenez-Salgado, T.; Fuentes-Ramirez, L.F.; Tapia-Hernandez, A.; Mascarua-Esparza, M.A.; Martinez-Romero, E.; Caballero-Mellado, J. *Coffea arabica* L., A new host plant for *Acetobacter diazotrophicus*, and isolation of other nitrogen-fixing acetobacteria. *Appl. Environ. Microbiol.* **1997**, *63*, 3676–3683.
19. Lopez, B.R.; Bashan, Y.; Bacilio, M. Endophytic bacteria of Mammillaria fraileana, an endemic rock-colonizing cactus of the southern Sonoran Desert. *Arch. Microbiol.* **2011**, *193*, 527–541.
20. Doty, S.L.; Dosher, M.R.; Singleton, G.L.; Moore, A.L.; Aken, B.; van Stettler, R.F.; Gordon, M.P. Identification of an endophytic Rhizobium in stems of Populus. *Symbiosis* **2005**, *39*, 27–35.
21. Doty, S.L.; Oakley, B.; Xin, G.; Kang, J.W.; Singleton, G.; Khan, Z.; Vajzovic, A.; Staley, J.T. Diazotrophic endophytes of native black cottonwood and willow. *Symbiosis* **2009**, *47*, 23–33.
22. Xin, G.; Zhang, G.; Kang, J.W.; Staley, J.T.; Doty, S.L. A diazotrophic, indole-3-acetic acid-producing endophyte from wild cottonwood. *Biol. Fert. Soils* **2009**, *45*, 669–674.
23. Bormann, B.T.; Bormann, W.B.; Bowden, R.S.; Piece, S.P.; Hamburg, D.; Wang, M.; Snyder, C.; Ingersoll, R.C. Rapid N_2 Fixation in Pines, Alder, and Locust: Evidence From the Sandbox Ecosystems Study. *Ecosystems* **1993**, *74*, 583–598.
24. Pankievicz, V.C.S.; do Amaral, F.P.; Santos, K.F.D.N.; Agtuca, B.; Xu, Y.; Schueller, M.J.; Arisi, A.C.M.; Steffens, M.B.R.; de Souza, E.M.; Pedrosa, F.O.; *et al*. Robust biological nitrogen fixation in a model grass-bacterial association. *Plant J.* **2015**, *81*, 907–919.
25. Mocali, S.; Bertelli, E.; Dicello, F.; Mengoni, A.; Falanga, A.; Villani, F.; Caciotti, A.; Tegli, S.; Surico, G.; Fani, R. Fluctuation of bacteria isolated from elm tissues during different seasons and from different plant organs. *Res. Micribiol.* **2003**, *154*, 105–114.
26. Bal, A.S.; Anand, R.; Berge, O.; Chanway, C.P. Isolation and identification of diazotrophic bacteria from internal tissues of *Pinus contorta* and *Thuja plicata*. *Can. J. For. Res.* **2012**, *42*, 807–813.
27. Izumi, H.; Anderson, I.C.; Killham, K.; Moore, E.R. Diversity of predominant endophytic bacteria in European deciduous and coniferous trees. *Can. J. Microbiol.* **2008**, *54*, 173–179.
28. Ulrich, K.; Ulrich, A.; Ewald, D. Diversity of endophytic bacterial communities in poplar grown under field conditions. *FEMS Microbiol.* **2008**, *63*, 169–180.
29. Pirttila, A.M.; Laukkanen, H.; Pospiech, H.; Myllyla, R.; Hohtola, A. Detection of intracellular bacteria in the buds of scotch pine (*Pinus sylvestris* L.) by insitu hybridization. *Appl. Environ. Microbiol.* **2000**, *66*, 3073–3077.
30. Bal, A.S.; Chanway, C.P. Evidence of nitrogen fixation in lodgepole pine inoculated with diazotrophic *Paenibacillus polymyxa*. *Botany* **2012**, *90*, 891–896.
31. Anand, R.; Grayston, S.; Chanway, C.P. N_2-fixation and seedling growth promotion of lodgepole pine by endophytic *Paenibacillus polymyxa*. *Microb. Ecol.* **2013**, *66*, 369–374.
32. Carrell, A.A.; Frank, A.C. *Pinus flexilis* and *Picea engelmannii* share a simple and consistent needle endophyte microbiota with a potential role in nitrogen fixation. *Front. Microbiol.* **2014**, doi:10.3389/fmicb.2014.00333.
33. Stack, R.W.; Sinclair, W.A. Protection of Douglas-fir seedlings against *Fusarium* root rot by a mycorrhizal fungus in the absence of mycorrhiza formation. *Phytopathology* **1975**, *65*, 468–472.
34. Horton, T.R.; Bruns, T.D.; Parker, V.T. Ectomycorrhizal fungi associated with *Arctostaphylos* contribute to *Pseudotsuga menziesii* establishment. *Can. J. Bot.* **1999**, *77*, 93–102.

35. Nara, K. Ectomycorhizal networks and seedling establishment during early primary succession. *New Phytol.* **2006**, *169*, 169–178.
36. Rodríguez, H.; Fraga, R. Phosphate solubilizing bacteria and their role in plant growth promotion. *Biotechnol. Adv.* **1999**, *17*, 319–339.
37. Neilands, J.B. Siderophores: Structure and function of microbial iron transport compounds. *J. Biol. Chem.* **1995**, *270*, 26723–26726.
38. Glick, B.R. Plant Growth-Promoting Bacteria: Mechanisms and Applications. *Scientifica* **2012**, *2012*, doi:10.6064/2012/963401.
39. Khan, Z.; Guelich, G.; Phan, H.; Redman, R.; Doty, S. Bacterial and Yeast Endophytes from Poplar and Willow Promote Growth in Crop Plants and Grasses. *ISRN Agric*. **2012**, *2012*, 890280:1–890280:11.
40. Knoth, J.L.; Kim, S.H.; Ettl, G.J.; Doty, S.L. Effects of cross host species inoculation of nitrogen-fixing endophytes on growth and leaf physiology of maize. *GCB Bioenergy* **2013**, *5*, 408–418.
41. Knoth, J.L.; Kim, S.H.; Ettl, G.J.; Doty, S.L. Biological nitrogen fixation and biomass accumulation within poplar clones as a result of inoculations with diazotrophic endophyte consortia. *New Phytol.* **2014**, *201*, 599–609.
42. Dumroese, K.R.; Wenny, L.D. Propagation protocol for production of Container (plug) *Pseudotsuga menziesii* Franco plants 66 mL (4cu.in) Ray Leach "Cone-tainers". 2009. Available online: http://www.NativePlantNetwork.org (accessed on 28 September 2015).
43. Khan, Z.; Rho, H.; Firrincieli, A.; Hung, S.H.; Luna, V.; Masciarelli, O.; Kim, S.H.; Doty, S.L. Growth improvement and drought tolerance of hybrid poplar by inoculation with endophyte consoria. **2015**, Submitted.
44. Cangelosi, G.A.; Best, E.A.; Martinetti, G.; Nester, E.W. Genetic analysis of *Agrobacterium*. *Meth. Enzymol.* **1991**, *204*, 384–397.
45. Nautiyal, C.S. An efficient microbiological growth medium for screening phosphate solubilizing microorganisms. *FEMS Microbiol. Lett.* **1999**, *170*, 265–270.
46. Alexander, D.B.; Zuberer, D.A. Use of chrome azurol S reagent to evaluate siderohore production by rhizosphere bacteria. *Biol. Fert. Soils* **1991**, *12*, 39–45.
47. Doty, S.; Chang, M.; Nester, E.W. The chromosomal virulence gene. ChvE of *Agrobacterium tumafaciens* is regulated by a LysR family member. *J. Bacteriol.* **1993**, *175*, 7880–7886.
48. Martin, W.; Lydiate, D.; Brinkmann, H.; Forkmann, G.; Saedle, R.H.; Cerff, R. Molecular phylogenies in angiosperm evolution. *Mol. Biol. Evol.* **1993**, *10*, 140–162.
49. Parker, A.K.; Dangerfield, J.A. Influence of bacterial inoculations on growth of containerized Douglas-fir seedlings. *Can. J. For. Res.* **1975**, *31*, 14–15.
50. Chanway, C.P.; Radley, R.A.; Holl, F.B. Bacterial Inoculation of Lodgepole Pine, White Spruce. In Proceedings of the Intermountain Forest Nursery Association Annual Meeting, Bismarck, ND, USA, 14–18 August 1989; Landis, T.D., Ed.; USDA Forest Service: Fort Collins, CO, USA, 1989.
51. Singh, P.; Kumar, V.; Agrawal, S. Evaluation of phytase producing bacteria for their plant growth promoting activities. *Int. J. Microbiol.* **2014**, *2014*, 7.

52. Horton, T.R.; Cazaras, E.; Bruns, T.D. Ectomycorrhizal, vesicular-arbuscular and dark septate fungal colonization of bishop pine (*Pinus muricata*) seedlings in the first 5 months of growth after wildfire. *Mycorrhiza* **1998**, *7*, 11–18.
53. Kazantseva, O.; Bingham, M.; Simard, S.W.; Berch, S.M. Effects of growth medium, nutrients, water, and aeration on mycorhization and biomass allocation of greenhouse-grown Douglas-fir seedlings. *Mycorrhiza* **2009**, *20*, 51–66.
54. Teste, F.P.; Schmidt, M.G.; Berch, S.M.; Bulmer, C.; Egger, K.N. Effects of ectomycorrhizal inoculants on survival and growth of interior Douglas-fir seedlings on reforestation sites and partially rehabilitated landings. *Can. J. For. Res.* **2004**, *4*, 2074–2088.
55. Rungin, S.; Indananda, C.; Suttiviriya, P.; Kruasuwan, W.; Jaemsaeng, R.; Thamchaipenet, A. Plant growth enhancing effects by a siderophore producing endophytic streptomycete isolated from a Thai jasmine rice plant (*Oryza sativa* L. KDML105). *Antonie Leeuwenhoek* **2012**, *102*, 463–472.
56. Thomashow, L.S. Biological control of plant root pathogens. *Curr. Opin. Biotech.* **1996**, *7*, 343–347.
57. Duijiff, B.J.; Alabouvette, C.; Lemanceau, P. Involvement of pseudobactin 358 in the inhibition of *Fusarium oxysporum* f.sp.lini in flax rhizosphere by the combination of *Pseudomas putida* WCS358 and nonpathogenic *Fusarium oxysporum* F047 and *Pseudomas putida* WCS358. *Phytopathology* **1997**, *89*, 1073–1079.
58. Khan, Z.; Ramos, D.; Matthew, A.; Ettl, J.G.; Doty, S.L. Impact of endophyte inoculation on health of Douglas-fir seedlings. Unpublished work, 2015.
59. Hunt, G.A. Effects of Mycorrhizal Fungi on Quality of Nursery Stock and Plantation Performance in the Southern Interior of British Columbia; FRDA Report ISSN 0835-0752; Natural Resources Canada: Victoria, BC, Canada, 1992; p. 185.
60. Marx, D.H.; Hatch, A.B.; Mendicino, J.F. High soil fertility decreases sucrose content and susceptibility of loblolly pine to ectomycorrhizal infection by *Pisolithis tinctorius*. *Can. J. Bot.* **1977**, *55*, 1569–1574.

10

Soil Nitrogen Transformations and Availability in Upland Pine and Bottomland Alder Forests

Tae Kyung Yoon [1,†], Nam Jin Noh [2,†], Haegeun Chung [3], A-Ram Yang [4] and Yowhan Son [5,6,*]

[1] Department of Environmental Science and Engineering, Ewha Womans University, Seoul 03760, Korea; E-Mail: yoon.ecology@gmail.com

[2] River Basin Research Center, Gifu University, Gifu 501-1193, Japan; E-Mail: treeworld@green.gifu-u.ac.jp

[3] Department of Environmental Engineering, Konkuk University, Seoul 05029, Korea; E-Mail: hchung@konkuk.ac.kr

[4] Forest Practice Research Center, Korea Forest Research Institute, Pocheon 11186, Korea; E-Mail: ramrami0130@gmail.com

[5] Department of Environmental Science and Ecological Engineering, Graduate School, Korea University, Seoul 02841, Korea

[6] Department of Biological and Environmental Science, Qatar University, Doha P.O. Box 2713, Qatar

† These authors contributed equally to this work.

* Author to whom correspondence should be addressed; E-Mail: yson@korea.ac.kr;

Academic Editor: Heinz Rennenberg

Abstract: Soil nitrogen (N) processes and inorganic N availability are closely coupled with ecosystem productivity and various ecological processes. Spatio-temporal variations and environmental effects on net N transformation rates and inorganic N concentrations in bulk soil and ion exchange resin were examined in an upland pine forest (UPF) and a bottomland alder forest (BAF), which were expected to have distinguishing N properties. The annual net N mineralization rate and nitrification rate (kg N·ha^{-1}·year^{-1}) were within the ranges of 66.05–84.01 and 56.26–77.61 in the UPF and −17.22–72.24 and 23.98–98.74 in the BAF, respectively. In the BAF, which were assumed as N-rich conditions, the net N mineralization rate was suppressed under NH_4^+ accumulated soils and was independent from soil temperature. On the other hand, in the UPF, which represent moderately fertile N conditions, net N transformation rates and N availability were dependent to the generally known regulation by soil temperature and soil water content. Stand density might indirectly affect the N transformations, N availability, and ecosystem productivity through different soil moisture conditions. The differing patterns of different inorganic N indices provide useful insight into the N availability in each forest and potential applicability of ion exchange resin assay.

Keywords: *Alnus japonica*; ion exchange resin; nitrogen mineralization; *Pinus densiflora*; soil moisture

---

1. Introduction

Soil nitrogen (N) has become a focus of terrestrial ecosystem studies for two main reasons: (1) N availability is a limiting factor for terrestrial ecosystem productivity [1–3] and (2) N losses from terrestrial ecosystems impact the atmosphere especially by greenhouse gas emission, aquatic ecosystems by eutrophication, and drinking water by nitrate contamination [4,5]. N mineralization, the microbial conversion of organic N to inorganic N, has been intensively studied because it is believed to be the principal control of N availability to plants in terrestrial ecosystems [6]. Recent progress in the knowledge of and techniques related to organic N and the priming of rhizospheres for N availability has shaken the classical assumption that N mineralization dominates as the primary source of available N, [7–10]; nevertheless, in field N studies, N mineralization and other inorganic N indices remain meaningful standard measures which provide useful insight.

Environmental controls of N mineralization and availability, particularly temperature and soil moisture, have been investigated through a theoretical model [11], literature review [9,12], laboratory experiments [13–15], and field manipulations [16–19]. For example, soil moisture varies significantly in space and time, and affects various biogeochemical processes in different ways [20–22]. Therefore, understanding soil moisture and its effect on N and nutrient processes is essential in soil ecology. Various studies have analyzed the effect of the soil moisture gradient on N mineralization and developed empirical models that describe the response of N mineralization to soil moisture based on laboratory incubation experiments. However, these laboratory studies may insufficiently reflect the effects of soil moisture on N mineralization in nature, where differing soil moisture ranges have indirect effects on N mineralization through other variables, such as the vegetation community, soil chemistry, and substrate availability [23–25]. To understand the complex interactions among such environmental factors and N availability, field studies incorporating the repeated measurement on various environmental conditions are necessary since they can better reflect the temporal and spatial variation of soil moisture and substrate gradients.

Numerous N indices techniques, each with respective advantages and limitations, have been used to trace the real availability of N under various conditions [6,26]. These techniques include ascertaining net N mineralization using a buried bag or resin core, net N mineralization potential using laboratory incubation, gross N mineralization and immobilization using a stable N isotope, and the total or inorganic N concentration of bulk soil. Although ion exchange resins (IERs) that efficiently adsorb "available" inorganic N from soil would provide an alternative useful measure of N availability by imitating the N uptake processes of plant roots [6,27,28], they have been rarely applied to soil N studies, and the knowledge and insight so far gained from this measure of N availability has been limited. Furthermore, because no single N index can give complete insight into N availability [6], a measurement of multiple N indices increases the understanding of N availability and its relationship with environmental variables. Therefore, this study uses multiple N indices, including IER assays.

The aim of the current study was to investigate net N transformation and N availability in an upland pine forest and a bottomland alder forest, which are expected to have distinguishing N properties due to the different species functional type and topography. Broad-leaved deciduous and coniferous forests are generally assumed to be developed on fertile and infertile soils, respectively [29–31], even though several recent syntheses failed to find the general pattern and harmonized empirical evidences supporting this assumption [32,33]. Consequently, distinctive litter N properties between broad-leaved deciduous and coniferous trees respectively affect soil N availability [34]. Topography also discriminates various soil processes coupled with hydrology. Therefore, the upland pine forest and the bottomland alder forest could be assumed to represent N-moderate and N-rich environments, respectively, and the forests would exhibit distinctive N cycle processes. Specifically, our objectives were: (1) to determine the spatio-temporal patterns of net N transformation rates and N availability and their relationship with environmental factors under different stand densities or soil moisture conditions; (2) to analyze the similarities and differences of multiple N indices, including IER assays, to better understand N availability; and (3) to characterize the soil N transformation, availability, and uptake in each forest.

2. Materials and Methods

2.1. Study Sites

This study was carried out in an upland pine forest (UPF) and a bottomland alder forest (BAF) in central Korea (Table 1). The UPF is located in a rural mountainous area, 30 km north of the city Seoul at Gwangneung Experimental Forest (37°47′01″ N, 127°10′37″ E, 410–440 m a.s.l.). The canopy is dominated by Japanese red pine (*Pinus densiflora* Sieb. et Zucc.) with deciduous middle and understory species (e.g., *Acer palmatum*, *Carpinus laxiflora*, *C. cordata*, *Fraxinus rhynchophylla*, *Quercus mongolica*, *Q. variabilis*, and *Styrax obassia*). The UPF area is typical of the upland pine forest that is the most common forest type in South Korea and represents 23.1% of the total national forest by area. The BAF is located at Heonilleung Ecosystem Landscape Conservation Area (37°27′52″ N, 127°04′53″ E, 40 m a.s.l.) in a southern suburb of Seoul. The canopy is dominated by Japanese alder (*Alnus japonica* Steud.) with several middle and understory species (e.g., *Euonymus oxyphyllus*, *F. rhynchophylla*, *Ginkgo biloba*, *Ligustrum obtusifolium*, *Q. aliena*, *Q. mongolica*, and *Rosa multiflora*). Japanese alder is a typical species tolerant of hydric conditions in East Asian bottomlands [35,36], although its cover is currently small due to land use change. Interestingly, both forests have been well preserved from a harvest for hundreds of years due to the preservation of the royal tombs established within them during the 15th century [37,38]. The forests were sheltered from disturbances during the first half of the 20th century, in which the rule of Japanese Imperialism

(1910–1945) and the Korean War (1950–1953) severely degraded nation-wide forests. The forests could provide reference status of the upland and bottomland forests in Korea, accompanied with abundant previous studies [25,39–42].

Table 1. Characteristics of the upland pine forest (UPF) and bottomland alder forest (BAF) study sites. Numbers in parentheses indicate the standard error of the means. Climate data was obtained from the nearest meteorological stations and site description data was taken from previous studies of the UPF [39,41] and BAF [25,42].

	Upland Pine Forest (UPF)		**Bottomland Alder Forest (BAF)**		
	Low Density (LD)	**High Density (HD)**	**Low Moisture (LM)**	**Middle Moisture (MM)**	**High Moisture (HM)**
Site (1980–2010 average for climate data)					
Location	37°47′01″ N, 127°10′37″ E Rural area 30 km north of Seoul		37°27′52″ N, 127°04′53″ E Southern suburb of Seoul		
Altitude (m)	410–440		40		
Aspect, slope (°)	West, 13–22		South, <1		
Temperature (°C)	11.4 (monthly range: −5.0–26.5)		13.5 (monthly range: −1.9–27.2)		
Annual precipitation (mm)	1471		1473		
Vegetation (in 2007 for the UPF and in 2009 for the BAF)					
Canopy species	*Pinus densiflora*		*Alnus japonica*		
Density ($trees \cdot ha^{-1}$)	983	1517	900	575	550
Diameter at breast height (cm)	20.5	17.6	32.6	27.1	24.2
Aboveground biomass ($Mg\ C \cdot ha^{-1}$)	88.1	86.7	187.9	78.9	61.8
Needle or leaf litter C:N ratio	72.7 (2.0)	73.2 (2.2)	21.8 (0.3)	20.8 (0.2)	18.6 (0.2)
Total N storage ($Mg\ N \cdot ha^{-1}$)	0.78	0.78	1.65	0.74	0.60
Soil (0–15 cm; during the study period)					
Soil texture	Silt loam	Silt loam	Sandy loam	Sandy loam	Silt loam
Bulk density ($g \cdot cm^{-3}$)	1.05 (0.04)	1.02 (0.03)	0.86 (0.06)	0.81 (0.03)	0.50 (0.14)
Soil water content ($kg \cdot kg^{-1}$)	0.25 (0.02)	0.20 (0.03)	0.35 (0.06)	0.72 (0.17)	1.23 (0.10)
pH	4.54 (0.03)	4.56 (0.01)	4.45 (0.07)	5.04 (0.10)	5.14 (0.04)
Total C concentration ($g \cdot kg^{-1}$)	33.4 (2.3)	29.7 (2.1)	20.8 (2.0)	31.2 (1.8)	61.7 (2.3)
C:N ratio	15.8 (0.4)	17.8 (0.4)	9.6 (0.3)	12.4 (0.2)	12.3 (0.1)
Total N storage (0–30 cm; $Mg\ N \cdot ha^{-1}$)	3.62 (0.24)	2.98 (0.12)	4.64 (0.52)	5.06 (0.70)	7.14 (0.20)

The sites are only 35 km apart and at the sites, precipitation was similar (Table 1). Temperature, however, differs and the UPF, due to its higher altitude, is slightly colder than the BAF. Their soils are classified into silt loam or sandy loam and are acidic (pH: 4.45–5.14) at both sites. The UPF had less soil water content (SWC) than the BAF during the all study period (Figure S1f), reflecting the topographical characteristics of each site. Moreover, the UPF and the BAF would be determined as moderately fertile (medium) and N-rich (high) conditions, respectively, based on the needle or leaf litter and soil C:N ratio, and total N (sum of organic and inorganic N) storage in vegetation and soil, using the nutrient availability criteria by Vicca *et al.* [43].

UPF stands were classified by stand density and BAF stands were classified by soil moisture, in order to characterize the spatial variability of stand structure and ecological processes in each type of forest (for UPF see [39,41] and for BAF see [25,42]). Stand density and soil hydrology are assumed to be the main environmental driver of upland forest and bottomland ecosystems, respectively; this assumption is generally applied to ecosystem management practices in each ecosystem. In the UPF, low density (LD; 983 trees·ha^{-1}) and high density (HD; 1517 trees·ha^{-1}) stands were selected and in the BAF, low moisture (LM), middle moisture (MM), and high moisture (HM) stands were selected with SWC of <0.50, 0.50–1.00, and 1.00–2.00 kg·kg^{-1}, respectively. Note that the MM and HM stands, which were underlain with poorly drained soils, can be classified as forested wetland. Three replicate plots, each covering 400 m^2 in the UPF and 100 m^2 in the BAF, were established in each stand.

2.2. Field Experiment and Laboratory Analysis

Three inorganic N indices of top mineral soils, net N transformation rates, inorganic N concentrations in bulk soil and IER assays, were investigated at five points, every 2 m along a single transect across each plot from March 2008 to April 2009. Net N transformation rates including net N mineralization and net nitrification were determined using a field incubation method [44]. Two soil cores from the 0–15 cm mineral layer were sampled from the five points in each plot. The soil sample for determining soil status before field incubation was taken immediately to the laboratory at a cool temperature (<4 °C) and analyzed for inorganic N concentrations in bulk soil. The other soil sample, the field incubation sample, was put into a gas permeable polyethylene bag and returned to its original location. After 45 days, the incubated soil core was collected and also taken to the laboratory for analysis. This soil sampling and incubation procedure was repeated every 45 days, except for a 90 day winter interval due to the difficulty of sampling and incubating frozen soil.

Two IER bags, which were made of stocking nylon and contained 14 mL of either anion or cation resins (Sybron IONAC ASB-1P OH^- and C-251 H^+, Sybron International, Birmingham, NJ, USA), were placed 10 cm below the surface of the mineral soil in each plot, near the five sampling points of the initial and incubated soils. The IER bags were retrieved and replaced every 90 days.

In the laboratory, the moist initial and incubated soil samples were subjected to an extraction process with 2 M KCl solution for 24 h, immediately upon arrival from the sites. The IERs were air-dried first, and then the inorganic N ions were extracted. The extracts were filtered through Whatman No. 42 filter paper and the NH_4^+ and NO_3^- concentrations were determined colorimetrically using a flow-injection auto analyzer (QuikChem AE, Lachat Instruments, Loveland, CO, USA). The NH_4^+ and NO_3^- concentrations extracted from moist initial and incubated soil samples were corrected to a dry soil basis using the gravimetric SWC ((moist soil weight – dry soil weight)/dry soil weight) of each soil.

The net N mineralization and net nitrification rates were calculated as the difference between the NH_4^+ and NO_3^- concentrations of the incubated and initial soil samples. Annual net N mineralization and annual net nitrification per unit area in the upper 15 cm mineral soil layer was calculated by summing the net N mineralization and net nitrification values of incubation periods spanning a year [6,45].

Annual N uptake was estimated from aboveground net primary productivity (ANPP) multiplied by the concentrations of N in woody tissues and leaves, determined using an elemental analyzer (Vario Macro CN analyzer, Elementar Analysensysteme GmbH, Hanau, Germany). ANPP was calculated as the sum of total aboveground biomass increment and litter production (See [41] for details in the UPF.). Aboveground biomass in the BAF was calculated using a DBH-based allometric equation [46] and C concentrations in each compartment determined by the elemental analyzer.

2.3. Statistical Analysis

The values of the five soil samples from each plot were averaged. Thus, plots, of which there were three in each stand, were considered as true replicates. Correlation and regression analyses were conducted on the data derived from individual plots to take into account the spatial variability within each stand. Significant differences in net N transformation rates, inorganic N concentrations in bulk soil and IER, and SWC across stands and seasons were tested using the two-way analysis of variance (ANOVA). Duncan's multiple range test was performed to determine the differences between the means of each stand, when the ANOVA model was significant ($p < 0.05$). Correlation and regression analyses were conducted among the N indices and soil environment variables. All statistical analyses were conducted using R [47].

3. Results

The mean NH_4^+, NO_3^-, and total inorganic N concentrations in bulk soils varied significantly between the stands (Figure 1, $p < 0.01$). NH_4^+ was abundant at the HM and the MM BAF stands (Figure 1a); whereas NO_3^- was rich at the LM BAF stand (Figure 1b). Total inorganic N concentration was lowest at the UPF stands (Figure 1c). As a proportion of total inorganic N, the N in NH_4^+ form represented 61% in the LD UPF stand, 63% in the HD UPF stand, and 36%, 69%, and 78% in the LM, MM, and HM BAF stands, respectively. Conversely, as a proportion of total inorganic N, the N in NO_3^- form represented 39% in the LD UPF stand and 37% in the HD UPF stand and 64%, 31%, and 22% in the LM, MM, and HM BAF stands, respectively. There were also differences in the seasonal patterns of NH_4^+ and NO_3^- levels and of total inorganic N concentrations in bulk soils between the sites. Whereas in the UPF, inorganic N concentrations increased during the growing season (May to August), in the BAF, they were lower (Figure 1a–c, Figure S1).

The total IER-inorganic N and IER-NO_3^- concentrations did not vary significantly (Figure 1e,f). However, IER-NH_4^+ concentration (mg N·bag^{-1}) did differ between the stands; it was highest at the HD UPF stand (5.17), followed by the LD UPF stand (4.29), and then the LM (3.01), HM (1.75), and MM (0.66) BAF stands in turn (Figure 1d). IER-NO_3^- accounted for a major portion of IER-inorganic N (61%–93%) and, conversely, IER-NH_4^+ accounted for a very small portion of IER-inorganic N, especially at the MM (7%) and HM (15%) BAF stands. In MM and HM, the majority proportion of IER inorganic NO_3^- strongly contrasted that of NH_4^+ in bulk soil.

The net N transformation rates were higher under the drier stands (LD and HD UPF and LM BAF) than the wetter ones (MM and HM BAF) (Figure 1g,h). Negative net N mineralization rate was observed in the MM and HM BAF stands (Figure 1g). The seasonal pattern of net N transformation rates at the UPF stands and the LM BAF stand closely followed that of soil temperature (Figure S1g,h), in that the rates increased in the growing season and decreased in the non-growing season. On the other hand, the seasonal pattern was not clear at MM and HM BAF stands; negative net N mineralization rates were observed during most seasons (Figure S1g).

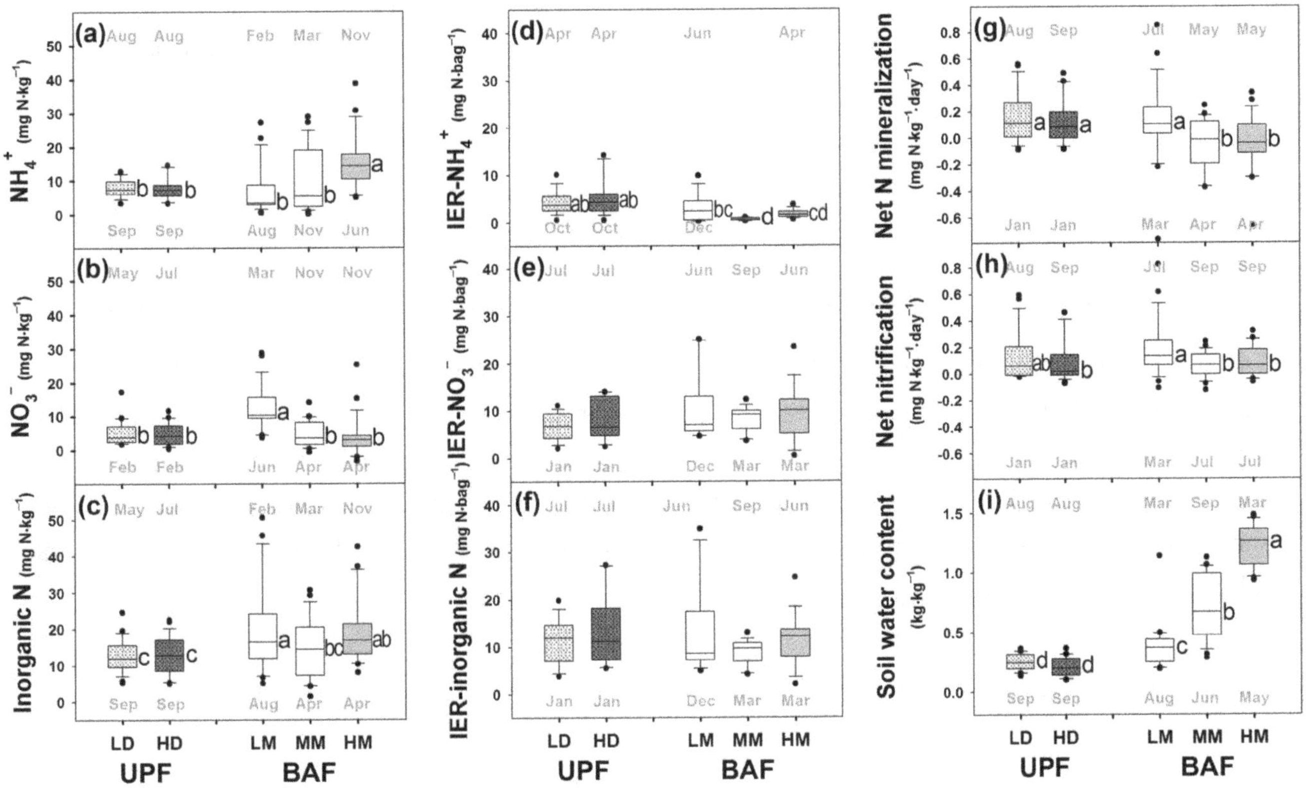

Figure 1. Inorganic N indices in an upland pine forest (UPF) with different stand densities and a bottomland alder forest (BAF) with different soil moisture content. (**a–c**) Inorganic N concentrations in bulk soil; (**d–f**) Inorganic N concentrations attached to ion exchange resin (IER); (**g,h**) Net N transformation rates; (**i**) Soil water content. Months written at the top and the bottom of each panel indicate the months corresponding to the highest and lowest values. Letters beside boxes indicate the significant difference, if there is one, among the means determined by Duncan's multiple range tests.

Table 2 shows the correlations between inorganic N indices and environmental variables. Generally, inorganic N indices in the UPF were considerably dependent to soil temperature and SWC; meanwhile, those in the BAF were less dependent to the environmental variables. IER-NO_3^-, IER-inorganic N, net N mineralization, and net nitrification were positively correlated with soil temperature in both the UPF and BAF. Correlation coefficients were higher in the UPF (r = 0.57–0.71) than the BAF (r = 0.24–0.47). NH_4^+ and inorganic N only in the UPF and IER-NH_4^+ only in the BAF were correlated with soil temperature. NH_4^+, inorganic N, net N mineralization and net nitrification in the UPF (r = 0.46–0.60) and NH_4^+ in the BAF (r = 0.57) were positively correlated with SWC. On the other hand, NO_3^- (r = −0.49), net N mineralization (r = −0.37), and net nitrification (r = −0.29) in the BAF were negatively correlated with SWC.

Table 2. Environmental dependence of inorganic N indices in an upland pine forest (UPF) and a bottomland alder forest (BAF). A correlation coefficient (r) that satisfies the statistical significance is presented ($p < 0.05$). A dash indicates insignificant.

	Soil Temperature		Soil Water Content	
	UPF	**BAF**	**UPF**	**BAF**
NH_4^+	0.31	–	0.54	0.57
NO_3^-	–	–	–	−0.48
Inorganic N	0.34	–	0.43	–
IER-NH_4^+	–	0.37	–	–
IER-NO_3^-	0.71	0.46	–	–
IER-inorganic N	0.57	0.47	–	–
Net N mineralization	0.58	0.24	0.56	−0.37
Net nitrification	0.67	0.32	0.60	−0.29

Spatio-temporal variations of net N mineralization and NH_4^+ were related, especially in the BAF (Figure 2). In the BAF, a significant negative relationship between net N mineralization and NH_4^+ was observed (Figure 2b, $p < 0.001$, $R^2 = 0.45$), whereas, in the UPF, the positive and weak relationship was only close to marginally significant (Figure 2a, $p = 0.10$, $R^2 = 0.06$). Seasonal patterns of the net N mineralization rate and of NH_4^+ concentration were coupled in the UPF (Figure 2c,d), but inversely coupled in the BAF (Figure 2e–g).

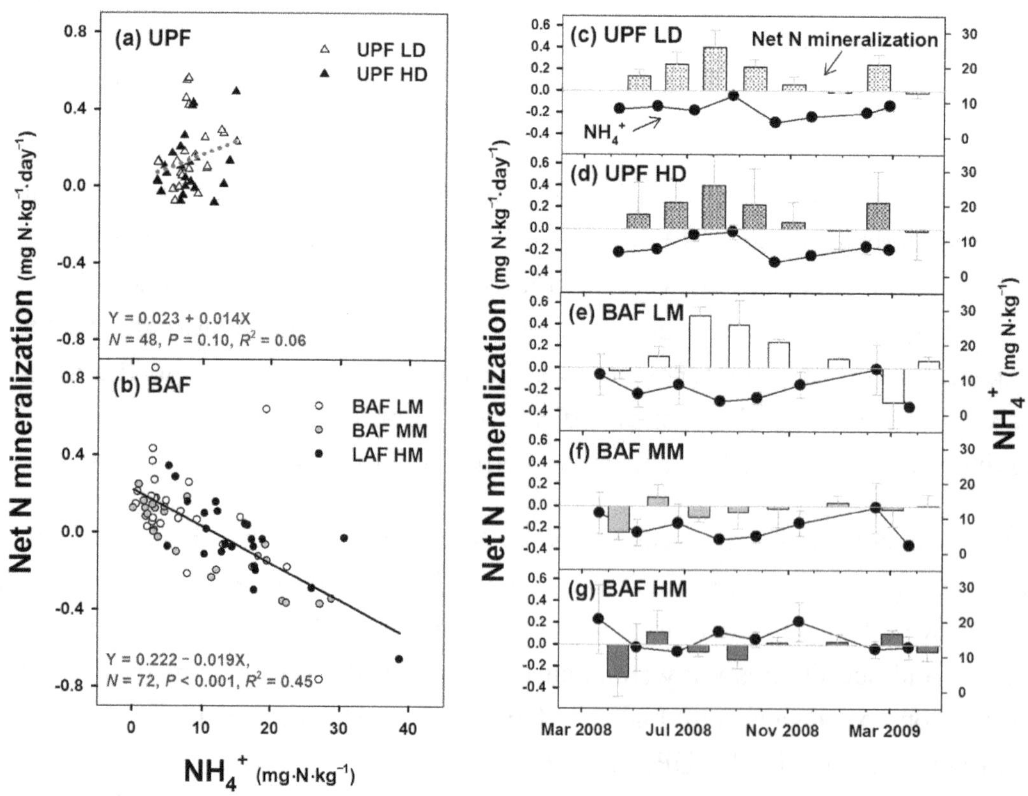

Figure 2. **(a,b)** Relationship between NH_4^+ concentrations and net N mineralization rates in an upland pine forest (UPF) and a bottomland alder forest (BAF); **(c–g)** Seasonal patterns of net N mineralization rates and NH_4^+ concentrations in the UPF and the BAF.

The annual net N mineralization rate (kg N·ha^{-1}·year^{-1}) was higher at the LD (84.01 ± 18.18) and HD (66.05 ± 11.25) UPF stands and the LM BAF stand (72.24 ± 11.13) than at the MM (−17.22 ± 43.10) and HM (−10.26 ± 17.60) BAF stands (Table 3, $p < 0.01$). The annual net N nitrification rate (kg N·ha^{-1}·year^{-1}), from highest to lowest, was LM (98.74 ± 16.01) ≥ LD (77.61 ± 18.26) = HD (56.26 ± 14.56) ≥ MM (32.69 ± 12.67) = HM (23.98 ± 9.80) (Table 3). Figure 3 describes the relationship between SWC and annual net N transformation rates. In the BAF, annual N transformation rates showed a linear inverse correlation with SWC. In contrast, in the UPF the correlation was close to being positive. Annual N uptake (kg N·ha^{-1}·year^{-1}) was estimated to be 73.19, 60.18, 191.77, 141.21, and 128.76 for LD, HD, LM, MM, and HM stands, respectively (Table 3).

Table 3. Summary of net N transformation rates, N availability, N uptake, and ANPP.

	Annual Net N Mineralization Rate †	Annual Net Nitrification Rate †	IER-NH_4^+ ‡	IER-NO_3^- ‡	IER-Total Inorganic N ‡	Annual N Uptake †	ANPP §
UPF							
LD	84.01 ± 18.18	77.61 ± 18.26	4.29 ± 0.49	6.81 ± 0.58	11.10 ± 0.97	73.19	6.17
HD	66.05 ± 11.25	56.26 ± 14.56	5.17 ± 0.79	8.17 ± 0.88	13.34 ± 1.53	60.18	5.03
BAF							
LM	72.24 ± 11.13	98.74 ± 16.01	3.03 ± 0.59	10.73 ± 1.43	13.76 ± 1.98	191.77	5.01
MM	−17.22 ± 43.10	32.69 ± 12.67	0.66 ± 0.05	8.14 ± 0.52	8.80 ± 0.54	141.21	6.06
HM	−10.26 ± 17.60	23.98 ± 9.80	1.75 ± 0.17	9.62 ± 1.13	11.37 ± 1.07	128.76	4.51

IER; ion exchange resin, ANPP; aboveground net primary productivity; Unit: † kg N·ha^{-1}·year^{-1}; ‡ mg N·bag^{-1}; § Mg C·ha^{-1}·year^{-1}.

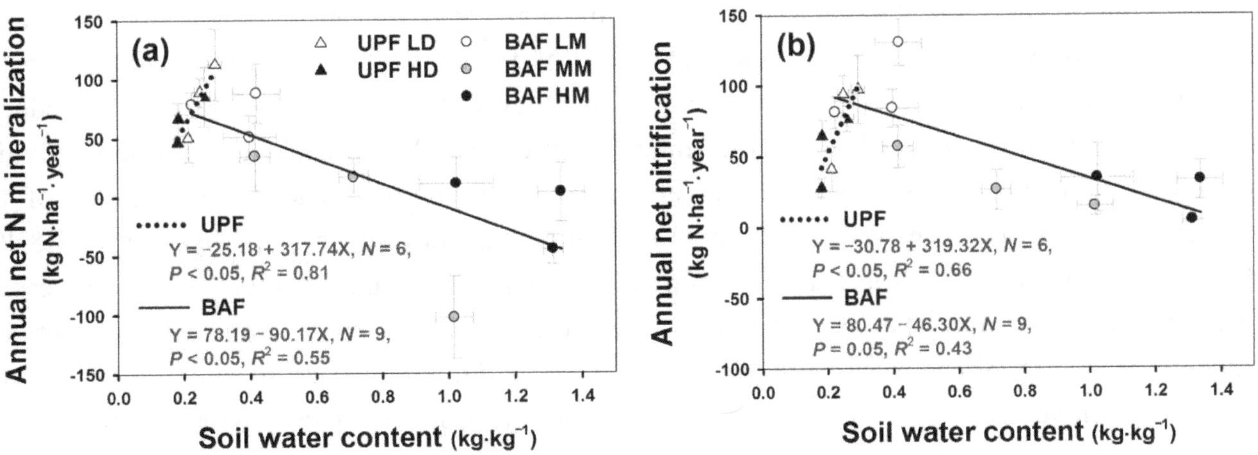

Figure 3. Relationship between gravimetric soil water content and (**a**) annual net N mineralization or (**b**) nitrification rates.

4. Discussion

The results in the UPF corresponded to the general established patterns of N availability in forest soils [12–16,18]. For example, net N transformation rates increased with soil temperature and SWC (Table 2, Figure 2). Net N transformation rates and inorganic N availability were higher during the growing season and lower during the non-growing season (Figure 1). Spatio-temporal patterns of three inorganic N indices, N transformation rates, IER-inorganic N, and bulk soil inorganic N concentrations, corresponded with each other (Figure 1). Results for the LM BAF stand were also similar to general established patterns.

However, some abnormal results, mainly from MM and HM BAF stands, led to some unanswered questions. Why was net N mineralization negatively related to bulk NH_4^+ concentration (Figure 2)? Why was IER-NH_4^+ low at the MM and HM BAF stands where bulk soil NH_4^+ concentration was high (Figure 1)? Moreover, could negative net N mineralization occur in ecosystems that require a substantial amount of N to sustain plant uptake (Figure 1, Table 3)? In addition, for the UPF stands, what was the effect of stand density on soil N transformation rates and N availability?

4.1. Restricted Net N Mineralization under NH_4^+ Rich Condition

Net N transformation rates were independent of soil temperature in the MM and HM BAF stands. Net N transformation rates in the wetter stands (MM and HM) were restricted by the high SWC as compared to the drier stand (LM) (Figures 1 and 3, Table 2). SWC could limitedly explain the spatio-temporal variations in soil N transformation rates and inorganic N indices (Table 2), unlike the soil carbon processes and soil properties which were strongly regulated by SWC in the previous studies [25,42]. Correlation between SWC and net N mineralization and net nitrification rates were much weaker in the BAF ($r = -0.29$ and -0.37) than in the UPF ($r = 0.56$ and 0.60) (Table 2). While soil moisture strongly regulated soil properties and soil C processes in the BAF [25,42], it only weakly affected N transformations processes.

Instead, the net N mineralization rate in the BAF was best explained by the negative relationship with NH_4^+ concentration (Figure 2b,f,g). Ideally, NH_4^+ will increase as a result of a net N mineralization process even though a proportion of mineralized NH_4^+ can be lost by plant uptake or nitrification. However, net N mineralization rate was low for soils with a high level of NH_4^+ in the BAF. This unexpected result can be speculated that a high level of NH_4^+ in bulk soil may have reduced the microbial activities involved in the mineralization process. Theoretically, the enzyme activities involved in mineralization could be negatively affected by nutrient availability because of the cost-efficiency strategy of microbes, known as the "microbial nitrogen mining" hypothesis [48,49]. When N is less available, microbes actively mine recalcitrant organic matter to acquire it; whereas, in N rich environments, microbes can assimilate inorganic N without costly mineralization activity. Although the relationship between inorganic N availability and microbial activities have been reported to be both positive [50,51] and negative [52–54] in empirical studies mainly based on fertilization experiments in forests, wetlands, or agricultural land, a meta-analysis of Treseder [55] suggested that N addition reduced microbial biomass by 15% and, consequently, also microbial activities such as soil respiration. Those articles could provide a theoretical proof and empirical evidence supporting the decrease of net N mineralization under high NH_4^+ concentrations in the BAF. Consequently, the intensive restricting effect of NH_4^+ on net N mineralization probably overspread the effect of soil temperature or SWC.

Meanwhile, it could be questioned how the high soil NH_4^+ concentration could be supported with a deficient inorganic N supply from extremely low net N mineralization. In the wet soils of the MM and HM stands, a major source of inorganic N was not N mineralization which was nearly zero year-round due to the inhibiting effect of high soil moisture on N mineralization. Instead, inorganic N could be supplied through atmospheric N deposition [56,57], accumulation of external N through run-off from upland forests [4,58], and symbiotic and non-symbiotic N fixation [59,60]. For instance, Shin *et al.* [57] reported notable atmospheric N deposition in Seoul (16.5 kg N·ha^{-1}·year^{-1} for NH_4^+ and NH_3 deposits and 18.1 kg N·ha^{-1}·year^{-1} for NO_3^-, NO_2 and HNO_3 deposits); the BAF is located within the range of the urban effect on atmospheric N deposition [61]. N fixation of pure alder forests was known to be 37–150 kg N·ha^{-1}·year^{-1} [62]. Even though lack of atmospheric N deposition and N fixation data in our study limits the certainty and depth of this implication, evidences from other studies suggest that there may be sufficient NH_4^+ sources without N mineralization. In addition, at MM and HM, the anaerobic conditions could have restricted the nitrification process, especially since the mean net nitrification rate at MM and HM was two to three times lower than that at LM (Figure 1), and this would tend to conserve NH_4^+ as the major form of inorganic N. The dominance of NH_4^+ in the inorganic N pool agreed with previous studies from alder forests [18,63] and wetlands [17,64].

4.2. Discrepancy among Inorganic N in Bulk Soil and IER, and Net N Transformation Rates: Which N Indices Could Describe the Real Status of N Availability?

The net N mineralization rate and IER-NH_4^+ were extremely low whereas soil NH_4^+ concentration was high at MM and HM BAF stands. Why was there a discrepancy between bulk soil and IER-NH_4^+ concentration? Understanding this discrepancy may elucidate the real status of N availability from the balance among microbial mineralization and immobilization, soil adsorption and desorption, and plant root uptake in the field. The low value of IER-NH_4^+ at MM and HM suggests that the rich NH_4^+ in bulk soil were not actually assimilated by plants. Due to the strong affinity of NH_4^+ for negatively charged soil organic matter, NH_4^+ may be strongly associated with soil organic matter and unavailable to plants [18] at the MM and HM stands where soil organic matter concentration was high (Table 1). This hypothesis is supported by the positive relationship between soil NH_4^+ and total carbon concentration ($p < 0.001$, $R^2 = 0.53$). On the other hand, the pattern of NO_3^- concentration in soil and IER showed an opposite pattern to that of NH_4^+; soil NO_3^- concentration was low and IER-NO_3^- was high. The low soil NO_3^- concentration is likely due to its high uptake by plants, as suggested by the high IER-NO_3^-. Abundant moisture at MM and HM could increase the availability of NO_3^- since it is mobilized by soil moisture.

This result implies that IER or ion exchange membranes could be comprehensive and practical indices for N availability in reality [27,65], even though the lack of standardized application remains the challenge [66]. The assumption that plant available N is mainly driven through N mineralization, was not applicable in the MM and HM BAF stands; net N transformation rates failed to meet the interest in N availability. On the other hand, IER assay data would be the best indirect measure of N availability unless the direct measure of plant uptake using stable isotope. An additional advantage of IER or ion exchange membrane assays is to determine availabilities of other nutrients such as phosphorus and cations simultaneously with inorganic N [66–68].

4.3. Stand Density Effects on Soil N Transformation and Availability

The annual net N mineralization rate of the UPF (84.01 and 66.05 kg N·ha^{-1}·year^{-1} for LD and HD stands, respectively) was within reasonable range of the generally accepted rate for temperate pine forests (43.6–87.0 kg N·ha^{-1}·year^{-1} [30,45,69,70]). The differences in net N transformation rates and N availability between LD and HD stands were insignificant according to the statistical tests (Figure 1). Nevertheless, the difference may not be negligible; annual net N mineralization rate, annual net nitrification rate, and IER-inorganic N concentration were 40%, 38%, and 20% higher, respectively, at LD than at HD (Table 3). The higher net N transformation rates and N availability at LD than at HD seem to correspond to the difference in SWC (t-test; $p < 0.05$). The higher SWC at LD, which was probably due to lower evaporation under the lower stand density and a less steep topography (Table 1), would promote soil N cycling. In addition, the difference of N processes matched that of ANPP in the both stands [41]. ANPP was 23% higher at the LD stands than the HD stands (Table 3). This result supports the relationship between N availability and primary production in forests [1–3]. It may provide an implication for stand density control (e.g., forest tending and thinning) with respect to nutrient availability. Stand density control as a management practice can be applied to improve nutrient use efficiency for tree production associated with C sequestration in unmanaged pine forests with high stand density in Korea [71,72]. In addition, this application may contribute to maintenance of stability in stand dynamics by reducing density-dependent tree mortality [73].

5. Conclusions

The UPF and BAF are presented, distinguishing net N transformation rates, availabilities, and environmental controls on those N processes. First, the UPF, which were assumed as moderately fertile environments, confirmed the generally accepted knowledge of net N transformation rates and N availability. Second, soil N transformation rates and N availability at the LM BAF stand represent active N cycling under an N-rich environment. The annual plant N uptake was the highest and the annual net N mineralization rate accounted for 37% of that uptake. The lower net N mineralization (72.24 kg N·ha^{-1}·year^{-1}) than net nitrification (98.74 kg N·ha^{-1}·year^{-1}) at the LM stand indicates that most of the mineralized NH_4^+ was converted to NO_3^- during the field incubation. Thus, the relatively low moisture condition at LM supported NO_3^- accumulation due to the high net nitrification rate and low NO_3^- loss via leaching and denitrification [19,74,75]. Third, the N-rich and wet MM and HM BAF stands exhibited restricted net N transformation rates and N availability unlike high inorganic N concentration in bulk soils and low C:N ratio in leaf litters and soils.

In the UPF, annual net N mineralization slightly exceeded the annual N uptake (Table 3). As taking account into the potential N input from atmospheric N deposition, plant N uptake did not consume the all N input. The surplus of inorganic N would not be retained in the stands; instead, it might leach to N runoff by a ground water and stream. The N saturation in the UPF agreed with other coniferous stands in this region [76]. Meanwhile, in the BAF annual net N mineralization failed to satisfy the annual N uptake (Table 3); the external N input might be retained in the stands. It suggests the importance of bottomland or riparian forests for N retention function [4,77].

The significant findings are summarized as follows. First, net N transformation rates and N availability were not regulated by SWC and soil temperature, instead regulated by initial soil NH_4^+ in the MM and HM BAF stands. The net N mineralization rate was inhibited under NH_4^+-rich soils. The

NH_4^+ accumulation and plant uptake-N might be supplied from external N sources. Second, net N transformation rates, inorganic N concentrations in bulk soil, and those on IER corresponded in the UPF, but did not in the BAF. Discrepancies between the patterns of N availability indices in the BAF may enhance our knowledge of the real status of N availability. Third, differences in soil moisture conditions between high and low density stands might affect N transformation and N availability, and, consequently, ecosystem productivity.

Acknowledgments

We appreciate the cooperation of Forest Practice Research Center, Korea Forest Research Institute and Heolleung Office, Culture Heritage Administration of Korea for allowing us access to the UPF and the BAF sites, respectively. We also appreciate Ah Reum Lee, Su Jin Heo, Kyung Won Seo, Sue Kyoung Lee, and Koong Yi for their assistance both in the laboratory and the field. This work was supported by research grants from the Korea Environmental Industry & Technology Institute (C314-00131-0408, G214-00181-0402).

Author Contributions

Tae Kyung Yoon and Nam Jin Noh equally led the field and laboratory work, data analyses, and manuscript writing. Tae Kyung Yoon and Nam Jin Noh have the main authorship of the results from the BAF and UPF, respectively. Haegeun Chung collaborated on discussion of the data and the manuscript writing, especially from the perspective of microbiology. A-Ram Yang contributed to the field and laboratory work and manuscript writing. Yowhan Son supervised the whole process of the study including the experimental design, discussion of the data, and manuscript writing.

References

1. Binkley, D.; Vitousek, P. Soil nutrient availability. In *Plant Physiological Ecology*; Pearcy, R., Ehleringer, J., Mooney, H., Rundel, P., Eds.; Chapman and Hall Ltd.: New York, NY, USA, 1989; pp. 75–96.
2. Vitousek, P.M.; Howarth, R.W. Nitrogen limitation on land and in the sea: How can it occur? *Biogeochemistry* **1991**, *13*, 87–115.
3. LeBauer, D.S.; Treseder, K.K. Nitrogen limitation of net primary productivity in terrestrial ecosystems is globally distributed. *Ecology* **2008**, *89*, 371–379.
4. Saunders, D.L.; Kalff, J. Nitrogen retention in wetlands, lakes and rivers. *Hydrobiologia* **2001**, *443*, 205–212.
5. Cameron, K.C.; Di, H.J.; Moir, J.L. Nitrogen losses from the soil/plant system: A review. *Ann. Appl. Biol.* **2013**, *162*, 145–173.
6. Binkley, D.; Hart, S.C. The components of nitrogen availability assessments in forest soils. In *Advances in Soil Science*; Stewart, B.A., Ed.; Springer: New York, NY, USA, 1989; Volume 10, pp. 57–112.

7. Schimel, J.P.; Bennett, J. Nitrogen mineralization: Challenges of a changing paradigm. *Ecology* **2004**, *85*, 591–602.
8. Frank, D.A.; Groffman, P.M. Plant rhizospheric N processes: What we don't know and why we should care. *Ecology* **2009**, *90*, 1512–1519.
9. Ros, G.H.; Temminghoff, E.J.M.; Hoffland, E. Nitrogen mineralization: A review and meta-analysis of the predictive value of soil tests. *Eur. J. Soil Sci.* **2011**, *62*, 162–173.
10. Zhu, B.; Gutknecht, J.L.M.; Herman, D.J.; Keck, D.C.; Firestone, M.K.; Cheng, W. Rhizosphere priming effects on soil carbon and nitrogen mineralization. *Soil Biol. Biochem.* **2014**, *76*, 183–192.
11. Manzoni, S.; Porporato, A. Soil carbon and nitrogen mineralization: Theory and models across scales. *Soil Biol. Biochem.* **2009**, *41*, 1355–1379.
12. Paul, K.I.; Polglase, P.J.; O'Connell, A.M.; Carlyle, J.C.; Smethurst, P.J.; Khanna, P.K. Defining the relation between soil water content and net nitrogen mineralization. *Eur. J. Soil Sci.* **2003**, *54*, 39–48.
13. Sierra, J. Temperature and soil moisture dependence of n mineralization in intact soil cores. *Soil Biol. Biochem.* **1997**, *29*, 1557–1563.
14. Guntiñas, M.E.; Leirós, M.C.; Trasar-Cepeda, C.; Gil-Sotres, F. Effects of moisture and temperature on net soil nitrogen mineralization: A laboratory study. *Eur. J. Soil Biol.* **2012**, *48*, 73–80.
15. Sleutel, S.; Moeskops, B.; Huybrechts, W.; Vandenbossche, A.; Salomez, J.; De Bolle, S.; Buchan, D.; de Neve, S. Modeling soil moisture effects on net nitrogen mineralization in loamy wetland soils. *Wetlands* **2008**, *28*, 724–734.
16. Zak, D.R.; Grigal, D.F. Nitrogen mineralization, nitrification and denitrification in upland and wetland ecosystems. *Oecologia* **1991**, *88*, 189–196.
17. Cartaxana, P.; Caçador, I.; Vale, C.; Falcão, M.; Catarino, F. Seasonal variation of inorganic nitrogen and net mineralization in a salt marsh ecosystem. *Mangroves Salt Marshes* **1999**, *3*, 127–134.
18. Uri, V.; Lõhmus, K.; Tullus, H. Annual net nitrogen mineralization in a grey alder (*Alnus incana* (L.) Moench) plantation on abandoned agricultural land. *For. Ecol. Manag.* **2003**, *184*, 167–176.
19. Hefting, M.; Clément, J.C.; Dowrick, D.; Cosandey, A.C.; Bernal, S.; Cimpian, C.; Tatur, A.; Burt, T.P.; Pinay, G. Water table elevation controls on soil nitrogen cycling in riparian wetlands along a European climatic gradient. *Biogeochemistry* **2004**, *67*, 113–134.
20. Western, A.W.; Grayson, R.B.; Bloschl, G. Scaling of soil moisture: A hydrologic perspective. *Annu. Rev. Earth Pl. Sc.* **2002**, *30*, 149–180.
21. Porporato, A.; D'Odorico, P.; Laio, F.; Rodriguez-Iturbe, I. Hydrologic controls on soil carbon and nitrogen cycles. I. Modeling scheme. *Adv. Water Resour.* **2003**, *26*, 45–58.
22. Robinson, D.A.; Campbell, C.S.; Hopmans, J.W.; Hornbuckle, B.K.; Jones, S.B.; Knight, R.; Ogden, F.; Selker, J.; Wendroth, O. Soil moisture measurement for ecological and hydrological watershed-scale observatories: A review. *Vadose Zone J.* **2008**, *7*, 358–389.
23. Updegraff, K.; Pastor, J.; Bridgham, S.D.; Johnston, C.A. Environmental and substrate controls over carbon and nitrogen mineralization in northern wetlands. *Ecol. Appl.* **1995**, *5*, 151–163.

24. Ehrenfeld, J.G.; Han, X.; Parsons, W.F.J.; Zhu, W. On the nature of environmental gradients: Temporal and spatial variability of soils and vegetation in the New Jersey pinelands. *J. Ecol.* **1997**, *85*, 785–798.
25. Yoon, T.K.; Noh, N.J.; Han, S.; Kwak, H.; Lee, W.-K.; Son, Y. Small-scale spatial variability of soil properties in a Korean swamp. *Landsc. Ecol. Eng.* **2015**, *11*, 723–734.
26. Khanna, P.K.; Raison, R.J. *In situ* core methods for estimating soil mineral-N fluxes: Re-evaluation based on 25 years of application and experience. *Soil Biol. Biochem.* **2013**, *64*, 203–210.
27. Friedel, J.K.; Herrmann, A.; Kleber, M. Ion exchange resin–soil mixtures as a tool in net nitrogen mineralisation studies. *Soil Biol. Biochem.* **2000**, *32*, 1529–1536.
28. Johnson, D.W.; Verburg, P.S.J.; Arnone, J.A. Soil extraction, ion exchange resin, and ion exchange membrane measures of soil mineral nitrogen during incubation of a tallgrass prairie soil. *Soil Sci. Soc. Am. J.* **2005**, *69*, 260–265.
29. Vitousek, P. Nutrient cycling and nutrient use efficiency. *Am. Nat.* **1982**, *119*, 553–572.
30. Gower, S.T.; Son, Y. Differences in soil and leaf litterfall nitrogen dynamics for five forest plantations. *Soil Sci. Soc. Am. J.* **1992**, *56*, 1959–1966.
31. Aerts, R. The advantages of being evergreen. *Trends Ecol. Evol.* **1995**, *10*, 402–407.
32. Reich, P.B.; Grigal, D.F.; Aber, J.D.; Gower, S.T. Nitrogen mineralization and productivity in 50 hardwood and conifer stands on diverse soils. *Ecology* **1997**, *78*, 335–347.
33. Mueller, K.E.; Hobbie, S.E.; Oleksyn, J.; Reich, P.B.; Eissenstat, D.M. Do evergreen and deciduous trees have different effects on net N mineralization in soil? *Ecology* **2012**, *93*, 1463–1472.
34. Binkley, D.; Giardina, C. Why do tree species affect soils? The warp and woof of tree-soil interactions. *Biogeochemistry* **1998**, *42*, 89–106.
35. Fujita, H.; Fujimura, Y. Distribution pattern and regeneration of swamp forest species with respect to site conditions. In *Ecology of Riparian Forests in Japan*; Sakio, H., Tamura, T., Eds.; Springer: Tokyo, Japan, 2008; pp. 225–236.
36. Eom, H.J.; Chang, K.S.; Kim, H.; Chang, C.-S. Notes on a new overlooked taxon of *Alnus* (*Betulaceae*) in Korea. *For. Sci. Technol.* **2011**, *7*, 42–46.
37. Whang, B.-C.; Lee, M.-W. Landscape ecology planning principles in korean Feng-Shui, Bi-Bo woodlands and ponds. *Landsc. Ecol. Eng.* **2006**, *2*, 147–162.
38. Yoon, T.K.; Noh, N.J.; Kim, R.-H.; Seo, K.W.; Lee, S.K.; Yi, K.; Lee, I.K.; Lim, J.-H.; Son, Y. Mass dynamics of coarse woody debris in an old-growth deciduous forest of Gwangneung, Korea. *For. Sci. Technol.* **2011**, *7*, 145–150.
39. Noh, N.J.; Son, Y.; Lee, S.K.; Yoon, T.K.; Seo, K.W.; Kim, C.; Lee, W.-K.; Bae, S.W.; Hwang, J. Influence of stand density on soil CO_2 efflux for a *Pinus densiflora* forest in Korea. *J. Plant Res.* **2010**, *123*, 411–419.
40. Noh, N.J.; Son, Y.; Jo, W.; Yi, K.; Park, C.W.; Han, S. Preliminary study on estimating fine root growth in a natural *Pinus densiflora* forest using a minirhizotron technique. *For. Sci. Technol.* **2012**, *8*, 47–50.

41. Noh, N.J.; Kim, C.; Bae, S.W.; Lee, W.K.; Yoon, T.K.; Muraoka, H.; Son, Y. Carbon and nitrogen dynamics in a *Pinus densiflora* forest with low and high stand densities. *J. Plant Ecol.* **2013**, *6*, 368–379.
42. Yoon, T.K.; Noh, N.J.; Han, S.; Lee, J.; Son, Y. Soil moisture effects on leaf litter decomposition and soil carbon dioxide efflux in wetland and upland forests. *Soil Sci. Soc. Am. J.* **2014**, *78*, 1804–1816.
43. Vicca, S.; Luyssaert, S.; Peñuelas, J.; Campioli, M.; Chapin, F.S.; Ciais, P.; Heinemeyer, A.; Högberg, P.; Kutsch, W.L.; Law, B.E.; *et al.* Fertile forests produce biomass more efficiently. *Ecol. Lett.* **2012**, *15*, 520–526.
44. Eno, C.F. Nitrate production in the field by incubating the soil in polyethylene bags. *Soil Sci. Soc. Am. J.* **1960**, *24*, 277–279.
45. Son, Y.; Lee, I.K. Soil nitrogen mineralization in adjacent stands of larch, pine and oak in central Korea. *Ann. For. Sci.* **1997**, *54*, 1–8.
46. Johansson, T. Biomass equations for determining fractions of common and grey alders growing on abandoned farmland and some practical implications. *Biomass Bioenerg.* **2000**, *18*, 147–159.
47. R Development Core Team. *R: A Language and Environment for Statistical Computing*; R Foundation for Statistical Computing: Vienna, Austria, 2011. Available online: http://www.r-project.org (accessed on 15 June 2015).
48. Craine, J.M.; Morrow, C.; Fierer, N. Microbial nitrogen limitation increases decomposition. *Ecology* **2007**, *88*, 2105–2113.
49. Shi, W. Agricultural and ecological significance of soil enzymes: Soil carbon sequestration and nutrient cycling. In *Soil Enzymology*; Shukla, G.; Varma, A., Eds.; Springer: Berlin-Heidelberg, Germany, 2011; Volume 22, pp. 43–60.
50. Regina, K.; Nykänen, H.; Maljanen, M.; Silvola, J.; Martikainen, P.J. Emissions of N_2O and NO and net nitrogen mineralization in a boreal forested peatland treated with different nitrogen compounds. *Can. J. For. Res.* **1998**, *28*, 132–140.
51. Höfferle, Š.; Nicol, G.W.; Pal, L.; Hacin, J.; Prosser, J.I.; Mandić-Mulec, I. Ammonium supply rate influences archaeal and bacterial ammonia oxidizers in a wetland soil vertical profile. *FEMS Microbiol. Ecol.* **2010**, *74*, 302–315.
52. Carpenter-Boggs, L.; Pikul, J.L.; Vigil, M.F.; Riedell, W.E. Soil nitrogen mineralization influenced by crop rotation and nitrogen fertilization. *Soil Sci. Soc. Am. J.* **2000**, *64*, 2038–2045.
53. Fisk, M.; Fahey, T. Microbial biomass and nitrogen cycling responses to fertilization and litter removal in young northern hardwood forests. *Biogeochemistry* **2001**, *53*, 201–223.
54. Bowden, R.D.; Davidson, E.; Savage, K.; Arabia, C.; Steudler, P. Chronic nitrogen additions reduce total soil respiration and microbial respiration in temperate forest soils at the Harvard forest. *For. Ecol. Manag.* **2004**, *196*, 43–56.
55. Treseder, K.K. Nitrogen additions and microbial biomass: A meta-analysis of ecosystem studies. *Ecol. Lett.* **2008**, *11*, 1111–1120.
56. Matson, P.; Lohse, K.A.; Hall, S.J. The globalization of nitrogen deposition: Consequences for terrestrial ecosystems. *AMBIO* **2002**, *31*, 113–119.

57. Shin, A.Y.; Sung, M.Y.; Choi, J.S.; On, J.S.; Roh, S.A.; Ahn, J.Y.; Han, J.S.; Lee, G.W. A study on the acid deposition of inorganic ions and flux in Seoul, Gangwha, Icheon, 2008. *J. Korean Soc. Urban Environ.* **2012**, *12*, 14–22. (In Korean)
58. Jacks, G.; Joelsson, A.; Fleischer, S. Nitrogen retention in forest wetlands. *AMBIO* **1994**, *23*, 358–362.
59. Son, Y. Non-symbiotic nitrogen fixation in forest ecosystems. *Ecol. Res.* **2001**, *16*, 183–196.
60. Vitousek, P.; Cassman, K.; Cleveland, C.; Crews, T.; Field, C.; Grimm, N.; Howarth, R.; Marino, R.; Martinelli, L.; Rastetter, E., *et al.* Towards an ecological understanding of biological nitrogen fixation. *Biogeochemistry* **2002**, *57–58*, 1–45.
61. Han, J.S.; Hong, Y.D.; Ahn, J.Y.; Chung, I.R.; Shin, A.Y.; Lim, B.K.; Noh, S.A.; Son, J.S.; Park, J.H. *Acid Deposition Monitoring and Impact Assessment (V): Centered on Forest Ecosystem*; National Institute of Environmental Research: Incheon, Korea, 2008.
62. Lee, Y.; Son, Y. Diurnal and seasonal patterns of nitrogen fixation in an *Alnus hirsuta* plantation of central Korea. *J. Plant Biol.* **2005**, *48*, 332–337.
63. Kim, D.K. Comparisons of Physico-Chemical Properties of Soil between the *Alnus japonica* Wetlands and Adjacent Slope. Master's Thesis, Kongju National University, Gongju, Korea, Febuary 2014. (In Korean)
64. Bedard-Haughn, A.; Matson, A.L.; Pennock, D.J. Land use effects on gross nitrogen mineralization, nitrification, and N_2O emissions in ephemeral wetlands. *Soil Biol. Biochem.* **2006**, *38*, 3398–3406.
65. Durán, J.; Delgado-Baquerizo, M.; Rodríguez, A.; Covelo, F.; Gallardo, A. Ionic exchange membranes (IEMs): A good indicator of soil inorganic N production. *Soil Biol. Biochem.* **2013**, *57*, 964–968.
66. Qian, P.; Schoenau, J.J. Practical applications of ion exchange resins in agricultural and environmental soil research. *Canad. J. Soil Sci.* **2002**, *82*, 9–21.
67. Van Raij, B. Bioavailable tests: Alternatives to standard soil extractions. *Commun. Soil Sci. Plant Anal.* **1998**, *29*, 1553–1570.
68. Meason, D.F.; Idol, T.W. Nutrient sorption dynamics of resin membranes and resin bags in a tropical forest. *Soil. Sci. Soc. Am. J.* **2008**, *72*, 1806–1814.
69. Nadelhoffer, K.J.; Aber, J.D.; Melillo, J.M. Leaf-litter production and soil organic matter dynamics along a nitrogen-availability gradient in southern Wisconsin (U.S.A.). *Can. J. For. Res.* **1983**, *13*, 12–21.
70. Binkley, D.; Valentine, D. Fifty-year biogeochemical effects of green ash, white pine, and Nrway spruce in a replicated experiment. *For. Ecol. Manag.* **1991**, *40*, 13–25.
71. Kim, C.; Son, Y.; Lee, W.-K.; Jeong, J.; Noh, N.-J.; Kim, S.-R.; Yang, A.R.; Ju, N.-G. Influence of forest tending (Soopkakkugi) works on litterfall and nutrient inputs in a *Pinus densiflora* stand. *For. Sci. Technol.* **2012**, *8*, 83–88.
72. Kang, J.-S.; Shibuya, M.; Shin, C.-S. The effect of forest-thinning works on tree growth and forest environment. *For. Sci. Technol.* **2014**, *10*, 33–39.

73. Franklin, J.F.; Spies, T.A.; van Pelt, R.; Carey, A.B.; Thornburgh, D.A.; Berg, D.R.; Lindenmayer, D.B.; Harmon, M.E.; Keeton, W.S.; Shaw, D.C.; *et al.* Disturbances and structural development of natural forest ecosystsems with silvicultural implications, using Douglas-fir forests as an example. *For. Ecol. Manag.* **2002**, *1–3*, 399–423.
74. Zak, D.R.; Groffman, P.M.; Pregitzer, K.S.; Christensen, S.; Tiedje, J.M. The vernal dam: Plant-microbe competition for nitrogen in northern hardwood forests. *Ecology* **1990**, *71*, 651–656.
75. Aulakh, M.S.; Singh, K.; Singh, B.; Doran, J.W. Kinetics of nitrification under upland and flooded soils of varying texture. *Commun. Soil Sci. Plant Anal.* **1996**, *27*, 2079–2089.
76. Yoo, J. Hydrological Properties and Nutrient Budgets in a Coniferous Forest Catchment in Gwangneung, Gyeonggi Province, Korea. Ph.D. Thesis, Korea University, Seoul, Korea, Febuary 2010. (In Korean)
77. Fisher, J.; Acreman, M.C. Wetland nutrient removal: A review of the evidence. *Hydrol. Earth Syst. Sci.* **2004**, *8*, 673–685.

11

Impact of Nitrogen Fertilization on Forest Carbon Sequestration and Water Loss in a Chronosequence of Three Douglas-Fir Stands in the Pacific Northwest

Xianming Dou [1,2], Baozhang Chen [1,3,*], T. Andrew Black [4], Rachhpal S. Jassal [4] and Mingliang Che [3,5]

1 School of Environment Science and Spatial Informatics, China University of Mining and Technology, Xuzhou 221116, China; E-Mail: douxianming88@163.com

2 School of Resources and Earth Sciences, China University of Mining and Technology, Xuzhou 221116, China

3 State Key Laboratory of Resources and Environmental Information System, Institute of Geographic Sciences and Nature Resources Research, University of Chinese Academy of Sciences, Beijing 100101, China; E-Mail: chemingliangs@163.com

4 Faculty of Land and Food Systems, University of British Columbia, Vancouver, BC V6T 1Z4, Canada; E-Mails: andrew.black@ubc.ca (T.A.B.); rachhpal.jassal@ubc.ca (R.S.J.)

5 University of Chinese Academy of Sciences, Beijing 100049, China

* Author to whom correspondence should be addressed; E-Mail: baozhang.chen@igsnrr.ac.cn;

Academic Editors: Heinz Rennenberg and Mark A. Adams

Abstract: To examine the effect of nitrogen (N) fertilization on forest carbon (C) sequestration and water loss, we used an artificial neural network model to estimate C fluxes and evapotranspiration (ET) in response to N fertilization during four post-fertilization years in a Pacific Northwest chronosequence of three Douglas-fir stands aged 61, 22 and 10 years old in 2010 (DF49, HDF88 and HDF00, respectively). Results showed that N fertilization increased gross primary productivity (GPP) for all three sites in all four years with the largest absolute increase at HDF00 followed by HDF88. Ecosystem respiration increased in all four years at HDF00, but decreased over the last three years at HDF88 and over all four years at DF49. As a result, fertilization increased the net ecosystem productivity of all three stands with the largest increase at HDF88, followed by DF49. Fertilization had no discernible effect on ET in any of the stands. Consequently, fertilization increased water use efficiency (WUE) in all four post-fertilization years at all three sites and also increased light use efficiency (LUE) of all the stands, especially HDF00. Our results suggest that the effects of fertilization on forest C sequestration and water loss may be associated with stand age and fertilization; the two younger stands appeared to be more efficient than the older stand with respect to GPP, WUE and LUE.

Keywords: forest fertilization; carbon sequestration; evapotranspiration; water use efficiency; light use efficiency; artificial neural network

1. Introduction

The factors affecting the physiological processes controlling the amount of terrestrial carbon (C) sequestered in terrestrial ecosystems mainly include atmospheric CO_2 concentration, climatic variability, land use change and nitrogen (N) fixation [1–3]. In terrestrial ecosystems, the important processes of determining whether an ecosystem is a C sink or source are C fixation and release through photosynthesis and respiration, respectively. For N-limited terrestrial ecosystems, additional N supply can affect these processes, and in turn affect the strength of C sinks or sources [4–9].

Research has shown that the response of C exchange to N fertilization in terrestrial ecosystems is positive or negative depending on the ecosystem N status [10–13]. Many researchers have studied the effects of N enrichment on net ecosystem productivity (NEP) in the forest ecosystem, but it remains uncertain as to how much N addition contributes to the magnitude of terrestrial NEP [1,14,15].

Many studies have found that N addition increases NEP [16–21]. Magnani *et al.* [22] found NEP was highly correlated with N deposition in forest ecosystems and the response of NEP to N in wet deposition was approximately 725 kg C kg^{-1} N. However, Sutton *et al.* [23] found that the response of NEP to N was about 50–75 kg C kg^{-1} N based on a model re-analysis across 22 European forest sites and considering the impacts of climatological differences among stands. de Vries *et al.* [24] also found that the response of NEP to N would be 30–70 kg C kg^{-1} N based on a multi-factor analysis of European forest measurements at nearly 400 intensively monitored forest plots in Europe. The results of both Sutton *et al.* [23] and de Vries *et al.* [24] are much smaller than the estimates of Magnani *et al.* [22]. Therefore, there is still some controversy over the magnitude and sustainability of NEP resulting from N enrichment and its potential mechanisms [25–27].

However, other studies have reported that N addition may cause a relatively small increase in NEP [12,28], have no effect [29], or have a negative effect by altering plant and microbial communities, including threatened and endangered species [30]. Nadelhoffer *et al.* [31] pointed out that the influence of N enrichment on ecosystem C storage should be minor in forest ecosystems in the northern temperate zone. Currie *et al.* [32] stated that small increases in C storage occurring primarily in living and dead wood may result from elevated N deposition over the next few decades. Using meta-analysis from global N addition experiments, Liu and Greaver [33] concluded that the increases in short-term C sequestration below ground caused by N enrichment are due to increased C storage in the surface organic layer, but it appeared to be difficult to predict the response of C storage to N enrichment through a long-term experiment. Harpole *et al.* [34] reported that the ability of grassland ecosystems to sequester C late in the growing season was decreased by N enrichment because of increased growing season water use and earlier leaf senescence.

Furthermore, water and C cycles in terrestrial ecosystems are closely coupled. A global sensitivity analysis for an integrated forest ecosystem model [35] also showed the intrinsic coupling between water and the C cycle through stomatal conductance, which affect both water and C fluxes between the biosphere and atmosphere. Nutrients may not only affect productivity and foliar biomass but are also associated with evapotranspiration (ET) in forests and other ecosystems [36]. Increased productivity and foliar biomass induced by increased nutrient availability are probably responsible for the increase in ET resulting from increased rainfall interception and transpiration. As a consequence, more attention should be paid to water use efficiency (WUE), defined as the ratio of C gain (usually gross primary productivity, GPP) to water loss, *i.e.*, ET [37,38]. Studies have reported that by increasing WUE, N addition could enhance plant productivity [39–41]. On the other hand, other researchers have found that N enrichment has no impact on WUE [42–44] or a negative effect on WUE [45,46]. Consequently, it is difficult to account for the differences in N-induced effects on WUE [47]. Furthermore, fertilization may increase NEP through an increase in GPP, resulting from an increase in resource capture and resource use efficiency, e.g., increase in absorbed photosynthetically active radiation (APAR), light use efficiency (LUE, the ratio of GPP to APAR), and/or ecosystem carbon-use efficiency (CUEe, the ratio of NEP to GPP) [5]. Although LUE models have been extensively used to estimate regional or global GPP through remote sensing techniques [48–50], the biophysical controls on LUE in terrestrial ecosystems, especially soil N, remain poorly understood, especially in Douglas-fir stands on the west coast of Canada.

The three different-aged Douglas-fir stands in this study (DF49, established in 1949, HDF88, established in 1988, and HDF00, established in 2000) are in the Pacific Northwest coastal forest region of Canada. Forests in this region between Oregon and Alaska cover approximately 10^5 km^2 and play an important role in the global C cycle. In this region, due to their considerable distances from heavy industrial activity, there is very little atmospheric N deposition, and N deficiency generally occurs in soils [51]. Fertilization using a single dose of urea in early 2007 in all three of these Douglas-fir stands and on-going long-term measurements of C fluxes using the eddy-covariance (EC) technique offer an opportunity to examine the effects of N fertilization on C and water fluxes. Previous studies on the effects of N fertilization have focused on C fluxes at DF49 in the first post-fertilization year using a process-based model [52], and C fluxes, ET and WUE in all three stands in the two post-fertilization years using an empirical model (multiple linear regression, MLR) [53].

In this study, we have used the measured C fluxes (viz., GPP, ecosystem respiration (*R*), NEP (= GPP − *R*)), ET and climatic variables in the three stands from 1998 to 2010. The objectives of this study are threefold: (1) to estimate the effects of N fertilization in the three stands on the C fluxes, ET, WUE and LUE using two common data-driven models, MLR and an artificial neural network (ANN); (2) to examine the variation of the N fertilization effects during the 4-year post-fertilization period; and (3) to gain insight into the mechanisms of N fertilization controlling the C fluxes, ET, WUE and LUE in the three stands.

2. Materials and Methods

2.1. Site Descriptions

Three stands were located on the east coast of Vancouver Island, BC, between Campbell River and Denman Island. In 2010, the ages of the stands were 61, 22 and 10 years old, which spanned the typical ages of stands in this area. The three stands were dominated by Douglas-fir (*Pseudotsuga menziesii* (Mirb.) Franco) with relatively similar stand densities, soil, topography, elevation, and biogeoclimatic classification [54]. On the whole, stand age and the corresponding stand structural characteristics were considered to be largely responsible for the differences among these stands in C uptake and ET. The oldest stand (49°52′7.8″ N, 125°20′6.3″ W, flux tower location), DF49, was planted with Douglas-fir seedlings in 1949 and occupied an area of 130 ha. The stand comprised 80% Douglas-fir, 17% western red cedar (*Thuja plicata*), and 3% western hemlock (*Tsuga heterophylla*). The mean annual temperature and precipitation measured during 1998–2010 in the stand are 8.1 °C and 1350 mm, respectively. The 22-year-old (pole-sapling) stand (49°32′10.49″ N, 124°54′7.18″ W), HDF88, 110 ha in size, was situated about 30 km southwest of DF49. The stand comprised 75% Douglas-fir, 21% western red cedar and 4% grand fir (*Abies grandis*). The mean annual temperature and rainfall in the stand, since measurements began in 2002, are 9.1 °C and 1579 mm, respectively. The youngest stand (49°52′1.08″ N, 125°16′43.80″ W), HDF00, was located about 2 km southeast of DF49. This 32-ha stand was harvested in the winter of 1999/2000 and in the following spring was planted with one-year-old seedlings (93% Douglas-fir, 7% western red cedar). Much of the leaf area at this site was due to the growth of pioneer and understory species from the previous stand. The mean annual temperature and rainfall at the site are 8.5 °C and 1321 mm, respectively. Additional details on stand history, vegetation and soil can be found in Humphreys *et al.* [54], Chen *et al.* [55], Krishnan *et al.* [56] and Morgenstern *et al.* [57].

2.2. Stand Fertilization

Urea fertilizer was spread aerially at 200 $kg \cdot N \cdot ha^{-1}$ over about 110 ha at DF49 and 20 ha at HDF88, which included most of the respective tower flux footprints, on 13 January and 17 February 2007, respectively. Because of the young age of the planted trees and competing understory at HDF00, 80 g urea per tree (~60 $kg \cdot N \cdot ha^{-1}$) was manually applied along the tree drip line of trees located in the 5-ha tower flux footprint on 13–14 February 2007. The fertilized areas around the respective towers accounted for more than 80% of the EC fluxes measured during unstable atmospheric conditions, which occurred during the daytime and nighttime [55].

2.3. EC and Climate Measurements

The EC technique has been used to measure continuous half-hourly CO_2 and water vapor fluxes since 2000, 2001 and 1997 at DF49, HDF88 and HDF00, respectively. Climate variables that were also measured included precipitation, net radiation (R_n), photosynthetically active radiation (PAR), wind speed, air temperature (T_a) and humidity, soil temperature (T_s) and volumetric soil moisture content (θ). Details on EC instrument characteristics, measurements and calculation procedures for these stands are described in Humphreys *et al.* [54], Chen *et al.* [55] and Morgenstern *et al.* [57]. Half-hourly

EC-measured CO_2 fluxes above the stand were corrected by adding the estimated rate of change in CO_2 storage in the air column below the EC sensor height to obtain the net ecosystem exchange (NEE). NEP was obtained as NEP = −NEE [54].

The methodology of gap-filling for EC-measured NEP and its separation into GPP and R was adopted from the Fluxnet-Canada Research Network protocol [55,58]. Missing nighttime values of R were obtained using an exponential relationship between NEE and T_s (Q_{10} model) using measured NEE values corresponding to friction velocity values greater than a threshold value. This model was also used to obtain daytime values of R. GPP during the growing season was obtained as the sum of daytime-measured NEP and daytime R and was set to zero during the cold season. An empirical light-response relationship was used to fill gaps in GPP. The details on NEP gap-filling and its separation into GPP and R as well as the uncertainties in the annual sums of NEP, GPP and R can be found in Chen *et al.* [59].

2.4. Artificial Neural Networks

An ANN model is characterized by flexible mathematical structures, which can be used to investigate complicated non-linear relationships between inputs and outputs [60]. For training of time series prediction, the ANN model is conducted by propagating the input data and then back-propagating the error by means of self-adjustment of the weights using least squares residuals so that the simulated outputs best approximate the target outputs (observed data). The performance of an ANN model as a purely empirical non-linear regression model is commonly affected by the quality of the training dataset, network architecture and network training [61]. Additional mathematical background on ANN models can be found in White [62], Jain *et al.* [63], Zhang *et al.* [64] and Basheer and Hajmeer [65]. The back-propagation (BP) algorithm used in this study for network training has been successfully applied in ecosystem C flux estimation [66–68] because it is regarded as the most widely applied algorithm in the ANN literature [69,70]. A three-layer BP neural network (BPNN) was employed with one hidden layer because Funahashi [71] and Cybenko [72] demonstrated that even with one hidden layer, a BPNN can approximate any continuous multivariate function with reasonable precision. In the calibration phase, the optimization approach applied was based on Levenberg–Marquardt [73], which is known to outperform the simple gradient descent and other conjugate gradient methods. In this three-layer BPNN, a hyperbolic tangent sigmoid transfer function (tansig) at hidden layer and a linear transfer function (purelin) at output layer was used. Aiming at obtaining the hidden node number and avoiding over-fitting during the training period, we used a trial and error method to select the optimal solution by altering the number of hidden nodes.

2.5. Comparison of ANN and MLR Models

Six environmental variables that included T_s at the 5-cm depth, T_a, downwelling PAR (Q), vapor pressure deficit (D), R_n, and θ in the 0–30 cm layer, and stand age were selected to train the ANN model using pre-fertilization flux measurements. The particular environmental variables that worked best for GPP, R and ET were chosen as inputs after trying several ANN networks. GPP, R and ET were simulated independently using different input variables for the ANN model. Specifically, three different variable groups, (T_a, θ, Q, D, stand age), (T_a, θ, T_s, stand age) and (T_a, θ, R_n, D, stand age), were then used to train the ANN model and predict monthly GPP, R and ET, respectively, for the three sites assuming the stands had not been fertilized. To ensure high precision in the period of model prediction, we used multi-year monthly values of climate variables, stand age and EC-measured C fluxes and ET before 2005 to train the ANN model and then verified the trained model with measurements in 2005 and 2006. When we were convinced that the optimized ANN model successfully simulated the multiyear seasonal variations in C fluxes and ET, the input values for the post-fertilization period were transferred into the trained model to predict the GPP, R and ET for 2007 to 2010. The resulting differences between the measurements and predictions were used to discern the impact of fertilization.

At the beginning of the training, inputs and outputs were normalized between 0 and 1, which is a common preprocessing method for variables with different values. Moreover, we also used an MLR model to simulate C fluxes and ET with the same variables to assess the advantages of the ANN modeling approach over traditional regression modeling techniques. Similar to the ANN model, we used a variable group (e.g., T_a, θ, Q, D, stand age) as the independent variables to establish an MLR equation and predict the dependent variable (e.g., GPP). Because of the differences in age among these three stands, it is important to note that both ANN and MLR models were run separately for each stand. The performances of the models were evaluated by comparing predicted with measured values in the calibration and verification periods. In this study, computer programming and data analysis were performed using MATLAB (version 7.13, R2011b).

3. Results

3.1. Environmental Variables

Since all three stands were exposed to similar weather conditions (warm dry summers and wet cool winters), we describe only the seasonal and annual variations in environmental variables at DF49. The seasonal patterns of T_a, T_s, θ, D, Q and R_n during 2007–2010 were similar to those measured before fertilization from 1998 to 2006 (Figure 1). Compared with the 9-year means during the pre-fertilization period, the annual average values of T_a during 2007–2010 were lower by about 0.7, 1.1, 0.4 and 0.8 °C, respectively, and the annual average values of T_s were higher by about 0.3, 0.1, 0.1 and 0.6 °C, respectively (Figure 1a,b). Annual values of total Q in 2007, 2008 and 2010 were lower by about 9%, 3% and 6%, respectively, suggesting possible growth limitation in those years, but were higher by about 8% in 2009 (Figure 1f). Annual average values of θ in the 0–30 cm soil layer in the growing season (May–September) in the post-fertilization period during 2007–2010 were 0.22, 0.20, 0.19 and 0.20 $m^3 \cdot m^{-3}$, respectively, which were higher, especially in the noticeably wet year of 2007, compared with the pre-fertilization 9-year mean (0.16 $m^3\ m^{-3}$) (Figure 1c).

3.2. Seasonal and Interannual Variations of Observed C Fluxes and ET before and after Fertilization

The seasonal variations in the monthly EC-measured C fluxes and ET in the three stands before and after fertilization are presented in Figure 2. Monthly GPP values during the 4 post-fertilization years were similar to the means during the 9 non-fertilized years (Figure 2a). The monthly values of R during the 4 post-fertilization years were close to the 9-year pre-fertilization means for Jan–May but lower for the later months except for June and July in 2009 (Figure 2b). Therefore, the monthly NEP values during the 4 post-fertilization years were greater than the 9-year pre-fertilization average values for March–December, especially in June, August and September (Figure 2c).

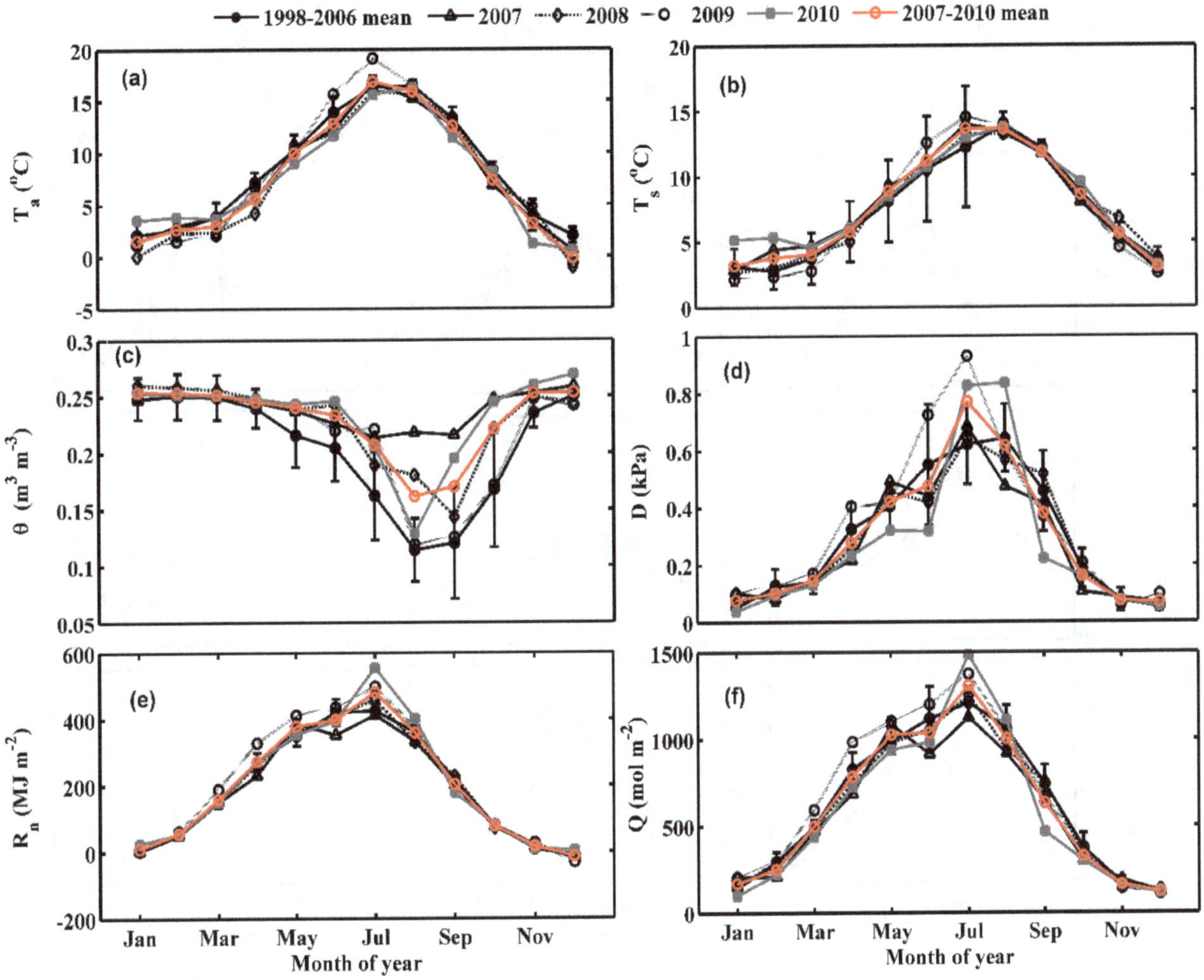

Figure 1. Comparisons of mean monthly values of environmental variables observed at DF49 for the pre-fertilization (1998–2006) and post-fertilization years (2007–2010). The error bars are ± 1 standard deviation (SD) for all the pre-fertilization data. **(a)** air temperature above the canopy (T_a); **(b)** 5-cm depth soil temperature (T_s); **(c)** 0–30-cm soil water content (θ); **(d)** water vapor pressure deficit (D); **(e)** total downwelling photosynthetically active radiation (Q) and **(f)** net radiation (R_n).

Similarly, for HDF88, monthly NEP values during the 4 post-fertilization years at HDF88 were generally greater than the pre-fertilization average values for 2002 to 2006 (Figure 2g). In the case of HDF00, compared with the 2001–2006 means, the largest increases in monthly values of NEP during 2007–2010 occurred in June, July and August (Figure 2k). The monthly GPP values during 2007–2010 at HDF88 and HDF00 were higher than the mean values for the pre-fertilization periods (Figure 2e,i). The monthly R values for HDF00 were also generally higher than the monthly means in the pre-fertilization period (Figure 2j). Furthermore, at DF49, the maximum NEP occurred in April and May (Figure 2c), while maximum NEP at HDF88 and HDF00 occurred in June and July (Figure 2g,k).

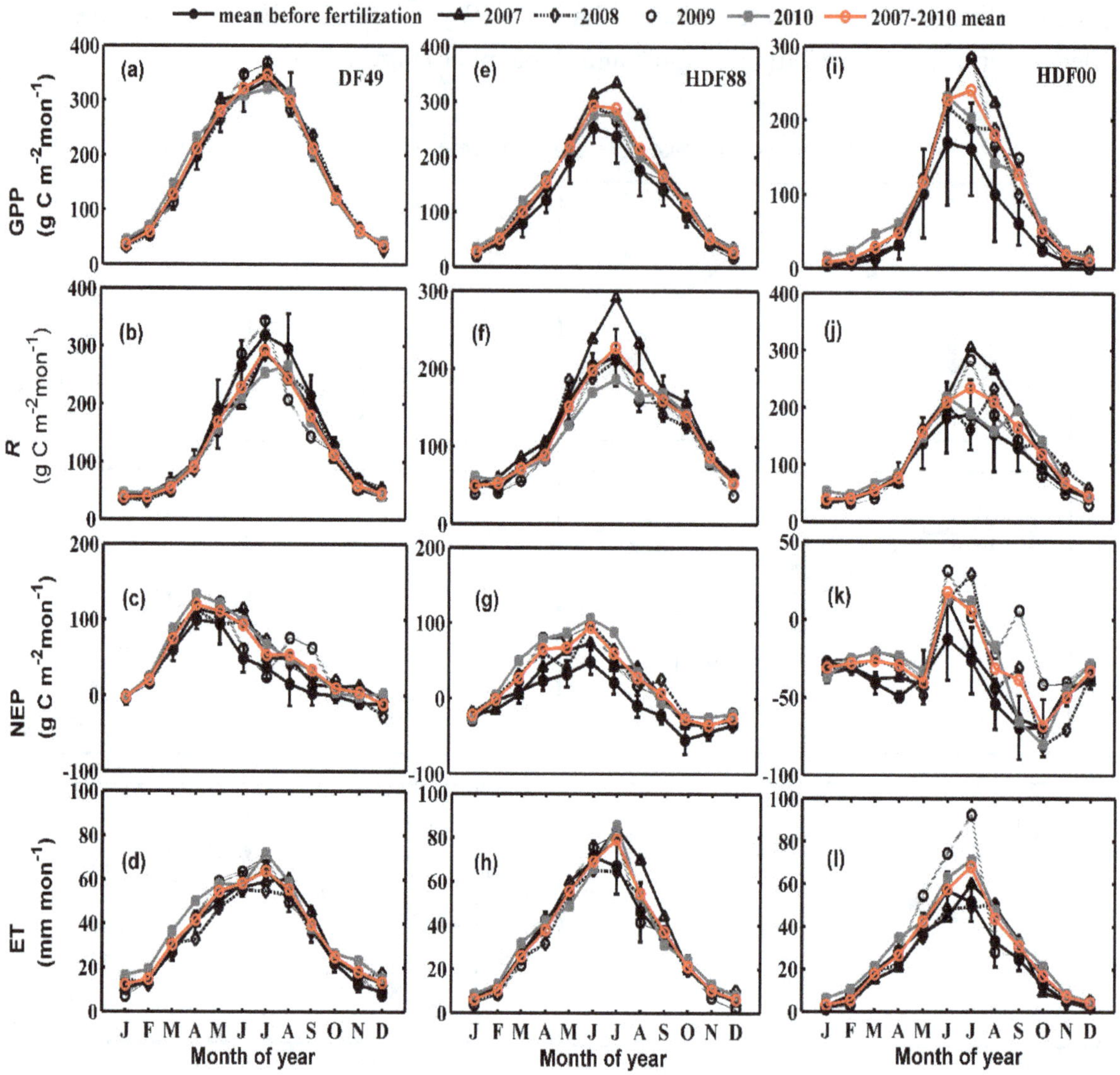

Figure 2. Comparisons of mean monthly values of eddy-covariance tower measured gross primary productivity (GPP), ecosystem respiration (*R*), net ecosystem productivity (NEP) and evapotranspiration (ET) in all three stands for the pre-fertilization and post-fertilization years (2007–2010). The means of measured carbon component fluxes and ET at DF49, HDF88 and HDF00 are shown for 1998 to 2006, 2002 to 2006 and 2001 to 2006, respectively. **(a–d)** for DF49; **(e–h)** for HDF88; and **(i–l)** for HDF00. The error bars are ± 1 SD for all the pre-fertilization data.

Most of the monthly ET values during 2007–2010 at DF49 and HDF88 were close to the monthly means in the pre-fertilization period (Figure 2d,h). However, at HDF00, monthly ET values for May–July during 2009 to 2010 were significantly higher than the means for the pre-fertilization period (Figure 2l). Moreover, for DF49, the monthly values of ET during January to April in 2010 were higher than the previous 9-year average values (Figure 2d).

To better examine the interannual variations of EC-measured C fluxes, ET, WUE and LUE, the annual measured values are compared in Figure 3. The results indicate that EC-measured NEP during the post-fertilization four years increased in all three stands (Figure 3c). Annual NEP at DF49 remained stable over post-fertilization years with values about 200 g C m^{-2} $year^{-1}$ higher than in the pre-fertilization period, despite the post-fertilization years being cooler and wetter than the pre-fertilization years (Figure 1a,c). However, at HDF88, annual NEP consistently increased with increasing stand age (Figure 3c).

The interannual variations in GPP and *R* at DF49 since 1998 show that *R* decreased while GPP showed little change during 2007–2010 (Figure 3a,b). However, for the other two stands, annual *R* increased in the first post-fertilization year and decreased in the next three years (Figure 3b). In addition, compared with the pre-fertilization (9-years) mean CUEe (17%) in DF49, annual average values of CUEe during 2007–2010 were higher by about 9%, 8%, 9% and 12%, respectively.

Annual ET during the post-fertilization period did not show an apparent difference from the previous 9-year means in all three stands (Figure 3d). Annual values of WUE varied greatly among the three stands (Figure 3e). Annual WUE of DF49 was the highest, ranging from 4.5 to 5.4 g·C (kg water)$^{-1}$, and steadily decreased during 2008–2010. Annual LUE values in DF49 were also the highest among the three stands, ranging from 0.0216 to 0.0254 mol·C·mol^{-1} photons (Figure 3f).

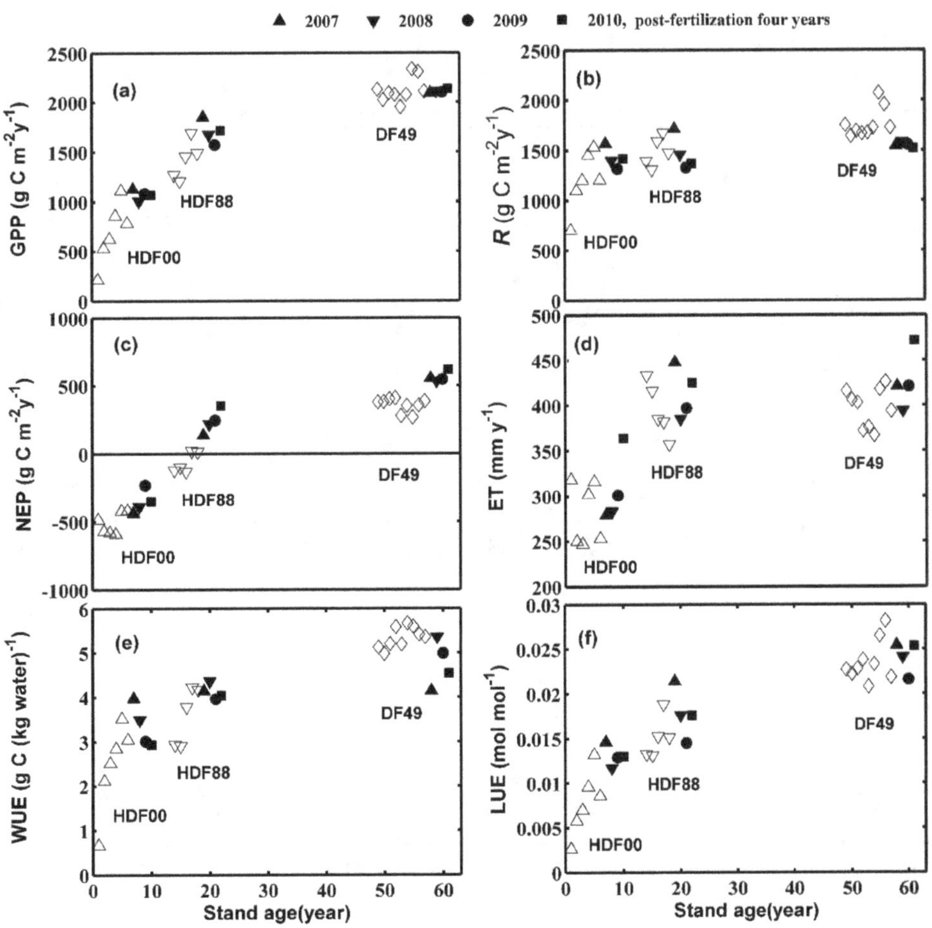

Figure 3. Interannual variations of eddy-covariance tower measured carbon fluxes, evapotranspiration (ET), water-use efficiency (WUE) and light-use efficiency (LUE) in the three Douglas-fir stands. (**a**) gross primary productivity (GPP); (**b**) ecosystem respiration (*R*); (**c**) net ecosystem productivity (NEP); (**d**) ET; (**e**) WUE and (**f**) LUE. Note: The open symbols in all three stands indicates the annual pre-fertilization values.

3.3. Verification of the ANN Model and Its Comparison to the MLR Model

The differences between EC-measured, ANN- and MLR-modeled monthly GPP, *R*, and NEP in the three stands are compared in Figure 4 and Tables 1–3. ANN simulations for the model calibration period (1998–2004) at DF49 showed strong agreement with the observed values and explained about 99%, 99% and 96% of the variance in monthly GPP, *R*, and NEP, respectively (Root mean square error (RMSE) = 2, 8, and 8 $g \cdot C \cdot m^{-2} \cdot month^{-1}$, respectively). These results indicate that by optimizing the ANN parameters, the trained ANN model achieved reasonable simulation of monthly GPP, *R* and NEP at DF49. The linear regression analysis between ANN-modeled and measured values for the model verification period (2005–2006) for DF49 showed R^2 = 0.99, 0.99 and 0.98 and RMSE = 3, 6 and 6 $g \cdot C \cdot m^{-2} \cdot month^{-1}$ for GPP, *R*, and NEP, respectively (Table 1).

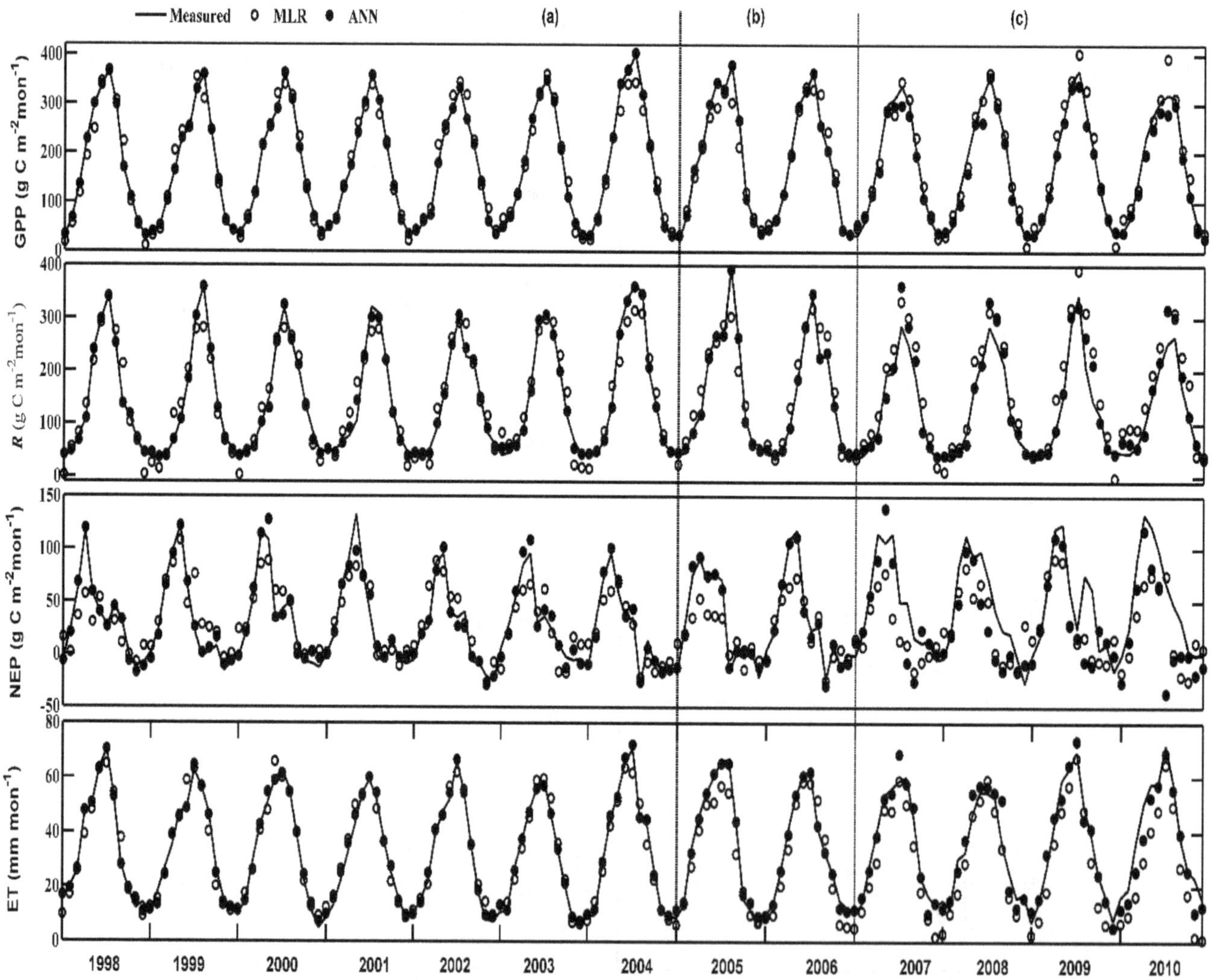

Figure 4. Artificial neural network (ANN) and multiple linear regression (MLR) simulated, and eddy-covariance tower measured monthly gross primary productivity (GPP), ecosystem respiration (*R*), net ecosystem productivity (NEP) and evapotranspiration (ET) at DF49 from 1998 to 2010. (**a**) for the ANN and MLR calibration years 1998–2004; (**b**) for the model validation years 2005–2006; and (**c**) for the four post-fertilization years 2007–2010.

MLR simulations for DF49 for the model calibration period, however, explained about 97%, 92% and 76% of the variance of monthly GPP, *R*, and NEP, respectively (Figure 4a and Table 1), which are somewhat lower than for the ANN simulations. The linear regression analysis showed RMSE = 21, 29, and 19 $g \cdot C \cdot m^{-2} \cdot month^{-1}$ for GPP, *R* and NEP, respectively, which are significantly higher than those for the comparison of ANN-simulated and EC-measured values. In addition, for the model verification period, linear regression between MLR-simulated and observed values showed RMSE = 30, 34 and 24 $g \cdot C \cdot m^{-2}\ month^{-1}$ for GPP, *R*, and NEP, respectively, which are all significantly higher than those for the calibration period.

For the other two stands, the performance of the ANN model was also superior to that of the MLR model in the estimates of GPP, *R* and NEP (Table 1). Regarding ET, similar results were observed in the calibration and verification periods in all three stands (Table 1). As a consequence, we focused on describing and discussing the results of ANN modeling in the prediction period to ascertain the responses of C fluxes and ET to fertilization.

3.4. Effects of N addition on C exchange Fluxes, ET, WUE and LUE in a Chronosequence

In order to provide insights into the influence of fertilization on the C fluxes, ET as well as WUE and LUE during the period of prediction, monthly measured and ANN-modeled values, and their differences are shown for the four post-fertilization years (2007–2010) in Figures 5 and 6. The annual effects of N fertilization during this period are summarized in Tables 2 and 3. Figure 7 compares the annual measured and ANN modeled values of the C fluxes and ET, and shows ± 1 standard deviation estimated using the optimized model parameters.

Over the four post-fertilization years at DF49, most of the ANN modeled monthly values of GPP were lower than the measured values (Table 1 and Figure 5a), but were higher for *R* (Table 1 and Figure 5d) and consequently, modeled monthly values of NEP were lower than the observed values (Table 1 and Figure 5g).

At the annual time scale over the four years, N fertilization at DF49 caused a 6% to 11% increase in annual GPP and a 6% to 10% decrease in *R*, resulting in an 84% to 164% increase in NEP (Table 2). For HDF88, N fertilization caused an 18% to 29% increase in annual GPP, an 11% increase in *R* in 2007, and a 6% to 14% decrease during 2008–2010, resulting in a 208% to 321% increase in NEP (Table 2). For HDF00, N fertilization caused a 43% to 70% increase in annual GPP and an 8% to 21% increase in *R*, resulting in a 27% to 50% increase in NEP (Table 2). Overall, N fertilization had a positive impact on GPP for all three stands with the response being the highest at HDF00, followed by HDF88 (Table 2).

No consistent effects on annual ET were evident in the four post-fertilization years in all three stands (Table 3). For annual WUE, N fertilization of DF49 caused an increase of 0.30 to 0.67 $g \cdot C\ (kg \cdot water)^{-1}$ during 2007 to 2009, but a decrease of 0.21 $g \cdot C\ (kg \cdot water)^{-1}$ in 2010. N fertilization caused increases of 0.59 to 1.05 $C\ (kg \cdot water)^{-1}$ and 0.76 to 1.75 $C\ (kg \cdot water)^{-1}$ in annual WUE during 2007 to 2010 at HDF88 and HDF00, respectively (Table 3). For annual LUE, N fertilization had positive impacts for all three stands, especially at HDF00 with the greatest response ranging from 0.0039 to 0.0057 $mol \cdot C \cdot mol^{-1}$ photons (Table 3).

Table 1. Comparisons of artificial neural network (ANN) and multiple linear regression (MLR) model performances among the calibration, verification and prediction periods for monthly gross primary productivity (GPP, $g \cdot C \cdot m^{-2} \cdot year^{-1}$), ecosystem respiration (*R*, $g \cdot C \cdot m^{-2} \cdot year^{-1}$) and evapotranspiration (ET, $mm \cdot year^{-1}$) in the three Douglas-fir stands.

Flux	Stand	Method	Calibration			Verification			Prediction		
			Regression equation	RMSE	R^2	Regression equation	RMSE	R^2	Regression equation	RMSE	R^2
GPP	DF49	MVMLR	$y = 0.97x + 5.94$	21	0.97	$y = 0.88x + 18.18$	30	0.94	$y = 1.01x + 6.1$	22	0.97
		MVANN	$y = x + 0.07$	2	0.99	$y = 0.99x - 0.35$	3	0.99	$y = 0.86x + 15.42$	27	0.97
	HDF88	MVMLR	$y = 0.92x + 8.65$	21	0.92	$y = 0.84x - 2.36$	36	0.91	$y = 0.82x + 16.47$	33	0.88
		MVANN	$y = x + 0.42$	5	0.99	$y = 0.97x + 2.22$	6	0.99	$y = 0.78x + 6.51$	40	0.90
	HDF00	MVMLR	$y = 0.82x + 8.23$	24	0.82	$y = 0.36x + 48.16$	62	0.55	$y = 0.43x + 68.56$	59	0.61
		MVANN	$y = 0.98x + 0.24$	7	0.98	$y = 0.98x - 1.25$	6	0.99	$y = 0.65x - 0.42$	43	0.96
R	DF49	MVMLR	$y = 0.92x + 11.45$	29	0.92	$y = 0.89x + 22.65$	34	0.90	$y = 1.2x - 17.4$	33	0.94
		MVANN	$y = 0.99x + 1.31$	8	0.99	$y = x + 0.03$	6	0.99	$y = 1.02x - 5.39$	25	0.96
	HDF88	MVMLR	$y = 0.92x + 9.71$	16	0.92	$y = 0.73x + 26.58$	25	0.92	$y = 0.59x + 43.8$	33	0.78
		MVANN	$y = x - 0.07$	3	0.99	$y = 0.98x + 1.63$	2	0.99	$y = 0.96x + 9.18$	23	0.89
	HDF00	MVMLR	$y = 0.84x + 14.57$	23	0.84	$y = 0.66x + 58.33$	40	0.80	$y = 0.81x + 59.39$	42	0.76
		MVANN	$y = 0.99x + 2.53$	3	0.99	$y = 1.01x - 0.98$	3	0.99	$y = 0.89x - 1.03$	28	0.91
NEP	DF49	MVMLR	$y = 0.71x + 8.56$	19	0.76	$y = 0.55x + 4.65$	24	0.82	$y = 0.59x - 2.14$	35	0.64
		MVANN	$y = 0.99x + 0.63$	8	0.96	$y = 0.99x - 0.54$	6	0.98	$y = 0.67x - 9.36$	39	0.54
	HDF88	MVMLR	$y = 0.93x - 0.65$	12	0.86	$y = 1.01x - 13.44$	23	0.81	$y = 1.08x - 5.52$	24	0.80
		MVANN	$y = 0.99x + 0.62$	4	0.98	$y = 0.96x - 1.33$	5	0.98	$y = 0.53x - 21$	41	0.51
	HDF00	MVMLR	$y = 0.67x - 15.4$	17	0.34	$y = -0.04x - 58.5$	42	0.01	$y = -0.21x - 55.3$	53	0.02
		MVANN	$y = 0.99x + 1.21$	8	0.82	$y = 0.89x - 6.47$	7	0.87	$y = 0.31x - 32.2$	42	0.06
ET	DF49	MVMLR	$y = 0.96x + 1.26$	4	0.96	$y = 0.91x - 0.82$	6	0.94	$y = 1.05x - 10.3$	5	0.93
		MVANN	$y = 0.99x + 0.34$	1	0.99	$y = 0.98x + 0.36$	1	0.99	$y = 1.04x - 2.1$	6	0.9
	HDF88	MVMLR	$y = 0.93x + 2.06$	6	0.93	$y = 1.24x - 14.9$	13	0.90	$y = 1.09x - 20.64$	20	0.91
		MVANN	$y = 0.99x + 0.21$	1	0.99	$y = 0.99x - 0.08$	1	0.99	$y = x + 0.56$	5	0.95
	HDF00	MVMLR	$y = 0.92x + 1.78$	5	0.92	$y = 0.91x - 2.6$	9	0.88	$y = 0.8x - 5.49$	16	0.76
		MVANN	$y = 0.99x - 0.03$	1	0.99	$y = x + 0.23$	1	0.99	$y = 0.83x + 3.62$	9	0.84

Note: The regressions among calibration, verification and prediction periods were obtained using the monthly EC-measured values for the C component fluxes as the independent variable and the monthly values estimated using the ANN and MLR models as the dependent variable. MVMLR indicates measured *versus* MLR modeled; MVANN indicates measured *versus* ANN modeled; three different-aged Douglas-fir stands: DF49, HDF88 and HDF00, established in 1949, 1988, and 2000, respectively.

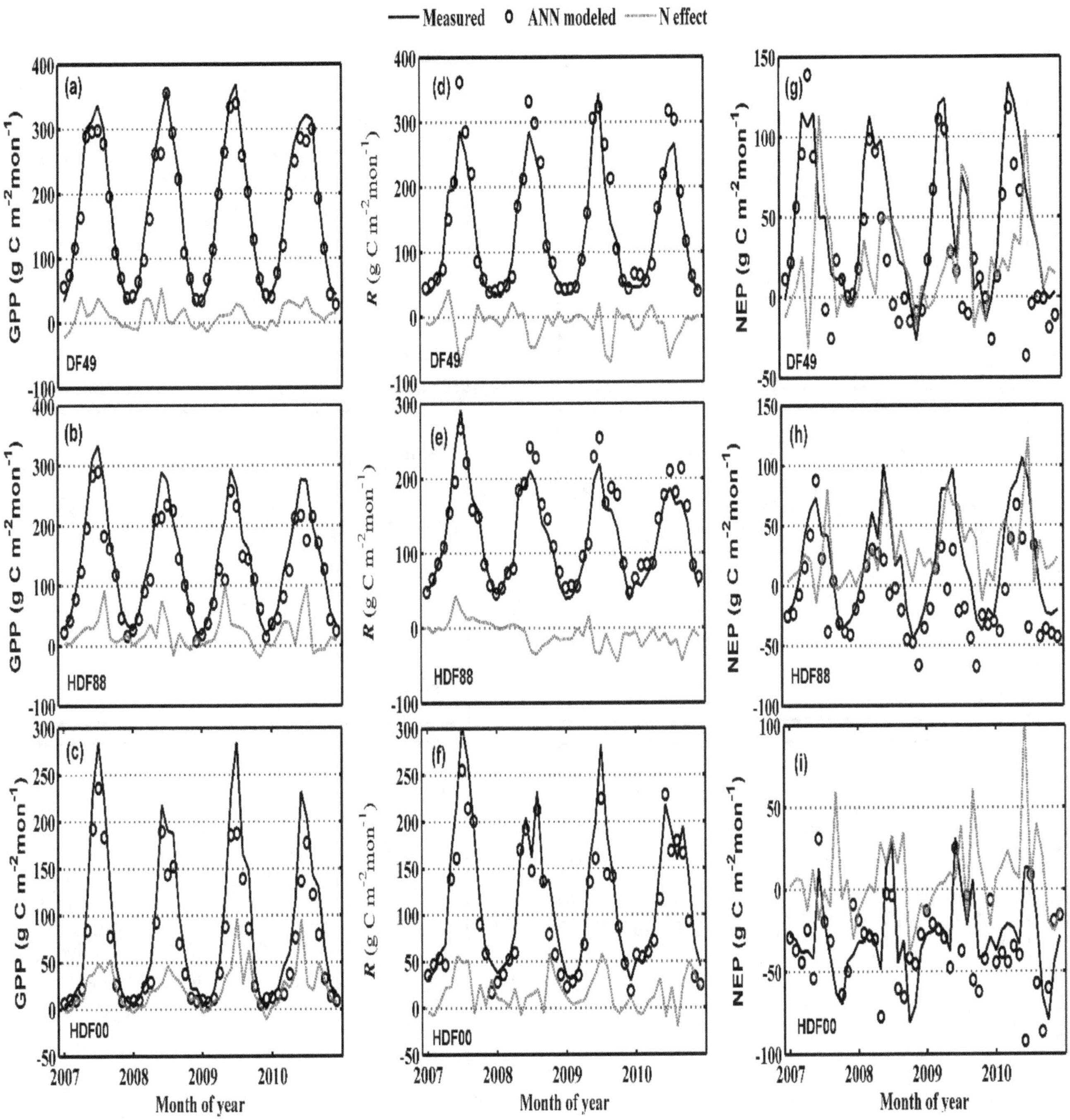

Figure 5. Artificial neural network (ANN) simulated and eddy-covariance tower measured monthly gross primary productivity (GPP), ecosystem respiration (*R*) and net ecosystem productivity (NEP) in all three stands for the four post-fertilization years 2007–2010. (**a–c**) for GPP; (**d–f**) for *R*; and (**g–i**) for NEP. The effects of N fertilization on the carbon (C) component fluxes were estimated as the differences between the measured C fluxes and their corresponding modeled values.

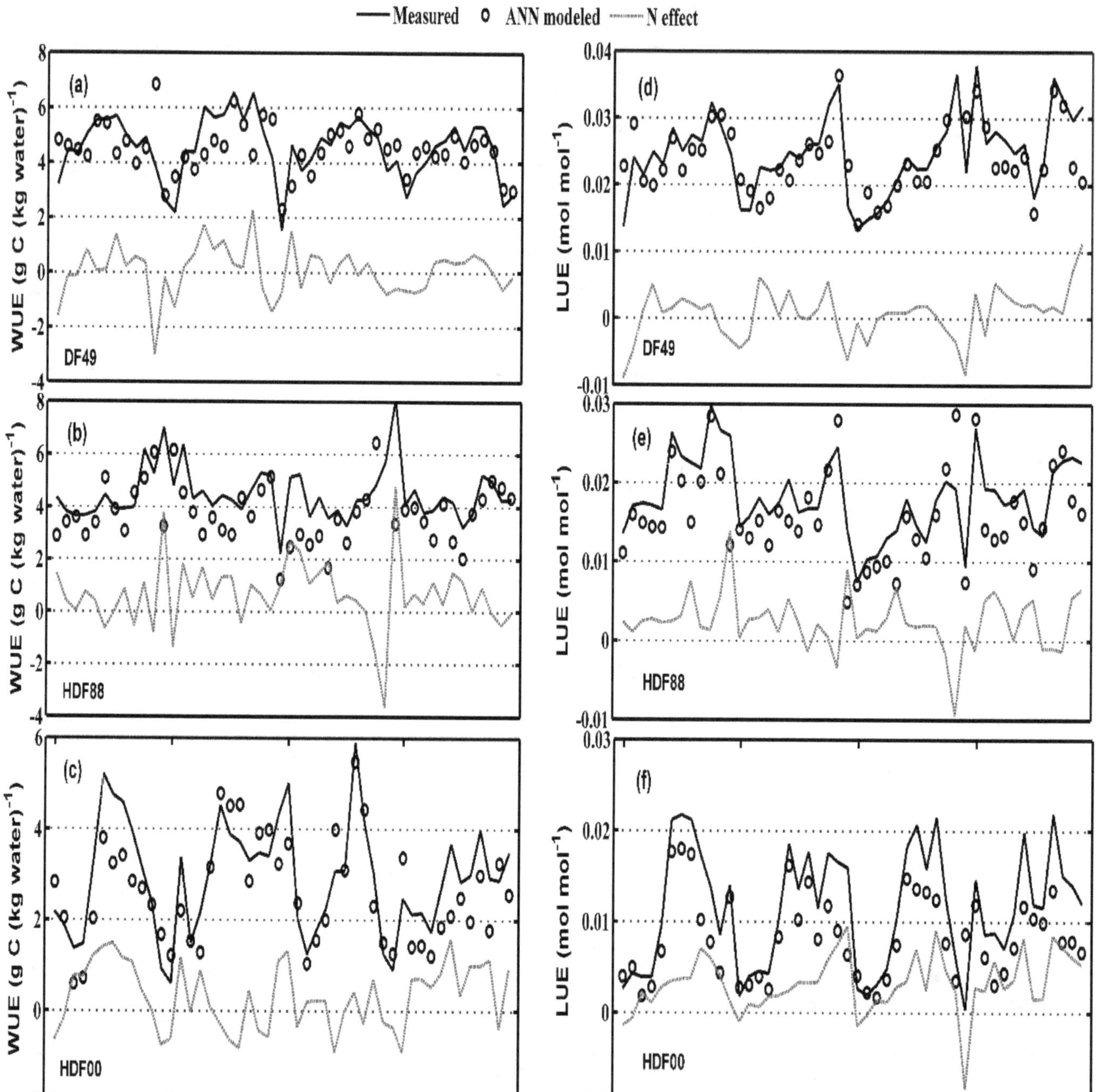

Figure 6. Artificial neural network (ANN) simulated and eddy-covariance tower measured monthly water-use efficiency (WUE) and light-use efficiency (LUE) in all three stands for the four post-fertilization years 2007–2010. **(a–c)** for WUE and **(d–f)** for LUE. The effects of N fertilization on WUE and LUE were estimated as the differences between the measured WUE and LUE values and their corresponding modeled values.

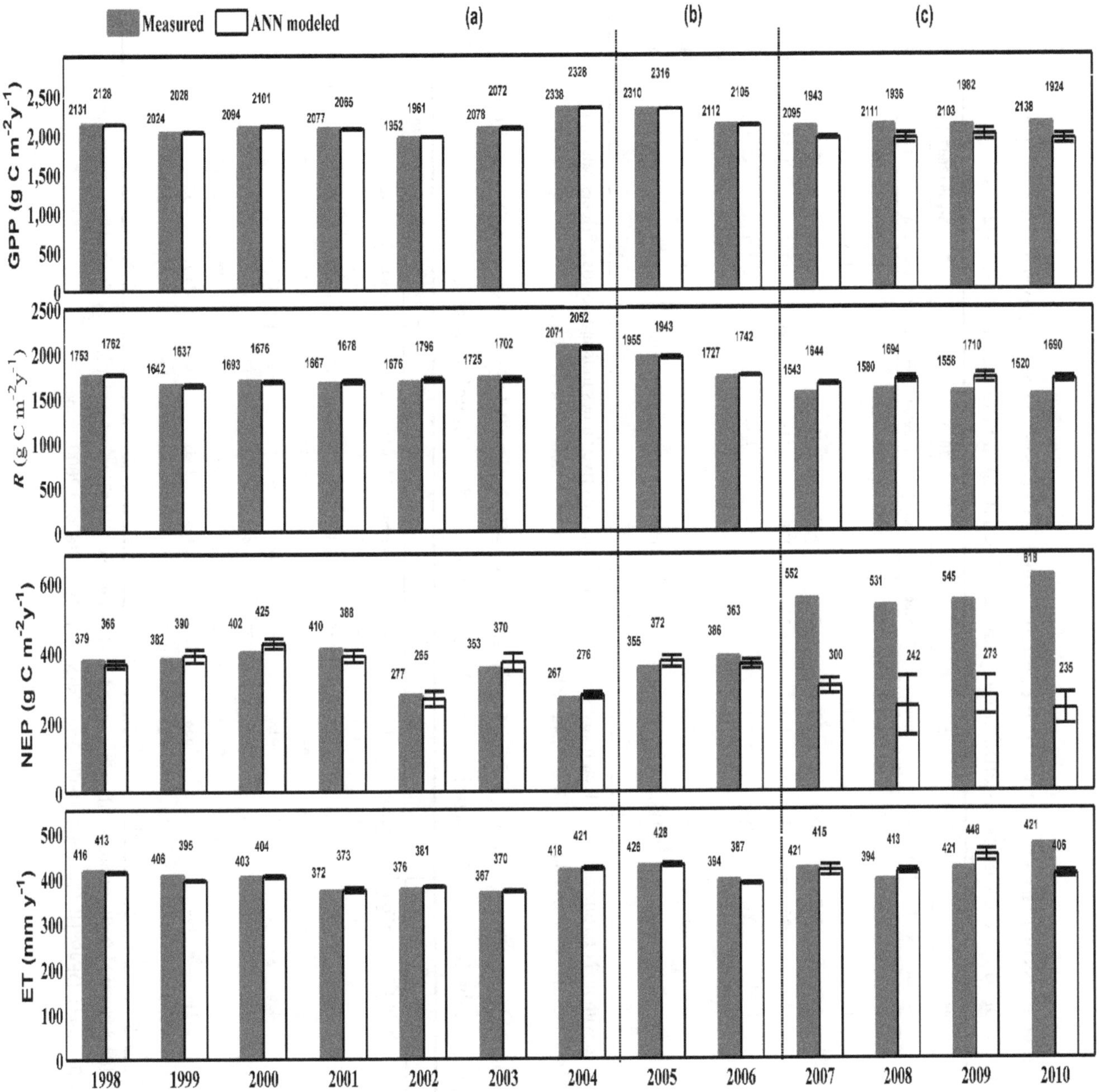

Figure 7. Comparisons of measured and artificial neural network (ANN) simulated annual gross primary productivity (GPP), ecosystem respiration (*R*), net ecosystem productivity (NEP) and evapotranspiration (ET) at DF49 from 1998 to 2010. (**a**) for the ANN calibration years 1998–2004; (**b**) for the model validation years 2005–2006; and (**c**) for the four post-fertilization years 2007–2010. The error bars with ± 1 SD are the standard errors for ANN-modeled annual values estimated using the optimized model parameters.

Table 2. Effects of nitrogen fertilization on annual gross primary productivity (GPP, g·C·m^{-2}·year^{-1}), ecosystem respiration (R, g·C·m^{-2}·year^{-1}) and net ecosystem productivity (NEP, g·C·m^{-2}·year^{-1}) for the four post-fertilization years 2007–2010 in the three Douglas-fir stands.

Stand	Flux	2007			2008			2009			2010			Mean (2007–2010)		
		Measured	Modeled	N effect	Measured	Modeled	N effect	Measured	Modeled	N effect	Measured	Modeled	N effect	Measured	Modeled	N effect
	GPP	2095	1943	+152	2111	1936	+175	2103	1982	+121	2138	1924	+214	2112	1924	+188
DF49	*R*	1543	1644	−101	1580	1694	−114	1558	1710	−152	1520	1690	−170	1550	1690	−140
	NEP	552	300	+253	531	242	+289	545	273	+272	618	235	+383	562	235	+327
	GPP	1851	1435	+416	1675	1406	+269	1572	1316	+256	1716	1456	+260	1704	1456	+248
HDF88	*R*	1714	1546	+168	1457	1548	-91	1326	1499	−173	1365	1583	−218	1466	1583	−117
	NEP	137	−112	+249	218	−142	+360	246	−183	+429	351	−159	+478	238	−159	+397
	GPP	1126	691	+435	1005	665	+340	1083	759	+324	1068	629	+439	1071	629	+442
HDF00	*R*	1566	1297	+269	1396	1278	+118	1314	1216	+98	1418	1258	+160	1424	1258	+166
	NEP	−440	−606	+166	−391	−613	+222	−231	−457	+226	−350	−629	+279	−353	−629	+276

Note: Modeled, artificial neural network (ANN) modeled without N fertilization; N effect, measured – ANN modeled. Positive values indicate positive effects caused by N fertilization, visa versa; three different-aged Douglas-fir stands: DF49, HDF88 and HDF00, established in 1949, 1988, and 2000, respectively.

Table 3. Effects of nitrogen fertilization on annual evapotranspiration (ET, mm·y^{-1}), water use efficiency (WUE, g·C (kg·water)$^{-1}$) and light use efficiency (LUE, mol·C·mol^{-1} photons) during the four post-fertilization years 2007–2010 in the three Douglas-fir stands.

Stand		2007			2008			2009			2010			Mean (2007-2010)		
		Measured	Modeled	N effect	Measured	Modeled	N effect	Measured	Modeled	N effect	Measured	Modeled	N effect	Measured	Modeled	N effect
	ET	421	415	+6	394	413	−19	421	448	−27	472	406	+66	427	420.5	+6.5
DF49	WUE	4.98	4.68	+0.3	5.36	4.69	+0.67	5	4.42	+0.58	4.53	4.74	−0.21	4.9675	4.6325	+0.335
	LUE	0.0254	0.0235	+0.0019	0.0241	0.0221	+0.002	0.0216	0.0204	+0.0012	0.0253	0.0228	+0.0025	0.0241	0.0222	+0.0019
	ET	448	405	+43	385	426	−41	397	413	−16	425	437	−12	413.75	420.25	−6.5
HDF88	WUE	4.13	3.54	+0.59	4.35	3.3	+1.05	3.96	3.19	+0.77	4.04	3.33	+0.71	4.12	3.34	+0.78
	LUE	0.0214	0.0166	+0.0048	0.0176	0.0148	+0.0028	0.0145	0.0121	+0.0024	0.0176	0.015	+0.0026	0.017775	0.014625	+0.00315
	ET	283	312	−29	288	244	+44	360	338	+22	364	431	−67	323.75	331.25	−7.5
HDF00	WUE	3.97	2.22	+1.75	3.49	2.73	+0.76	3.01	2.24	+0.77	2.93	1.46	+1.47	3.35	2.1625	+1.1875
	LUE	0.0146	0.0089	+0.0057	0.0117	0.0077	+0.004	0.0129	0.009	+0.0039	0.013	0.0077	+0.0053	0.01305	0.008325	+0.004725

Note: Modeled, artificial neural network (ANN) modeled without N fertilization; N effect, measured – ANN modeled. Positive values indicate positive effects caused by N fertilization, visa versa; three different-aged Douglas-fir stands: DF49, HDF88 and HDF00, established in 1949, 1988, and 2000, respectively.

4. Discussion

4.1. Effects of N Fertilization on Gross Primary Productivity

N fertilization generally results in GPP and net primary productivity (NPP) increases in terrestrial ecosystems by promoting growth and biomass accumulation in terrestrial plants [7,21,25,74,75]. Our results showed that N fertilization generally increased GPP in all three stands, especially in the younger stands, HDF00 and HDF88. N addition resulted in a GPP increase of 8% at DF49 in the first post-fertilization year, which is in excellent agreement with an 8% increase obtained by Chen *et al.* [52] using a model-data synthesis approach and a 10% increase calculated by Jassal *et al.* [53] using a simple empirical model (Table 4). This is lower than the 14% increase estimated by Grant *et al.* [76] using a process-based model. In addition, our results, which showed that N fertilization led to an increase in GPP in the three stands during the first two years after fertilization, are consistent in sign with the findings by Jassal *et al.* [53], although the magnitude of the increases in GPP are somewhat different. This difference may partly result from the reported uncertainty due to the use of the empirical relationships used for gap-filling, resulting in an uncertainty of ±50 $g \cdot C \cdot m^{-2} \cdot year^{-1}$ in EC-measured GPP, and ±75 and ±25 $g \cdot C \cdot m^{-2} \cdot year^{-1}$ for measured *R* and NEP, respectively [54]. Overall, N fertilization had a positive effect on GPP and the effectiveness of fertilization varied with stand age.

4.2. Effects of N Fertilization on Ecosystem Respiration

Our results indicated that N fertilization of DF49 reduced *R* by approximately 101 $g \cdot C \cdot m^{-2} \cdot year^{-1}$ in the first post-fertilization year, which is in good agreement with the 93 $g \cdot C \cdot m^{-2} \cdot year^{-1}$ decrease calculated by Chen, *et al.* [52] (Table 4). During the next three years (2008–2010), modeled *R* remained greater than measured *R* by 114 to 170 $g \cdot C \cdot m^{-2} \cdot year^{-1}$, implying that N fertilization suppressed *R*. However, the increase in *R* induced by N fertilization at HDF00 during the first two years after fertilization agrees with the findings reported in Jassal *et al.* [53]. Our ANN modeling indicated that this positive effect on *R* at HDF00 continued for the third and fourth post-fertilization years (2009 to 2010) (Table 2). It should be noted that Jassal *et al* [53] partitioned NEE by estimating daytime *R* using the relationship between daytime NEP and PAR rather than by using the relationship between nighttime NEE and T_s. Such variable effects of N addition on *R* in different stands suggest the need to investigate the mechanisms.

4.3. Effects of N fertilization on Net Ecosystem Production

N fertilization significantly increased NEP during the four post-fertilization years in all the three stands, especially HDF88, followed by DF49, largely due to N-induced GPP increases and *R* decreases. These results are consistent with those reported by Chen, *et al.* [52] in the first post-fertilization year, and are also consistent in sign with the results obtained by Jassal *et al.* [53] both at DF49 and HDF88 in the first two post-fertilization years (Table 4), although the magnitude of the increase in NEP was less than that in our study. However, at HDF00, a substantial increase in NEP in the four post-fertilization years was primarily due to GPP increasing more than *R* (Table 2).

The C:N response [kg C (sequestered) kg^{-1} N added] has been widely used to estimate the response of C sequestration to N addition [22–24,77]. Our results suggest that C:N response varied with stand age. In the first post-fertilization year, DF49 and HDF88 had approximately the same C:N response (13 kg C kg^{-1} N, followed by HDF00 (8 kg C kg^{-1} N). Our results are consistent with those of Jassal *et al.* [53], de Vries *et al.* [24], Högberg [26] and Sutton *et al.* [23], which are significantly smaller than the assessment made by Magnani *et al.* [22], with an extremely high C:N response (approximately 725 kg C kg^{-1} N) to wet deposition. Furthermore, it seems that the 60 $kg \cdot N \cdot ha^{-1}$ application to individual trees at HDF00 was less efficient with respect to NEP than the larger application (200 kg N ha^{-1}) at DF49 and HDF88.

Table 4. N-induced changes (= measured – modeled) in annual gross primary productivity (GPP, $g \cdot C \cdot m^{-2} \cdot year^{-1}$), ecosystem respiration (*R*, $g \cdot C \cdot m^{-2} \cdot year^{-1}$) and net ecosystem productivity (NEP, $g \cdot C \cdot m^{-2} \cdot year^{-1}$) from different study results in the three West Coast Douglas-fir stands during 2007 to 2010.

Stand	Flux	Study	2007	2008	2009	2010
DF49	NEP	This study (ANN)	253	289	272	383
		Jassal *et al.* [53]	168	78		
		Chen *et al.* [52]	250			
	GPP	This study (ANN)	152	175	121	214
		Jassal *et al.* [53]	203	139		
		Chen *et al.* [52]	157			
	R	This study (ANN)	−101	−114	−152	−170
		Jassal *et al.* [53]	35	61		
		Chen *et al.* [52]	−93			
HDF88	NEP	This study (ANN)	249	360	429	478
		Jassal *et al.* [53]	182	179		
	GPP	This study (ANN)	416	269	256	260
		Jassal *et al.* [53]	615	366		
	R	This study (ANN)	168	−91	−173	−218
		Jassal *et al.* [53]	433	187		
HDF00	NEP	This study (ANN)	166	222	226	279
		Jassal *et al.* [53]	−3	−82		
	GPP	This study (ANN)	435	340	324	439
		Jassal *et al.* [53]	305	329		
	R	This study (ANN)	269	118	98	160
		Jassal *et al.* [53]	308	411		

Three different-aged Douglas-fir stands: DF49, HDF88 and HDF00, established in 1949, 1988, and 2000, respectively.

4.4. Effects of N fertilization on ET, WUE and LUE

N application showed no consistent effect on ET among the three stands. N fertilization slightly decreased ET at HDF88 in the last three years and had no clear effect on ET at HDF00 over the four post-fertilization years (Table 3). On the other hand, increased nutrient availability (e.g., N fertilization) had significantly positive effects on WUE at HDF88 and HDF00 due to its strong positive effect on GPP compared with its small effect on ET, but had no significant effect at DF49. Ripullone *et al.* [47] found that N addition increased both WUE and biomass production. They concluded that this was mainly due to the positive response of photosynthesis to N availability with it having no impact on either stomatal conductance or transpiration. Consequently, improved tree nutrition (e.g., N fertilization) could cause an increase in forest productivity [78] without leading to an increase in demand for water.

Furthermore, N fertilization caused an increase in the 4-year mean WUE of 0.34 g·C (kg·water)$^{-1}$ for DF49, implying a relatively conservative response to N fertilization, and increases of 0.78 and 1.19 g·C (kg·water)$^{-1}$ for HDF88 and HDF00, respectively, likely due to their younger stand ages and less developed root system [56]. Wu *et al.* [79] suggested that low or appropriate N application might improve the survivability of *Sophora davidii* seedlings under water deficit conditions, resulting in an increase in WUE and stimulating growth and biomass production. Our results indicated that the supply of 60 kg N ha^{-1} applied to individual trees at HDF00 may be more efficient in terms of WUE compared with the higher application of 200 kg N ha^{-1} at HDF88 and DF49, especially in the latter stand.

N fertilization in our study had positive impacts on LUE in all three stands in the four post-fertilization years, especially at HDF00 followed by HDF88 with the smallest increase occurring in 2009 in all three stands accompanied by the highest annual PAR. This suggests that there was a reduction in the increase in LUE induced by N fertilization under high PAR, which has also been reported by Ibrom *et al.* [80] and Schwalm *et al.* [81]. Furthermore, N fertilization caused a 0.002·mol C·mol^{-1} photon increase in the 4-year mean LUE over the post-fertilization period for DF49, implying a relatively conservative plant productivity response to N fertilization, while higher increases of 0.0032 and 0.0047 mol·C·mol^{-1} photons for HDF88 and HDF00, respectively, were likely due to their younger stand age and rapid stand development. Overall, stand development had a significant impact on plant productivity, LUE and WUE as well as nutrient demand (e.g., N fertilization) [82].

4.5. Modeling Uncertainty and Limitations

Several studies have found that there is observable interannual variability of C component fluxes and ET in these stands, which to a large extent, results from interannual variability in climate [42,54,56,59]. We used both ANN and MLR approaches for removing the impacts of climatic variability on C fluxes and ET in the three stands during the post-fertilization period. In addition, a significant effect of stand development on C fluxes in the two younger stands (HDF88 and HDF00) was demonstrated in Figure 3. Our results show that pre-fertilization monthly C fluxes and ET affected by climate and stand age were better described using the ANN approach than the MLR approach.

The determination of BPNN model parameters including the hidden layer node number, momentum coefficient and learning rate and its shortcomings, such as the slow rate of convergence and local minimum, have considerably restricted its application and have led to uncertainties in the estimate of the N effect in this study. Nevertheless, we believe this study contributes significantly to quantifying the effects of N fertilization on C fluxes and ET and reducing the uncertainty in model simulation. Considering the limitations of the BPNN model, in order to further improve the estimate of the N effect, more importance should be attached to a hybrid BPNN model with optimized algorithms such as the genetic algorithm and particle swarm optimization.

5. Conclusions

1. Pre-fertilization monthly C fluxes and ET were better described using the ANN approach than the MLR approach.

2. N fertilization increased GPP in all three Douglas-fir stands over the four post-fertilization years with the greatest increase in the youngest stand (HDF00) followed by HDF88, suggesting that the effectiveness of fertilization on GPP may be associated with stand age. In addition, fertilization decreased *R* over the four years after fertilization at DF49, and over the last three years at HDF88, but increased *R* over the four years at HDF00. As a result, N fertilization increased NEP in all three stands, with the greatest increase in HDF88 and followed by DF49.

3. N fertilization had no discernible effects on ET among the three stands. This result led to a significant increase in WUE in the two younger stands (HDF88 and HDF00) but a small increase in only the first three years for the oldest stand (DF49).

4. N fertilization increased LUE in all three stands, especially the youngest stand (HDF00).

Acknowledgments

This research was supported a research grant (41271116) funded by the National Science Foundation of China and an NSERC Discovery Grant (6123-2012) to TAB. Thanks to TimberWest Forest Company and Island Timberlands LP for logistical support and to Agrium Inc. for providing financial support and the urea fertilizer. We greatly appreciate the assistance of Dominic Lessard, Zoran Nesic and Rick Ketler with the climate and EC observations, and Christian Brümmer, Nick Grant and Praveena Krishnan for data preprocessing.

Author Contributions

Baozhang Chen and T. Andrew Black, conceived and designed the paper. Xianming Dou performed the data analyses and ANN modeling. Xianming Dou and Baozhang Chen wrote the paper. Rachhpal S. Jassal conducted the fertilization experiments. Mingliang Che prepared the data for ANN modeling. All coauthors contributed to manuscript revision and approved the final manuscript.

References

1. Houghton, R.A. Terrestrial carbon sinks–uncertain. *Biologist* **2002**, *49*, 155–160.
2. Piao, S.; Ciais, P.; Friedlingstein, P.; de Noblet-Ducoudré, N.; Cadule, P.; Viovy, N.; Wang, T. Spatiotemporal patterns of terrestrial carbon cycle during the 20th century. *Glob. Biogeochem. Cycles* **2009**, *23*, doi:10.1029/2008GB003339.
3. Tian, H.; Melillo, J.; Lu, C.; Kicklighter, D.; Liu, M.; Ren, W.; Xu, X.; Chen, G.; Zhang, C.; Pan, S. China's terrestrial carbon balance: Contributions from multiple global change factors. *Glob. Biogeochem. Cycles* **2011**, *25*, doi:10.1029/2010GB003838.

4. Cameron, D.R.; Oijen, M.V.; Werner, C.; Butterbach-Bahl, K.; Grote, R.; Haas, E.; Heuvelink, G.B.M.; Kiese, R.; Kros, J.; Kuhnert, M. Environmental change impacts on the C-and N-cycle of european forests: A model comparison study. *Biogeosciences* **2013**, *10*, 1751–1773.
5. Fernández-Martínez, M.; Vicca, S.; Janssens, I.A.; Sardans, J.; Luyssaert, S.; Campioli, M.; Chapin, F.S., III; Ciais, P.; Malhi, Y.; Obersteiner, M. Nutrient availability as the key regulator of global forest carbon balance. *Nat. Clim. Chang.* **2014**, *4*, 471–476.
6. Hilker, T.; Lepine, L.; Coops, N.C.; Jassal, R.S.; Black, T.A.; Wulder, M.A.; Ollinger, S.; Tsui, O.; Day, M. Assessing the impact of *N*-fertilization on biochemical composition and biomass of a Douglas-fir canopy—A remote sensing approach. *Agric. For. Meteorol.* **2012**, *153*, 124–133.
7. Högberg, P. Environmental science: Nitrogen impacts on forest carbon. *Nature* **2007**, *447*, 781–782.
8. Reich, P.B.; Hobbie, S.E.; Lee, T.; Ellsworth, D.S.; West, J.B.; Tilman, D.; Knops, J.M.H.; Naeem, S.; Trost, J. Nitrogen limitation constrains sustainability of ecosystem response to CO_2. *Nature* **2006**, *440*, 922–925.
9. Yang, Y.; Luo, Y.; Finzi, A.C. Carbon and nitrogen dynamics during forest stand development: A global synthesis. *New Phytol.* **2011**, *190*, 977–989.
10. Aber, J.; McDowell, W.; Nadelhoffer, K.; Magill, A.; Berntson, G.; Kamakea, M.; McNulty, S.; Currie, W.; Rustad, L.; Fernandez, I. Nitrogen saturation in temperate forest ecosystems. *Bioscience* **1998**, *48*, 921–934.
11. Du, Z.; Wang, W.; Zeng, W.; Zeng, H. Nitrogen deposition enhances carbon sequestration by plantations in northern China. *PLoS ONE* **2014**, *9*, doi:10.1371/journal.pone.0087975.
12. Krause, K.; Cherubini, P.; Bugmann, H.; Schleppi, P. Growth enhancement of picea abies trees under long-term, low-dose n addition is due to morphological more than to physiological changes. *Tree Physiol.* **2012**, *32*, 1471–1481.
13. Magill, A.H.; Aber, J.D.; Currie, W.S.; Nadelhoffer, K.J.; Martin, M.E.; McDowell, W.H.; Melillo, J.M.; Steudler, P. Ecosystem response to 15 years of chronic nitrogen additions at the harvard forest lter, Massachusetts, USA. *For. Ecol. Manag.* **2004**, *196*, 7–28.
14. Reay, D.S.; Dentener, F.; Smith, P.; Grace, J.; Feely, R.A. Global nitrogen deposition and carbon sinks. *Nat. Geosci.* **2008**, *1*, 430–437.
15. Fleischer, K.; Rebel, K.T.; Molen, V.D.M.K.; Erisman, J.W.; Wassen, M.J.; Loon, E.E.; Montagnani, L.; Gough, C.M.; Herbst, M.; Janssens, I.A. The contribution of nitrogen deposition to the photosynthetic capacity of forests. *Glob. Biogeochem. Cycles* **2013**, *27*, 187–199.
16. De Vries, W.; Du, E.; Butterbach-Bahl, K. Short and long-term impacts of nitrogen deposition on carbon sequestration by forest ecosystems. *Curr. Opin. Environ. Sust.* **2014**, *9*, 90–104.
17. Berg, B.; Matzner, E. Effect of n deposition on decomposition of plant litter and soil organic matter in forest systems. *Environ. Rev.* **1997**, *5*, 1–25.
18. Franklin, O.; Högberg, P.; Ekblad, A.; Ågren, G.I. Pine forest floor carbon accumulation in response to N and PK additions: Bomb ^{14}C modelling and respiration studies. *Ecosystems* **2003**, *6*, 644–658.
19. Leggett, Z.H.; Kelting, D.L. Fertilization effects on carbon pools in loblolly pine plantations on two upland sites. *Soil Sci. Soc. Am. J.* **2006**, *70*, 279–286.
20. Olsson, P.; Linder, S.; Giesler, R.; Högberg, P. Fertilization of boreal forest reduces both autotrophic and heterotrophic soil respiration. *Glob. Chang. Biol.* **2005**, *11*, 1745–1753.

21. Xia, J.; Wan, S. Global response patterns of terrestrial plant species to nitrogen addition. *New Phytol.* **2008**, *179*, 428–439.
22. Magnani, F.; Mencuccini, M.; Borghetti, M.; Berbigier, P.; Berninger, F.; Delzon, S.; Grelle, A.; Hari, P.; Jarvis, P.G.; Kolari, P. The human footprint in the carbon cycle of temperate and boreal forests. *Nature* **2007**, *447*, 849–851.
23. Sutton, M.A.; Simpson, D.; Levy, P.E.; Smith, R.I.; Reis, S.; van Oijen, M.; de Vries, W.I.M. Uncertainties in the relationship between atmospheric nitrogen deposition and forest carbon sequestration. *Glob. Chang. Biol.* **2008**, *14*, 2057–2063.
24. de Vries, W.; Solberg, S.; Dobbertin, M.; Sterba, H.; Laubhahn, D.; Reinds, G.J.; Nabuurs, G.J.; Gundersen, P.; Sutton, M.A. Ecologically implausible carbon response? *Nature* **2008**, *451*, E1–E3.
25. Dezi, S.; Medlyn, B.E.; Tonon, G.; Magnani, F. The effect of nitrogen deposition on forest carbon sequestration: A model-based analysis. *Glob. Chang. Biol.* **2010**, *16*, 1470–1486.
26. Högberg, P. What is the quantitative relation between nitrogen deposition and forest carbon sequestration? *Glob. Chang. Biol.* **2012**, *18*, 1–2.
27. Janssens, I.A.; Dieleman, W.; Luyssaert, S.; Subke, J.A.; Reichstein, M.; Ceulemans, R.; Ciais, P.; Dolman, A.J.; Grace, J.; Matteucci, G.; *et al.* Reduction of forest soil respiration in response to nitrogen deposition. *Nat. Geosci.* **2010**, *3*, 315–322.
28. Morell, F.J.; Lampurlanés, J.; Álvaro-Fuentes, J.; Cantero-Martínez, C. Yield and water use efficiency of barley in a semiarid mediterranean agroecosystem: Long-term effects of tillage and N fertilization. *Soil Till. Res.* **2011**, *117*, 76–84
29. Körner, C. Biosphere responses to CO_2 enrichment. *Ecol. Appl.* **2000**, *10*, 1590–1619.
30. Fenn, M.E.; Baron, J.S.; Allen, E.B.; Rueth, H.M.; Nydick, K.R.; Geiser, L.; Bowman, W.D.; Sickman, J.O.; Meixner, T.; Johnson, D.W. Ecological effects of nitrogen deposition in the western united states. *Bioscience* **2003**, *53*, 404–420.
31. Nadelhoffer, K.J.; Emmett, B.A.; Gundersen, P.; Kjønaas, O.J.; Koopmans, C.J.; Schleppi, P.; Tietema, A.; Wright, R.F. Nitrogen deposition makes a minor contribution to carbon sequestration in temperate forests. *Nature* **1999**, *398*, 145–148.
32. Currie, W.S.; Nadelhoffer, K.J.; Aber, J.D. Redistributions of highlight turnover and replenishment of mineral soil organic n as a long-term control on forest c balance. *For. Ecol. Manag.* **2004**, *196*, 109–127.
33. Liu, L.; Greaver, T.L. A global perspective on belowground carbon dynamics under nitrogen enrichment. *Ecol. Lett.* **2010**, *13*, 819–828.
34. Harpole, W.S.; Potts, D.L.; Suding, K.N. Ecosystem responses to water and nitrogen amendment in a California grassland. *Glob. Chang. Biol.* **2007**, *13*, 2341–2348.
35. Tian, S.; Youssef, M.A.; Amatya, D.M.; Vance, E.D. Global sensitivity analysis of drainmod-forest, an integrated forest ecosystem model. *Hydrol. Process.* **2014**, *28*, 4389–4410.
36. Felzer, B.S.; Cronin, T.W.; Melillo, J.M.; Kicklighter, D.W.; Schlosser, C.A.; Dangal, S.R.S. Nitrogen effect on carbon-water coupling in forests, grasslands, and shrublands in the arid western united states. *J. Geophys. Res.* **2011**, *116*, doi:10.1029/2010jg001621.
37. Hu, Z.; Yu, G.; Fu, Y.; Sun, X.; Li, Y.; Shi, P.; Wang, Y.; Zheng, Z. Effects of vegetation control on ecosystem water use efficiency within and among four grassland ecosystems in China. *Glob. Chang. Biol.* **2008**, *14*, 1609–1619.

38. Ito, A.; Inatomi, M. Water-use efficiency of the terrestrial biosphere: A model analysis focusing on interactions between the global carbon and water cycles. *J. Hydrometeorol.* **2012**, *13*, 681–694.
39. Brix, H.; Mitchell, A.K. Thinning and nitrogen fertilization effects on soil and tree water stress in a douglas-fir stand. *Can. J. For. Res.* **1986**, *16*, 1334–1338.
40. Dordas, C.A.; Sioulas, C. Safflower yield, chlorophyll content, photosynthesis, and water use efficiency response to nitrogen fertilization under rainfed conditions. *Ind. Crop. Prod.* **2008**, *27*, 75–85.
41. Livingston, N.J.; Guy, R.D.; Sun, Z.J.; Ethier, G.J. The effects of nitrogen stress on the stable carbon isotope composition, productivity and water use efficiency of white spruce (*Picea glauca* (moench) voss) seedlings. *Plant Cell Environ.* **1999**, *22*, 281–289.
42. Jassal, R.S.; Black, T.A.; Spittlehouse, D.L.; Brümmer, C.; Nesic, Z. Evapotranspiration and water use efficiency in different-aged Pacific Northwest Douglas-fir stands. *Agric. For. Meteorol.* **2009**, *149*, 1168–1178.
43. Korol, R.L.; Kirschbaum, M.U.F.; Farquhar, G.D.; Jeffreys, M. Effects of water status and soil fertility on the c-isotope signature in *Pinus radiata*. *Tree Physiol.* **1999**, *19*, 551–562.
44. Mitchell, A.K.; Hinckley, T.M. Effects of foliar nitrogen concentration on photosynthesis and water use efficiency in Douglas-fir. *Tree Physiol.* **1993**, *12*, 403–410.
45. Castellanos, M.T.; Tarquis, A.M.; Ribas, F.; Cabello, M.J.; Arce, A.; Cartagena, M.C. Nitrogen fertigation: An integrated agronomic and environmental study. *Agric. Water Manag.* **2013**, *120*, 46–55.
46. Claussen, W. Growth, water use efficiency, and proline content of hydroponically grown tomato plants as affected by nitrogen source and nutrient concentration. *Plant Soil* **2002**, *247*, 199–209.
47. Ripullone, F.; Lauteri, M.; Grassi, G.; Amato, M.; Borghetti, M. Variation in nitrogen supply changes water-use efficiency of pseudotsuga menziesii and populus x euroamericana; a comparison of three approaches to determine water-use efficiency. *Tree Physiol.* **2004**, *24*, 671–679.
48. Running, S.W.; Nemani, R.R.; Heinsch, F.A.; Zhao, M.; Reeves, M.; Hashimoto, H. A continuous satellite-derived measure of global terrestrial primary production. *Bioscience* **2004**, *54*, 547–560.
49. Yuan, W.; Liu, S.; Yu, G.; Bonnefond, J.M.; Chen, J.; Davis, K.; Desai, A.R.; Goldstein, A.H.; Gianelle, D.; Rossi, F. Global estimates of evapotranspiration and gross primary production based on modis and global meteorology data. *Remote Sens. Environ.* **2010**, *114*, 1416–1431.
50. Zhao, M.; Heinsch, F.A.; Nemani, R.R.; Running, S.W. Improvements of the modis terrestrial gross and net primary production global data set. *Remote Sens. Environ.* **2005**, *95*, 164–176.
51. Hanley, D.P.; Chappell, H.N.; Nadelhoffer, E.H. *Fertilizing Coastal Douglas-fir Forests*; Washington State University: Pullman, WA, USA, 1991.
52. Chen, B.; Coops, N.C.; Black, T.A.; Jassal, R.S.; Chen, J.M.; Johnson, M. Modeling to discern nitrogen fertilization impacts on carbon sequestration in a Pacific Northwest Douglas-fir forest in the first post-fertilization year. *Glob. Chang. Biol.* **2011**, *17*, 1442–1460.
53. Jassal, R.S.; Black, T.A.; Cai, T.; Ethier, G.; Pepin, S.; Brümmer, C.; Nesic, Z.; Spittlehouse, D.L.; Trofymow, J.A. Impact of nitrogen fertilization on carbon and water balances in a chronosequence of three Douglas-fir stands in the pacific northwest. *Agric. For. Meteorol.* **2010**, *150*, 208–218.

54. Humphreys, E.R.; Black, T.A.; Morgenstern, K.; Cai, T.; Drewitt, G.B.; Nesic, Z.; Trofymow, J.A. Carbon dioxide fluxes in coastal Douglas-fir stands at different stages of development after clearcut harvesting. *Agric. For. Meteorol.* **2006**, *140*, 6–22.
55. Chen, B.; Black, T.A.; Coops, N.C.; Hilker, T.; Trofymow, J.A.; Morgenstern, K. Assessing tower flux footprint climatology and scaling between remotely sensed and eddy covariance measurements. *Bound. Layer Meteorol.* **2009**, *130*, 137–167.
56. Krishnan, P.; Black, T.A.; Jassal, R.S.; Chen, B.; Nesic, Z. Interannual variability of the carbon balance of three different-aged Douglas-fir stands in the pacific northwest. *J. Geophys. Res.* **2009**, *114*, doi:10.1029/2008JG000912.
57. Morgenstern, K.; Black, T.A.; Humphreys, E.R.; Griffis, T.J.; Drewitt, G.B.; Cai, T.; Nesic, Z.; Spittlehouse, D.L.; Livingston, N.J. Sensitivity and uncertainty of the carbon balance of a pacific northwest Douglas-fir forest during an el niño/la niña cycle. *Agric. For. Meteorol.* **2004**, *123*, 201–219.
58. Barr, A.G.; Black, T.A.; Hogg, E.H.; Griffis, T.J.; Morgenstern, K.; Kljun, N.; Theede, A.; Nesic, Z. Climatic controls on the carbon and water balances of a boreal aspen forest, 1994–2003. *Glob. Chang. Biol.* **2007**, *13*, 561–576.
59. Chen, B.; Black, T.A.; Coops, N.C.; Krishnan, P.; Jassal, R.; Bruemmer, C.; Nesic, Z. Seasonal controls on interannual variability in carbon dioxide exchange of a near-end-of rotation Douglas-fir stand in the pacific northwest, 1997–2006. *Glob. Chang. Biol.* **2009**, *15*, 1962–1981.
60. Melesse, A.M.; Hanley, R.S. Artificial neural network application for multi-ecosystem carbon flux simulation. *Ecol. Model.* **2005**, *189*, 305–314.
61. Moffat, A.M.; Papale, D.; Reichstein, M.; Hollinger, D.Y.; Richardson, A.D.; Barr, A.G.; Beckstein, C.; Braswell, B.H.; Churkina, G.; Desai, A.R. Comprehensive comparison of gap-filling techniques for eddy covariance net carbon fluxes. *Agric. For. Meteorol.* **2007**, *147*, 209–232.
62. White, H. Learning in artificial neural networks: A statistical perspective. *Neural Comput.* **1989**, *1*, 425–464.
63. Jain, A.K.; Mao, J.; Mohiuddin, K.M. Artificial neural networks: A tutorial. *Computer* **1996**, *29*, 31–44.
64. Zhang, G.; Patuwo, B.E.; Hu, M.Y. Forecasting with artificial neural networks: The state of the art. *Int. J. Forecast.* **1998**, *14*, 35–62.
65. Basheer, I.A.; Hajmeer, M. Artificial neural networks: Fundamentals, computing, design, and application. *J. Microbiol. Methods* **2000**, *43*, 3–31.
66. Papale, D.; Valentini, R. A new assessment of european forests carbon exchanges by eddy fluxes and artificial neural network spatialization. *Glob. Chang. Biol.* **2003**, *9*, 525–535.
67. Ooba, M.; Hirano, T.; Mogami, J.I.; Hirata, R.; Fujinuma, Y. Comparisons of gap-filling methods for carbon flux dataset: A combination of a genetic algorithm and an artificial neural network. *Ecol. Model.* **2006**, *198*, 473–486.
68. Richardson, A.D.; Mahecha, M.D.; Falge, E.; Kattge, J.; Moffat, A.M.; Papale, D.; Reichstein, M.; Stauch, V.J.; Braswell, B.H.; Churkina, G. Statistical properties of random CO_2 flux measurement uncertainty inferred from model residuals. *Agric. For. Meteorol.* **2008**, *148*, 38–50.
69. Rumelhart, D.E.; Hintont, G.E.; Williams, R.J. Learning representations by back-propagating errors. *Nature* **1986**, *323*, 533–536.

70. Maier, H.R.; Dandy, G.C. Neural networks for the prediction and forecasting of water resources variables: A review of modelling issues and applications. *Environ. Model. Softw.* **2000**, *15*, 101–124.
71. Funahashi, K.I. On the approximate realization of continuous mappings by neural networks. *Neural Netw.* **1989**, *2*, 183–192.
72. Cybenko, G. Approximation by superpositions of a sigmoidal function. *Math. Control Signal Syst.* **1989**, *2*, 303–314.
73. Moré, J. The levenberg-marquardt algorithm: Implementation and theory. In *Numerical Analysis*; Watson, G.A., Ed.; Springer: Berlin Heidelberg, Germany, 1978; pp. 105–116.
74. Thomas, R.Q.; Canham, C.D.; Weathers, K.C.; Goodale, C.L. Increased tree carbon storage in response to nitrogen deposition in the us. *Nat. Geosci.* **2010**, *3*, 13–17.
75. Niu, S.; Wu, M.; Han, Y.I.; Xia, J.; Zhang, Z.; Yang, H.; Wan, S. Nitrogen effects on net ecosystem carbon exchange in a temperate steppe. *Glob. Chang. Biol.* **2010**, *16*, 144–155.
76. Grant, R.F.; Black, T.A.; Jassal, R.S.; Bruemmer, C. Changes in net ecosystem productivity and greenhouse gas exchange with fertilization of Douglas-fir: Mathematical modeling in ecosys. *J. Geophys. Res.* **2010**, *115*, doi:10.1029/2009jg001094.
77. Hyvönen, R.; Persson, T.; Andersson, S.; Olsson, B.; Ågren, G.I.; Linder, S. Impact of long-term nitrogen addition on carbon stocks in trees and soils in northern Europe. *Biogeochemistry* **2008**, *89*, 121–137.
78. Vicca, S.; Luyssaert, S.; Penuelas, J.; Campioli, M.; Chapin, F.S.; Ciais, P.; Heinemeyer, A.; Högberg, P.; Kutsch, W.L.; Law, B.E. Fertile forests produce biomass more efficiently. *Ecol. Lett.* **2012**, *15*, 520–526.
79. Wu, F.; Bao, W.; Li, F.; Wu, N. Effects of drought stress and n supply on the growth, biomass partitioning and water-use efficiency of sophora davidii seedlings. *Environ. Exp. Bot.* **2008**, *63*, 248–255.
80. Ibrom, A.; Oltchev, A.; June, T.; Kreilein, H.; Rakkibu, G.; Ross, T.; Panferov, O.; Gravenhorst, G. Variation in photosynthetic light-use efficiency in a mountainous tropical rain forest in Indonesia. *Tree Physiol.* **2008**, *28*, 499–508.
81. Schwalm, C.R.; Black, T.A.; Amiro, B.D.; Arain, M.A.; Barr, A.G.; Bourque, C.P.A.; Dunn, A.L.; Flanagan, L.B.; Giasson, M.-A.; Lafleur, P.M. Photosynthetic light use efficiency of three biomes across an east–west continental-scale transect in Canada. *Agric. For. Meteorol.* **2006**, *140*, 269–286.
82. Binkley, D.; Stape, J.L.; Bauerle, W.L.; Ryan, M.G. Explaining growth of individual trees: Light interception and efficiency of light use by eucalyptus at four sites in Brazil. *For. Ecol. Manag.* **2010**, *259*, 1704–1713.

12

Mid-Rotation Silviculture Timing Influences Nitrogen Mineralization of Loblolly Pine Plantations in the Mid-South USA

Michael A. Blazier [1,*], D. Andrew Scott [2] and Ryan Coleman [1]

[1] Hill Farm Research Station, Louisiana State University Agricultural Center, Homer, LA 71040, USA; E-Mail: rpcoleman2@gmail.com

[2] Southern Research Station, USDA Forest Service, P.O. Box 1927, Normal, AL 35762, USA; E-Mail: andyscott@fs.fed.us

* Author to whom correspondence should be addressed; E-Mail: mblazier@agcenter.lsu.edu;

Academic Editor: Heinz Rennenberg

Abstract: Intensively managed loblolly pine (*Pinus taeda* L.) plantations often develop nutrient deficiencies near mid-rotation. Common silvicultural treatments for improving stand nutrition at this stage include thinning, fertilization, and vegetation control. It is important to better understand the influence of timing fertilization and vegetation control in relation to thinning as part of improving the efficiency of these practices. The objective of this study was to determine the effects of fertilization and vegetation control conducted within a year prior to thinning and within a year after thinning on soil N supply in mid-rotation loblolly pine plantations on a gradient of soil textures. Net N mineralization (N_{min}) and exchangeable N were measured monthly. Fertilization increased annual N_{min} at all sites irrespective of timing relative to thinning, with the increase more pronounced when combined with vegetation control. This finding suggests some management flexibility in the timing of mid-rotation fertilization relative to thinning for increasing soil N supply. However, the site with the highest total soil N and the lowest C:N ratio was more prone to NO_3-N increases after fertilization conducted pre- and post-thinning. At all sites, fertilization with vegetation control promoted increases in NO_3-N when done after thinning, which may indicate that this practice increased soil N supply to levels that exceeded stand N demand.

Keywords: nitrogen; herbicide; thinning; fertilization timing

1. Introduction

As intensively managed loblolly pine plantations in the southeastern United States approach mid-rotation, stand nutrient demand increases and soil nutrient supply decreases [1,2]. To synchronize nutrient supply with plant demand, increase leaf area, and improve productivity, plantations are often fertilized, typically after thinning [3–5]. Nitrogen is typically the limiting nutrient in these plantations at this stage because it is needed in relatively high quantities for photosynthesis, carbohydrate assimilation, and foliage growth [6]. Herbicides are also sometimes applied at this stage to control competing vegetation and improve site resource availability for the crop tree [7]. Further improvement of loblolly pine production efficiency via mid-rotation silviculture will likely require more than simply increasing resource inputs; silvicultural practices that better manage the interactions among crop trees, non-crop vegetation, soil properties, and applied nutrients will likely be necessary [8]. Altering the timing of fertilization and vegetation control treatments relative to mid-rotation thinning is one potential method of influencing resource use efficiency. Fertilizing mid-rotation stands prior to thinning may promote higher stand-level fertilizer uptake, which could in turn reduce risks of N leaching on well-drained soils and denitrification on poorly drained soils. With pre-thinning fertilization, nutrient levels of fertilized trees remaining after thinning may be elevated, which could enable trees to more rapidly exploit increased growing space. However, increased intraspecific competition in stands at canopy closure can temper response to fertilization, with loblolly pine stands generally exhibiting declining potential response to fertilization as they approach basal areas greater than 35 $m^2 \cdot ha^{-1}$ [8].

A vital component of understanding loblolly pine plantation response to timing of fertilization relative to thinning is observing its effects on soil N supply. Soil N supply has been defined as net N mineralization [9]. Fertilization of mid-rotation loblolly pine plantations can increase N mineralization [2,10]. Non-crop vegetation suppression can likewise increase N mineralization in mid-rotation loblolly pine plantations, particularly when combined with fertilization [2]. Nitrogen mineralization rates may be relatively high in the year of thinning because stand conditions could be similar to those of complete harvesting, with increased radiation to the forest floor and decreased inputs of soluble organic C because of tree harvesting. Such conditions have been shown to result in relatively high N mineralization in the initial years of loblolly pine plantations following clearcuts [11].

Microbial biomass and activity are closely associated with soil N mineralization, and they are affected by fertilization and vegetation control in forests [12]. As such, it is appropriate to explore these variables in tandem with soil N mineralization trends. Fertilization can lead to a "priming effect" in which microbial N immobilization potential is reduced, leading to increases in N mineralization that lead to increases in soil N availability beyond that applied in fertilizer [9]. Microbial biomass and activity increased [13], decreased [14], or had no response [15] to forest fertilization in previous studies. Combining non-crop vegetation control with fertilization reduced microbial biomass and activity to a greater extent than to fertilization alone in a juvenile loblolly pine plantation, which was attributed to reduction in labile C sources for microbes due to vegetation control [16].

The influence of altering the timing of fertilization and vegetation control relative to thinning at mid-rotation loblolly pine plantations on soil N dynamics and its associated microbial parameters is relatively unstudied. Accordingly, the objectives of this study were to determine the effects of fertilization and vegetation control conducted within a year prior to thinning and within a year after thinning on soil N mineralization, exchangeable N, soil microbial biomass C, and dehydrogenase activity in mid-rotation loblolly pine plantations on a gradient of soil textures in the mid-South USA. Study sites were selected along a gradient of soil textures because soil texture influences nutrient cycling; mineral particle size distribution can affect organic matter retention, N mineralization, microbial biomass, and other soil properties [17].

2. Materials and Methods

2.1. Study Sites

In September and October 2003, three study sites were established in mid-rotation loblolly pine plantations in north central and southwest Louisiana. The sites were selected for their similarities in age and management history and the range of soil drainage classes they provided (Table 1). All sites had chop and burn site preparation; the Oakdale site had been fertilized with 50 kg·N·ha^{-1} and 56 kg·P·ha^{-1} as diammonium phosphate in 1997.

Table 1. Location, stand age and basal area in 2003, and U.S. Department of Agriculture (USDA) Natural Resource Conservation Service soil type classification of study sites established in loblolly pine plantations in north central and southwest Louisiana, USA.

Nearest Town	Geographic Coordinates	Stand Age (Years)	Stand Basal Area (m^2·ha^{-1})	Soil Type
Dodson	32.0727° N, 92.6697° W	13	23.5	Bowie (moderately well-drained, fine-loamy, siliceous, thermic Plinthic Paleudult)
Lucky	32.3029° N, 92.9240° W	12	19.9	Betis (excessively-drained, sandy siliceous, thermic Lamellic Paleudult)
Oakdale	30.8164° N, 92.6914° W	15	20.5	Caddo (poorly-drained, fine-silty, siliceous, thermic Typic Glossaqualf)

Understory and mid-story of the sites at the initiation of the study were dominated by hardwood tree species at the Dodson and Lucky sites and shrub species at the Oakdale site [7]. Red maple (*Acer rubrum* L.) and sweetgum (*Liquidambar styraciflua* L.) were frequently observed species at the Dodson and Lucky sites. Southern red oak (*Quercus falcata* Michx.) was present in high numbers at the Lucky site. The most common understory and mid-story species at the Oakdale site were yaupon holly (*Ilex vomitoria* Sol. Ex Aiton), wax myrtle (*Morella cerifera* L. (Small)), and Chinese tallow tree (*Triadica sebifera* L.).

2.2. Study Design and Treatments

Study design was a randomized complete block with soil texture differences to a 40-cm depth used as a blocking factor at each site. Experimental units were 0.1-ha plots, and all treatment combinations were replicated three times at each study site. Plots were separated by at least 13 m to insure independence of treatments. Measurement zones were established within the central 0.08 ha of each plot. Blocks, which contained one replicate of each treatment combination and buffer spaces between plots, were 1 ha in size.

Treatments conducted in this study were: (1) no fertilization or herbicide (CONTROL), (2) fertilization (FERT), and (3) fertilization and vegetation suppression (FERTVS). The FERT and FERTVS treatments were conducted prior to thinning or after thinning to separate sets of plots. The CONTROL treatment was applied to plots observed in pre-thin and post-thin condition, *i.e.*, the same CONTROL plots were observed pre- and post-thinning. The pre-thin FERT and FERTVS plots were sampled in only the year prior to thinning, and the post-thin FERT and FERTVS plots were observed in only the year after thinning. There were three replications for each treatment. With this treatment structure there were three CONTROL plots, three pre-thin FERT plots, three post-thin FERT plots, three pre-thin FERTVS plots, and three post-thin FERTVS plots.

Fertilization was conducted pre-thin in April 2004 and post-thin in April 2005. A mixture of urea and diammonium phosphate supplying 135 kg·N and 13 kg·P·ha^{-1} was applied using shoulder-mounted spreaders. Fertilizer was applied at each site within one week of predicted rainfall in order to minimize variability in the dissolution of granules [18].

Non-crop vegetation suppression treatments (herbicides) were applied as part of the FERTVS treatments by backpack sprayers in March 2004 (pre-thin) and March 2005 (post-thin). A 20% solution of triclopyr mixed with diesel as a penetrant was applied as a basal bark spray to suppress hardwood saplings and shrubs. A 5% solution of glyphosate mixed with water was used to suppress herbaceous vegetation and hardwood sprouts. These herbicide mixtures were re-applied as necessary for the remainder of each year in order to prevent regrowth of competing vegetation. Understory biomass was reduced by 95.4, 51.4, and 71.4% at the Dodson, Lucky, and Oakdale sites, respectively, in the FERTVS treatment relative to the CONTROL and FERT treatments [7].

Thinning was conducted operationally using feller-bunchers and skidders in October 2004 for the Dodson and Oakdale sites and in November 2004 for the Lucky site. Thinning was conducted as a fifth-row removal thinning from below at each site, with crown biomass returned to down rows as conventionally conducted in operational thinning. Crown biomass was separated from boles by gate de-limber at the logging deck, and skidders were used to transport and deposit the crown biomass along down rows as the skidders returned into the stand to continue tree felling. A target residual density of 494 trees·ha^{-1} was created for each plot by selecting and marking residual trees after down rows were cut in the plantations. Residual trees were selected on the basis of dominant or codominant height and DBH, good form, and spacing. Spacing between trees within rows was approximately 6 m, although this spacing varied depending on availability of trees of appropriate height, DBH, and form within 6 m of each tree. After marking residual trees between down rows, the areas between down rows were thinned by feller-bunchers. Achieved average residual densities at the Dodson, Lucky, and Oakdale sites were 421, 420, and 466 trees·ha^{-1}, respectively. Post-thinning basal areas were respectively 11.2, 8.5, and 8.9 m^{2}·ha^{-1} for the Dodson, Lucky, and Oakdale sites.

2.3. Nitrogen Mineralization and Exchangeable N

Net N mineralization (N_{min}) was measured using the *in situ* buried bag method [19] as described for a sweetgum (*Liquidambar styraciflua* L.) plantation by Scott *et al.* [9]. Soil samples for N_{min} were collected to a depth of 20 cm using punch augers at two randomly located subsample points within the central 0.08 ha of each plot. Sampling occurred every four to six weeks, with the six-week incubation durations occurring in fall and winter months from January 2004 through November 2005. For each sample period, two composite samples were taken at each sample point. Each composite sample was comprised of three punch auger samples; organic matter was removed from the soil surface immediately prior to sampling. One composite sample at each sample point was used for *in situ* incubation, and the other was used for pre-incubation analysis. Each composited sample for incubation was used to fill sterile 14 cm × 23 cm sampling bags with round-wire closure. The bags had a 0.08 mm thickness to allow gas exchange while retaining moisture. After bags were filled with soil, any surface organic matter was moved aside and a small hole was dug to a depth of 20 cm in the center of the three sampling points used to create the composite sample. The incubation bag was then placed in the hole, and any surface organic matter was replaced. Samples were incubated until the subsequent sample period. All pre- and post-incubation samples were stored in a cooler for transport to the laboratory, and the samples were then air-dried to constant weight and sieved to pass a 2 mm mesh for preparation of extraction of NH_4^+-N and NO_3^--N to be used for determination of N_{min}. Soil NH_4^+-N and NO_3^--N were extracted using a 2 M solution of KCl at a 10:1 ratio of solution to soil [20]. Concentrations of NH_4^+-N and NO_3^--N were quantified by the microdiffusion-conductivity method [21] using a Bran Luebbe AutoAnalyzer 3 (SPX Equipment, Delavan, WI, USA).

Net N mineralization was calculated as the difference in mineral N (mg·N·kg^{-1} soil) following incubation and prior to incubation $(NH_4\text{-N} + NO_3\text{-N})_{t+1} - (NH_4\text{-N} + NO_3\text{-N})_t$ [22]. The N_{min} rates for each incubation period were adjusted to a one-month (30-day) basis by dividing net N_{min} by the number of days in the incubation period and then multiplying by 30. Annual N_{min} was estimated by summing N_{min} measurements for all incubation periods for each plot, dividing by the total days of incubation for the year, and then multiplying by 365. Bulk density samples were collected to a 20-cm depth in all plots with a 5-cm diameter soil core sampler in spring 2004, and these measurements were used to convert annualized N_{min} to a kg·ha^{-1} basis.

The NH_4-N and NO_3-N concentrations of the pre-incubation bags were used to determine changes in exchangeable N within and between the years of this study. In addition to NH_4-N and NO_3-N concentrations, total exchangeable N was assessed as the sum of NH_4-N and NO_3-N. The ratio of NO_3-N to total exchangeable N was calculated for the pre-incubation bag samples as well.

2.4. Soil Microbial Biomass C, Dehydrogenase Activity

Soil samples used for assays of soil microbial biomass C (C_{mic}) and microbial activity were collected to a 20 cm depth with punch augers. Sampling for these assays was conducted in tandem with some dates at which N_{min} samples were collected using sampling and compositing protocol similar to that described above for N_{min}. In 2004, soil samples for which C_{mic} and microbial activity assays were conducted were collected in April, May, July, and September. In 2005, samples for which C_{mic} was assayed were collected monthly from March through September. Microbial activity assays were also conducted for the May, June, and September 2005 soil samples. All samples were refrigerated at 5 °C until analyses, which was initiated within 14 days of sampling.

The chloroform fumigation-incubation method was used to determine C_{mic} [23,24]. A 10-day pre-incubation was carried out prior to fumigation of soil samples in order to allow the influence of soil disturbance to subside and living fragments of roots to die [25,26]. Microbial biomass C was determined by fumigating soil samples with alcohol-free $CHCl_3$ vapor for 24 h, incubating soil samples at 25 °C for 10 days, collecting respired CO_2 in 2 M NaOH, measuring CO_2-C by titration of the NaOH with 0.1 N HCl, and using a proportionality constant of 0.45 to convert CO_2-C to C_{mic} after subtracting the CO_2-C produced by a non-fumigated control [23,24]. All results are expressed on an oven-dry soil basis (105 °C, 24 h).

Microbial activity was estimated by determining dehydrogenase activity [27,28]. Dehydrogenase, which is only active in viable living cells, serves as an indicator of total microbial metabolic activity [29,30]. To quantify dehydrogenase, triphenyltetrazolium chloride (TTC) was used as an artificial electron acceptor. Dehydrogenase reduces TTC to red-colored triphenyl formazan (TPF) that can be extracted with methanol and quantified colorimetrically [29]. Results are expressed as μg TPF kg^{-1} soil on an oven-dry soil basis (105 °C, 24 h).

2.5. Soil Labile C

Labile C was measured using the sequential fumigation-incubation procedure [31] for soil samples collected in September 2005 as described above for C_{mic}. The September sample was selected for this assay because above- and below-ground plant biomass is near its seasonal peak at that time. Soil samples were fumigated with alcohol-free $CHCl_3$ vapor for 24 h, and respired CO_2 was measured as described above for the chloroform fumigation-incubation method for C_{mic} measurement for a series of eight 10-day fumigation-incubation cycles. The method is based on the assumption that soil labile C is decayed during these cycles according to the negative exponential equation as widely reported in plant decomposition studies, so labile C was estimated from CO_2 measurements using a first-order kinetics model [31].

2.6. Soil Carbon and Nitrogen

In April 2005, five (for the Lucky site) and six (for the Dodson and Oakdale sites) months after thinning, soil samples were collected from all plots to characterize C and N concentrations of surface soil. Ten samples were randomly sampled in the central 0.08 ha of each plot to a 20 cm depth using punch augers; the ten samples were composited for each plot. Organic matter was removed from the soil surface prior to sampling. Carbon and nitrogen concentrations were measured by combustion using a LECO CNS 2000 analyzer (Leco, Inc., St. Joseph, MI, USA).

2.7. Soil Temperature and Volumetric Water Content

Soil temperature and volumetric water content (VWC) measurements in the top 20 cm of soil were collected concurrently with all soil sampling for soil N and microbial assays using a K-type thermocouple thermometer (Hanna Instruments, Inc., Woonsocket, RI, USA) and a TDR 200 time-domain reflectometry probe (Spectrum Technologies, Inc., Paxinos, PA, USA), respectively. Soil temperature and water content was measured within 10 cm of all samples collected for soil N and microbial assays.

2.8. Environmental Conditions

Precipitation and temperature trends for the regions in which all study sites were located were obtained from the Historical Climate Trends Tool of the Southern Climate Impacts Planning Program (SCIPP) [32]. This database facilitates comparison of average monthly precipitation and temperature trends relative to long-term (1971–2000) average trends for each month. The Lucky and Dodson sites were in the North Central Louisiana SCIPP climate region, and the Oakdale site was in the Southwest Louisiana SCIPP climate region.

2.9. Statistical Analysis

All treatment effects were analyzed for variance (ANOVA) at $\alpha = 0.05$ using the MIXED procedure of the SAS System (SAS Institute, Inc., Cary, NC, USA). When the null model likelihood ratio test revealed heterogeneous variances in a dataset, the GROUP option of MIXED was utilized to perform ANOVA using different variances for all treatment combinations. When an ANOVA indicated significant treatment effects, treatment means were calculated and separated by the DIFF and SLICE options of the LSMEANS procedure. The DIFF option provided multiple comparisons of treatment means by invoking *t*-tests to determine significant differences between all possible treatment combinations. The SLICE option provided *t*-tests of treatment means in which the effect of one treatment is evaluated at each level of another treatment. The SLICE option was used to investigate treatment main effects when significant two-way interactions were observed.

Models used in analyses of parameters measured in this study differed depending on sampling frequency. Analyses of N_{min}, NH_4-N, NO_3-N, total exchangeable N, ratio of NO_3-N to total exchangeable N, soil volumetric water content, and soil temperature were each performed with a model that included these parameters as a dependent variable and block, treatment, sampling date, and the interaction of treatment and date as independent variables. All independent variables were considered fixed effects. The model was a repeated measures model with an autoregressive correlation structure. These analyses were performed by site and year because of differences in sampling dates for each site and year. There were two numerator degrees of freedom for block and treatment associated with the model for all sites and both years. Due to sampling date differences for each site and year, there were differences in date and treatment × date numerator degrees of freedom associated with the model. For the Dodson and Oakdale sites in 2004 there were seven numerator degrees of freedom and 14 numerator degrees of freedom for date and treatment × date, respectively. For the Lucky site in 2004, the model was associated with six numerator degrees of freedom for date and 12 numerator degrees of freedom for treatment × date. For the Lucky and Oakdale sites in 2005, the model was associated with eight numerator degrees of freedom for date and 16 numerator degrees of freedom for treatment × date. The model had nine and 18 numerator degrees of freedom, respectively, for date and treatment × date for the Dodson site in 2005.

Microbial biomass C and dehydrogenase activity were sampled at the same dates for each site within both years of the study, but there were differences in the number of sampling dates between 2004 and 2005. As such, analyses of these parameters were performed by year. The model used in analyses included C_{mic} and dehydrogenase as dependent variables and block, treatment, site, sampling date, and all possible interactions of treatment, site, and sampling date as independent variables. All independent variables were fixed effects. The model was a repeated measures model with an autoregressive correlation structure. For C_{mic} and dehydrogenase in both years, numerator degrees of freedom associated with the model were two, two, two, and four for block, site, treatment, and site × treatment, respectively. For C_{mic} and dehydrogenase activity in 2004, numerator degrees of freedom associated with the model were six, six, and 12 for site × sampling date, treatment × sampling date, and site × sampling date × treatment, respectively. There were eight, eight, and 16 numerator degrees of freedom associated with site × sampling date, treatment × sampling date, and site × sampling date × treatment, respectively, for the model for C_{mic} in 2005. For dehydrogenase activity in 2005 there were four, four, and eight numerator degrees of freedom for site × month, treatment × sampling date, and site × sampling date × treatment, respectively, associated with the model.

Annualized N_{min} was measured over the same sample period for each site in both years of the study. As such, the model used in its analysis included annualized N_{min} as a dependent variable and block, site, treatment, year, and all possible interactions between site, treatment, and year as independent variables. All independent variables were fixed effects, and the model was a repeated measures model with an autoregressive correlation structure. Block, site, treatment, site × year, and treatment × year were each associated with two numerator degrees of freedom in analysis. Site × treatment and site × treatment × year were each associated with four numerator degrees of freedom, and year was associated with one numerator degree of freedom in analysis.

Labile C, soil C concentrations, soil N concentrations, and soil C:N ratios were each measured a single time at each site in this study. The model used in their analyses therefore included block, site, treatment, and site × treatment as independent variables, which were all fixed effects. For labile C (which was sampled only in CONTROL and post-thin FERT and FERTVS plots), block, site, and treatment each had two numerator degrees of freedom in analysis, and site × treatment had four numerator degrees of freedom. For C, N, and C:N ratios (which was sampled post-thinning in all plots) block and site each had two numerator degrees of freedom in analysis, treatment had four numerator degrees of freedom, and site × treatment had eight numerator degrees of freedom.

To provide greater insight into how thinning affected N_{min}, exchangeable N, C_{mic}, dehydrogenase activity, soil temperature, and soil moisture, cross-year comparisons were performed using data from CONTROL plots. CONTROL plot data from common sample dates for 2004 and 2005 of all sites were pooled. Analyses were performed using a repeated measures model with an autoregressive correlation structure that included block, site, year, and the interaction of site and year as fixed effects. Block, site, and site × year each had two numerator degrees of freedom in analysis, and year was associated with one numerator degree of freedom.

Correlations among N_{min}, volumetric soil moisture, and soil temperature were quantified for each site using the PROC CORR procedure of the SAS System, which produced Pearson correlation coefficients and the probabilities associated with these statistics. For each site, correlations were also determined between N_{min}, C_{mic}, and dehydrogenase activity for months in which all variables were measured.

3. Results

3.1. Environmental and Edaphic Conditions

Temperature trends for the regions of Louisiana in which the study sites were located were similar to 30-year trends in both years of the study. In both regions, 2004 precipitation exceeded the 30-year average by approximately 25% whereas precipitation for 2005 was 40 to 50% below the 30-year average. Deviations from the long-term trend were greatest in mid-summer and fall each year, with 2004 precipitation higher than the 30-year average and 2005 precipitation lower than the 30-year average during these months.

Soil temperature was similar among treatments at all sites prior to thinning in 2004. After thinning, the CONTROL treatment soils were about 2–3 °C warmer than soils in the FERT and FERTVS treatments throughout most of the summer and early autumn at the Dodson ($P = 0.001$) and Lucky sites ($P = 0.002$) (Figure 1). Soil temperatures were similar ($P = 0.88$) among all treatments at Oakdale. In the cross-year comparison of CONTROL plot data, across all sites soil temperature was significantly ($P = 0.03$) warmer (2–3 °C) in the CONTROL plots post-thinning than pre-thinning.

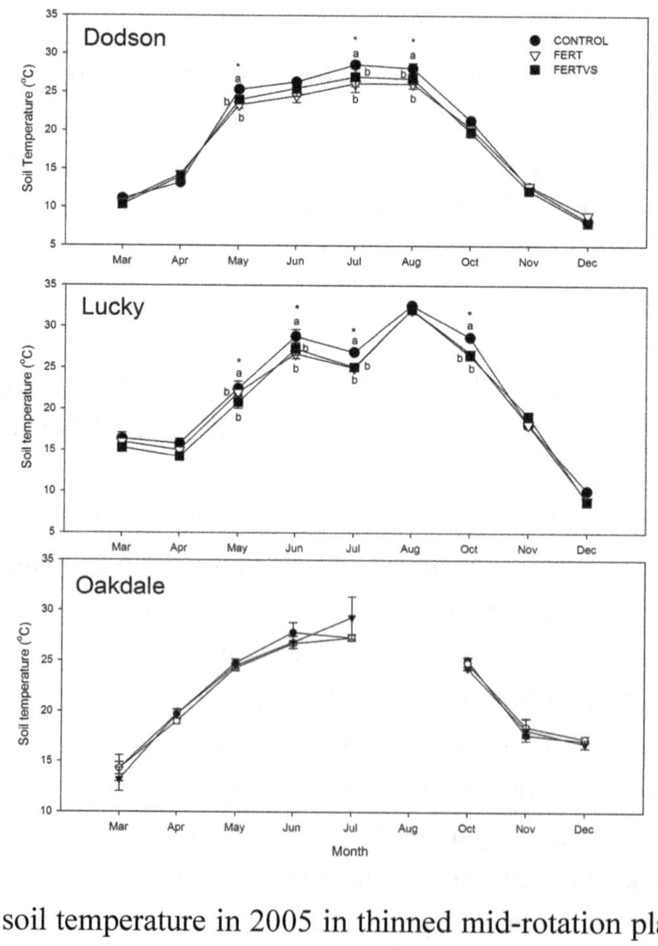

Figure 1. Monthly soil temperature in 2005 in thinned mid-rotation plantations (identified by the name of the nearest city) in north central (Dodson, Lucky) and southwest (Oakdale) Louisiana, USA, in response to no fertilization or herbicide (CONTROL), fertilization (FERT), and fertilization and vegetation suppression (FERTVS) treatments. The plantations were thinned in October and November 2004. For months headed by an asterisk, symbols noted by different letters differ at $P < 0.05$.

Fertilized soils were generally drier at the Dodson and Lucky sites but unaffected at the Oakdale site. In 2004, there was a significant (P = 0.01) treatment effect at the Lucky site, at which VWC of the CONTROL treatment (mean 11.5% v/v) was 40% wetter than that of the FERT (mean 6.4% v/v) and FERTVS (mean 7.7% v/v) treatments. Similarly, the CONTROL soils at Dodson were wetter than the FERT and FERTVS treatments in 2004, but only significantly in March and December (*i.e.*, there was a significant (P = 0.04) treatment × date interaction). At the Dodson site in 2005 there was a significant (P = 0.03) treatment effect; the CONTROL (mean 16.9% v/v) and FERTVS (mean 17.2% v/v) treatments had greater VWC than the FERT (mean 13.5% v/v) treatment. On average, soils were the wettest at Oakdale (25.1% v/v), moderate at Dodson (18.9% v/v), and driest at Lucky (12.0% v/v).

3.2. Nitrogen Mineralization and Exchangeable N

Monthly N_{min} differed among treatments at one or more sites in both years of the study (Figure 2). At Dodson in 2004, N_{min} in the FERTVS soils was more than twice that of the CONTROL and FERT soils (P = 0.04). No differences in monthly N_{min} were observed at the Lucky (P = 0.71) and Oakdale (P = 0.57) sites in 2004. In 2005, the FERT and FERTVS treatment soils had nearly double the average monthly N_{min} than the CONTROL treatment soils at the Dodson (P = 0.02) and Oakdale (P = 0.02) sites. No differences (P = 0.66) in monthly N_{min} among treatments were observed at the Lucky site in 2005.

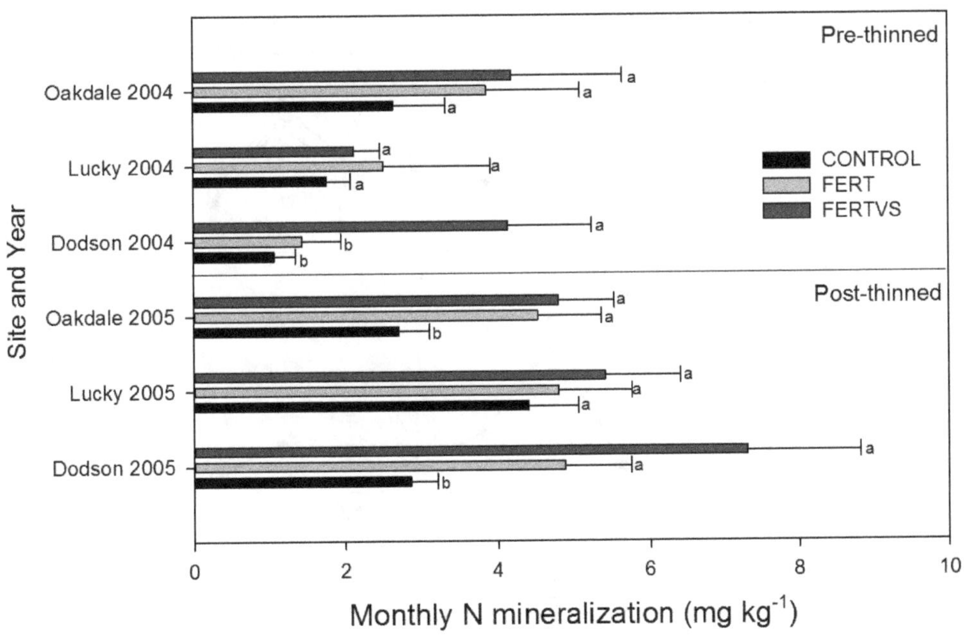

Figure 2. Average monthly N mineralization of non-thinned and thinned mid-rotation loblolly pine plantations in north central (Dodson, Lucky) and southwest Louisiana (Oakdale), USA, in response to no fertilization or herbicide (CONTROL), fertilization (FERT), and fertilization and vegetation suppression (FERTVS) treatments. Plantations were thinned in October and November 2004. For each site and year, columns headed by a different letter differ at $P < 0.05$.

Across all sites and years of the study, FERT (mean 113.5 kg·ha^{-1}·year^{-1}) and FERTVS (mean 146.3 kg·ha^{-1}·year^{-1}) treatments had greater annualized N_{min} than the CONTROL (mean

81.1 kg·ha^{-1}·year^{-1}) treatment; annualized N_{min} of the FERTVS treatment was also greater than that of the FERT treatment ($P = 0.004$). Annualized N_{min} was similar ($P = 0.98$) among sites. Across all sites, annual N_{min} significantly differed ($P < 0.0001$) between years of the study, with annual N_{min} nearly twice as high after thinning (mean 142.3 kg·ha^{-1}·year^{-1}) as prior to thinning (mean 85.0 kg·ha^{-1}·year^{-1}).

Soil NO_3-N concentrations were different among treatments at one site or more in both years of the study. Concentrations of NO_3-N differed at only the Oakdale site in 2004, for which there was a significant ($P = 0.004$) treatment × date interaction (Figure 3). Soil NO_3-N concentrations were greater for FERTVS treatment than the other treatments in May, June, and August. The FERT treatment had greater NO_3-N than that of the CONTROL treatment in May and June. A significant treatment × date effect ($P = 0.01$) was also observed for the Oakdale site for 2005 (Figure 4). In 2005, NO_3-N of the FERTVS treatment exceeded that of the CONTROL and FERT treatments in all months observed after April. The FERT treatment had greater NO_3-N concentrations than the CONTROL treatment in all months observed after April. Significant treatment effects were found for the Dodson ($P = 0.0003$) and Lucky ($P = 0.004$) sites in analyses of NO_3-N for 2005 (Table 2). At the Dodson site the FERTVS treatment had the highest NO_3-N concentrations, and the FERT treatment had greater NO_3-N concentrations than the CONTROL treatment. The FERT and FERTVS treatments had greater NO_3-N concentrations than the CONTROL treatment at the Lucky site.

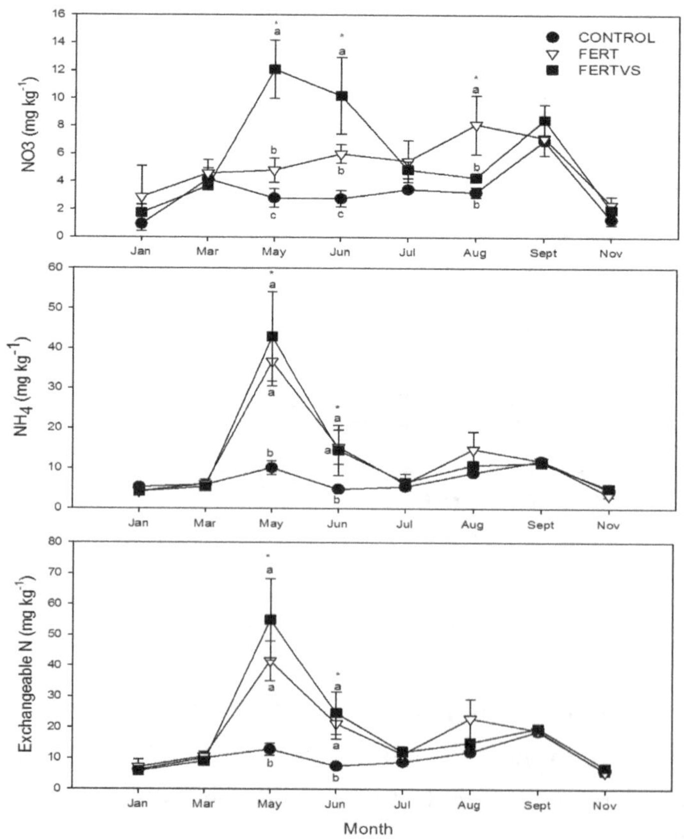

Figure 3. Monthly NO_3-N, NH_4-N, and total exchangeable N concentrations of a non-thinned mid-rotation loblolly pine plantation in 2004 at the Oakdale site in southwest Louisiana, USA, in response to no fertilization or herbicide (CONTROL), fertilization (FERT), and fertilization and vegetation suppression (FERTVS) treatments. For months headed by an asterisk, symbols noted by different letters differ at $P < 0.05$.

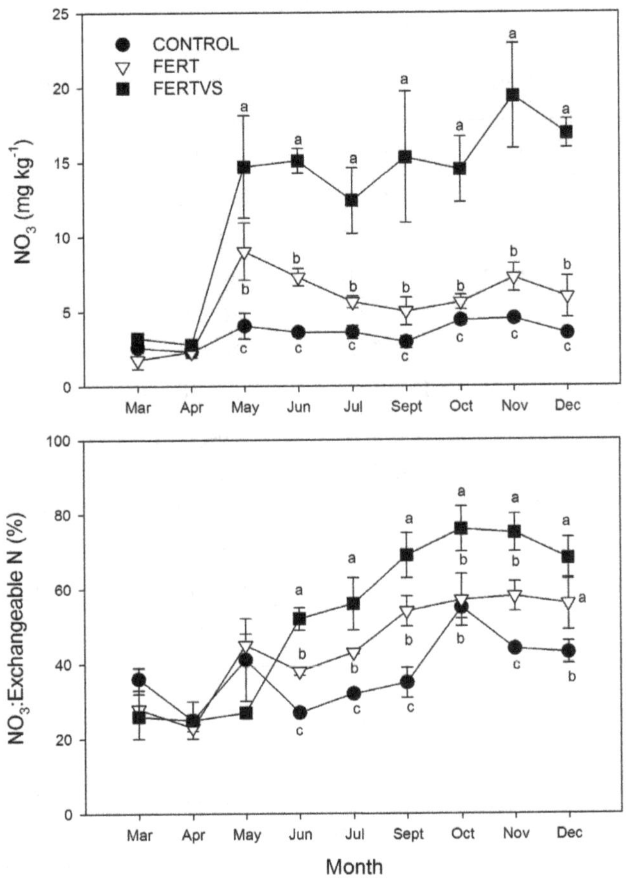

Figure 4. Monthly NO_3-N concentrations and ratios of NO_3 to total exchangeable N of a thinned mid-rotation loblolly pine plantation in 2005 at the Oakdale site in southwest Louisiana, USA, in response to no fertilization or herbicide (CONTROL), fertilization (FERT), and fertilization and vegetation suppression (FERTVS) treatments. For months headed by an asterisk, symbols noted by different letters differ at $P < 0.05$.

Table 2. Soil exchangeable N of thinned mid-rotation loblolly pine plantations in 2004 and 2005 in north central (Dodson, Lucky) and southwest Louisiana (Oakdale), USA in response to no fertilization or herbicide (CONTROL), fertilization (FERT), and fertilization and vegetation suppression (FERTVS) treatments. Means within a row followed by a different letter differ at $P < 0.05$; rows in which no letters are provided had a significant interaction between treatment and another parameter in analyses of variance. Values in parentheses are standard errors.

Exchangeable N	Year	Site	Treatment		
			CONTROL	FERT	FERTVS
NO_3-N ($mg \cdot kg^{-1}$)	2004	Dodson	3.4 (0.6) *a*	3.7 (0.7) *a*	3.9 (0.8) *a*
		Lucky	2.5 (0.3) *a*	2.9 (0.5) *a*	2.5 (0.3) *a*
		Oakdale	3.2 (0.4)	5.2 (0.6)	5.7 (0.8)
	2005	Dodson	3.3 (0.2) *c*	4.4 (0.4) *b*	8.4 (0.8) *a*
		Lucky	2.9 (0.1) *b*	4.2 (0.4) *a*	5.6 (0.6) *a*
		Oakdale	10.0 (0.6)	12.5 (1.0)	22.2 (1.8)

Table 2. *Cont.*

Exchangeable N	Year	Site	Treatment		
			CONTROL	FERT	FERTVS
NH_4-N (mg·kg^{-1})	2004	Dodson	5.4 (0.6) *a*	7.3 (1.3) *a*	7.1 (1.3) *a*
		Lucky	6.5 (1.5)	16.5 (6.9)	11.0 (3.0)
		Oakdale	7.3 (0.7)	11.9 (2.2)	11.1 (2.5)
	2005	Dodson	7.6 (1.0) *b*	11.5 (2.5) *ab*	12.7 (1.5) *a*
		Lucky	5.6 (0.6)	8.3 (1.3)	8.5 (1.4)
		Oakdale	6.6 (0.6) *c*	7.0 (0.8) *b*	9.5 (1.1) *a*
Total exchangeable N (mg·kg^{-1})	2004	Dodson	8.8 (1.0) *a*	10.9 (1.7) *a*	11.0 (1.8) *a*
		Lucky	9.0 (1.7)	19.4 (7.0)	13.4 (3.1)
		Oakdale	10.5 (1.0)	17.1 (2.4)	16.8 (3.1)
	2005	Dodson	10.9 (1.0) *b*	15.9 (2.7) *ab*	21.0 (1.8) *a*
		Lucky	8.6 (0.6)	12.5 (1.4)	14.1 (1.7)
		Oakdale	10.0 (0.7) *c*	12.5 (1.0) *b*	22.2 (1.8) *a*
NO_3: total exchangeable N (%)	2004	Dodson	37.9 (2.9) *a*	35.6 (2.9) *a*	35.7 (2.8) *a*
		Lucky	33.9 (2.2) *a*	28.4 (3.9) *a*	27.3 (3.0) *a*
		Oakdale	28.8 (2.3) *a*	34.9 (2.8) *a*	36.3 (2.9) *a*
	2005	Dodson	35.5 (2.6)	36.3 (2.6)	42.5 (3.2)
		Lucky	37.0 (2.4) *b*	37.5 (2.1) *b*	43.3 (2.4) *a*
		Oakdale	35.8 (1.9)	44.2 (2.9)	54.8 (3.8)

Ratios of NO_3-N to total exchangeable N were different among treatments in both years at the Oakdale site and at the Lucky site in 2005. There was a significant treatment effect for the Oakdale site in 2004, with the FERT (mean 34.0%) and FERTVS (mean 35.0%) treatments had a higher (P = 0.04) NO_3:exchangeable N ratio than the CONTROL treatment (mean 29.7%). In 2005, there was a significant treatment × date effect (P = 0.01) for the Oakdale site (Figure 4). In all months observed after May, the FERTVS treatment had a higher ratio than the CONTROL. The FERTVS treatment also had a higher ratio than the FERT treatment in all months observed between June and November except October. The FERT treatment had a greater NO_3:exchangeable N ratio than the CONTROL treatment in June, July, September, and November. At the Lucky site there was a significant treatment effect (P = 0.02), with the FERTVS treatment having a higher ratio than the FERT and CONTROL treatments (Table 2).

In 2004, there were significant treatment × date interactions for the Oakdale ($P < 0.0001$) and Lucky ($P < 0.0001$) sites in analyses of NH_4-N (Figures 3 and 5). At Oakdale, the FERT and FERTVS treatments had greater NH_4-N concentrations than the CONTROL treatment in May and June (Figure 3). At the Lucky site differences were observed among treatments in only April, in which the FERT treatment had NH_4-N concentrations that exceeded those of the CONTROL and FERTVS treatments (Figure 5). Soil NH_4-N concentrations of the FERTVS treatment were significantly greater than those of the CONTROL treatment in that month as well.

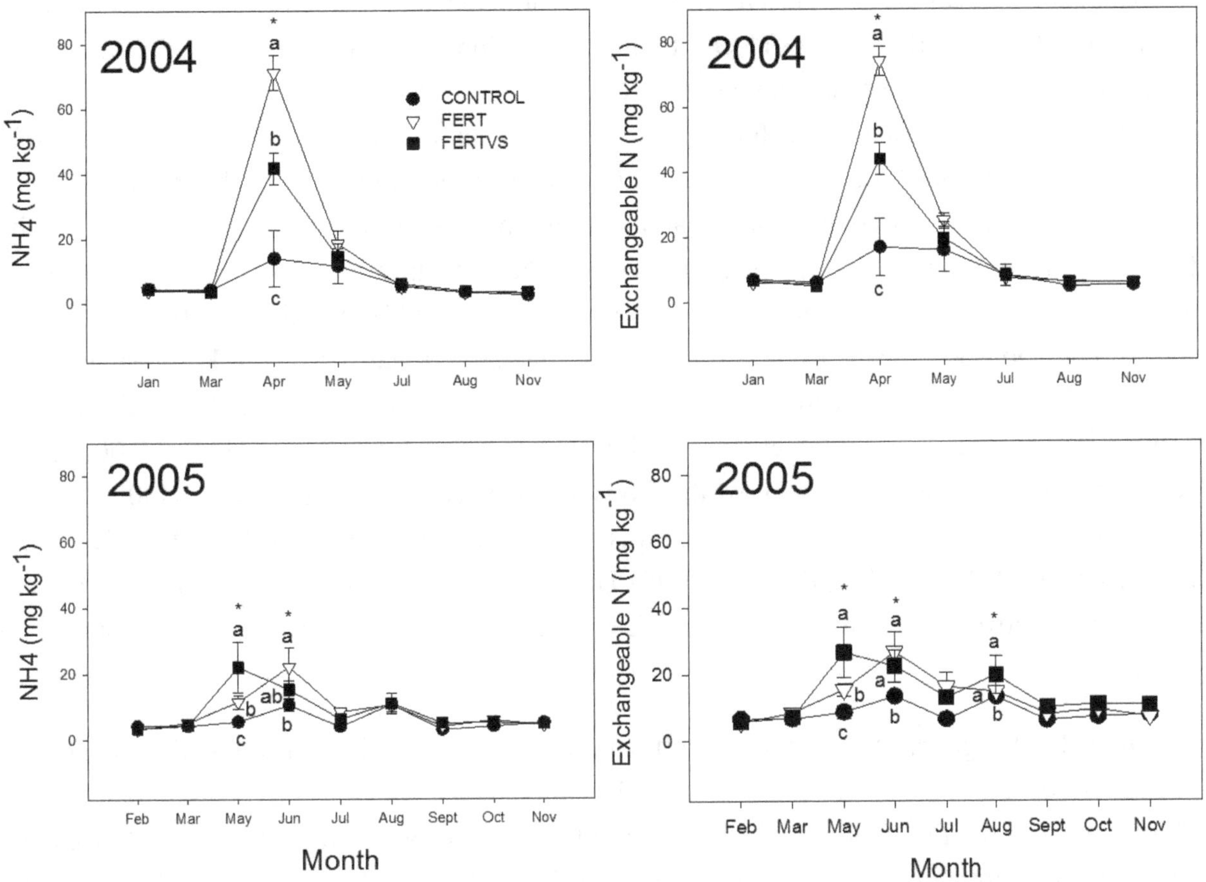

Figure 5. Monthly NH_4-N and total exchangeable N concentrations of a mid-rotation loblolly pine plantation at the Lucky site in north central Louisiana, USA, that was thinned in November 2004 in response to no fertilization or herbicide (CONTROL), fertilization (FERT), and fertilization and vegetation suppression (FERTVS) treatments. For months headed by an asterisk, symbols noted by different letters differ at $P < 0.05$.

Soil NH_4-N concentrations differed among treatments at all sites in 2005. The treatment effect was significant for the Dodson ($P = 0.003$) and Oakdale ($P = 0.03$) sites (Table 2), and the treatment × date interaction was significant ($P = 0.03$) for the Lucky site (Figure 5). At the Dodson site, the FERTVS treatment had greater NH_4-N concentrations than the CONTROL treatment (Table 2). At the Oakdale site, NH_4-N concentrations of the FERTVS treatment were higher than those of the other two treatments and NH_4-N of the FERT treatment greater than that of the CONTROL treatment (Table 2). This treatment ranking was also observed for the Lucky site in May, and in June the FERTVS treatment had greater NH_4-N than the CONTROL treatment (Figure 5).

Significant treatment × date interactions were observed in analyses of total exchangeable N for the Lucky ($P < 0.0001$) and Oakdale ($P < 0.0001$) sites for 2004 (Figures 3 and 5). At the Lucky site, treatments differed as described above for NH_4-N of 2004 (Figure 5). Differences among treatments were similarly identical to those described above for NH_4-N of 2004 at the Oakdale site (Figure 3).

For the Dodson ($P = 0.0002$) and Oakdale ($P < 0.0001$) sites, significant treatment effects were found in analyses of total exchangeable N for 2005 (Table 2). Total exchangeable N differences among treatments at both sites were identical to those described above for NH_4-N. At the Lucky site there was a significant ($P = 0.03$) treatment × date interaction (Figure 5). In May the FERTVS treatment had greater total exchangeable N than the CONTROL and FERT treatments, and total exchangeable N of the FERT treatment exceeded that of the CONTROL treatment. In June and August the FERT and FERTVS treatments had greater total exchangeable N than the CONTROL treatment.

All exchangeable N variables differed between years in the CONTROL treatment (Table 3). Soil NO_3-N concentrations of 2005 were greater ($P = 0.002$) than those of 2004 across all sites. Sites differed ($P = 0.03$) in NO_3-N concentrations across years in the CONTROL treatment as well, with the Oakdale site (mean 5.6 $mg \cdot kg^{-1}$) having greater NO_3-N concentrations than the Lucky site (mean 3.4 $mg \cdot kg^{-1}$). Total exchangeable N in 2005 was greater ($P = 0.005$) than that of 2004 across all sites (Table 3). The Dodson (mean 14.1 $mg \cdot kg^{-1}$) and Oakdale (mean 15.1 $mg \cdot kg^{-1}$) sites had total exchangeable N concentrations across years greater ($P = 0.04$) than those of the Lucky site (mean 11.1 $mg \cdot kg^{-1}$). The site × year interaction was significant in analyses of NH_4-N ($P = 0.0001$) and the NO_3: exchangeable N ratio ($P = 0.002$). At the Dodson site NH_4-N concentrations of 2005 were greater than those of 2004, but an opposite trend was observed for NH_4-N of the Oakdale site, with concentrations of 2004 exceeding those of 2005. The NO_3:exchangeable N ratio was different at only the Oakdale site between years in the CONTROL treatment. At that site, the ratio was greater for 2005 than 2004.

Table 3. Exchangeable N in 2004 and 2005 of mid-rotation loblolly pine plantations in north central (Dodson, Lucky) and southwest Louisiana (Oakdale) in CONTROL treatment of no fertilizer or herbicide. Plantations were thinned in October and November 2004. For each site, means within a row followed by a different letter differ at $P < 0.05$. All values except NO_3:exchangeable N ratio are in $mg \cdot kg^{-1}$.

	Site					
	Dodson		Lucky		Oakdale	
	2004	2005	2004	2005	2004	2005
NO_3-N	3.9 *b*	5.3 *a*	2.5 *b*	4.2 *a*	4.1 *b*	7.2 *a*
NH_4-N	6.5 *b*	12.4 *a*	6.7 *a*	8.5 *a*	11.0 *a*	7.6 *b*
Total exchangeable N	10.6 *b*	17.5 *a*	9.3 *b*	12.8 *a*	15.0 *b*	15.1 *a*
NO_3:exchangeable N	0.40 *a*	0.36 *a*	0.33 *a*	0.37 *a*	0.34 *b*	0.47 *a*

3.3. Soil Microbial Biomass C, Dehydrogenase Activity, and Labile C

Overall, soil microbial biomass C and dehydrogenase activity were relatively unaffected by site, treatment, or year. In only April 2004 ($P = 0.0004$), the FERT (mean 443.5 $mg \cdot kg^{-1}$) and FERTVS (mean 464.8 $mg \cdot kg^{-1}$) treatments had greater C_{mic} than the CONTROL treatment (mean 373.1 $mg \cdot kg^{-1}$). The FERT (mean 30.1 $mg \cdot kg^{-1}$) and FERTVS (mean 37.8 $mg \cdot kg^{-1}$) treatments had greater ($P = 0.003$) dehydrogenase activity than the CONTROL treatment (mean 13.6 $mg \cdot kg^{-1}$) in only April 2004 as well. Monthly average C_{mic} ranged from 126.9 to 460.8 $mg \cdot kg^{-1}$ over the years of the study, and monthly average dehydrogenase activity ranged from 13.4 to 77.1 $mg \cdot kg^{-1}$.

The site × treatment interaction was significant ($P = 0.02$) for the labile C analysis (Table 4). The CONTROL treatment had 6 and 10% greater labile C than the FERTVS treatment at the Dodson and Lucky sites, respectively. At the Oakdale site the FERTVS soils had about 6% greater labile C than the CONTROL and FERT soils.

Table 4. Labile C ($mg \cdot kg^{-1}$) of thinned mid-rotation loblolly pine plantations during 2005 in north central (Dodson, Lucky) and southwest Louisiana (Oakdale), USA in response to no fertilization or herbicide (CONTROL), fertilization (FERT), and fertilization and vegetation suppression (FERTVS) treatments. Means within a row followed by a different letter differ at $P < 0.05$; values in parentheses are standard errors.

Site	CONTROL	FERT	FERTVS
Dodson	6.23 (0.07) *a*	6.18 (0.31) *ab*	5.88 (0.05) *b*
Lucky	6.33 (0.09) *a*	6.16 (0.14) *ab*	5.78 (0.13) *b*
Oakdale	5.97 (0.14) *b*	5.89 (0.08) *b*	6.30 (0.05) *a*

3.4. Soil C and N

No differences were observed among treatments in soil C, N, and C:N ratios, but differences ($P = 0.04$, 0.0002, <0.0001, respectively) were observed among sites for each of these parameters (Table 5). The Dodson site had higher C concentrations than the Oakdale site. The Oakdale site had greater N concentrations than both other sites, and the Dodson site had greater N concentrations than the Lucky site. The Lucky site had the highest C:N ratio of the sites, while the Oakdale site had the lowest C:N ratio.

Table 5. Soil C and N concentrations and C:N ratios of thinned mid-rotation loblolly pine plantations in 2005 in north central (Dodson, Lucky) and southwest Louisiana (Oakdale), USA. Means within a row differ at $P < 0.05$; values in parentheses are standard errors.

	Site		
Soil Attribute	**Dodson**	**Lucky**	**Oakdale**
C ($mg \cdot kg^{-1}$)	13,138 (676) *a*	12,153 (813) *ab*	10,554 (676) *b*
N ($mg \cdot kg^{-1}$)	557 (23) *b*	465 (29) *c*	658 (23) *a*
C:N ratio	23.5 (0.66) *b*	26.5 (0.79) *a*	16.1 (0.66) *c*

3.5. Correlations

Monthly N_{min} was significantly positively correlated ($r = 0.18$, $P = 0.03$) with soil temperature. Soil N_{min}, C_{mic}, and dehydrogenase activity were uncorrelated. Microbial biomass C was significantly positively correlated ($r = 0.21$, $P < 0.0001$) with VWC and negatively correlated ($r = -0.21$, $P < 0.0001$) with soil temperature. Dehydrogenase activity was positively correlated with soil temperature ($r = 0.33$, $P < 0.0001$) and C_{mic} ($r = 0.19$, $P = 0.002$).

4. Discussion

This study yielded N_{min} values within the range reported in previous studies of loblolly pine that used similar buried bag procedures. The 81 $kg \cdot ha^{-1}$ annual N_{min} mean of the CONTROL treatment in this study was similar to the 75 $kg \cdot ha^{-1}$ annual N_{min} reported for a 7-year-old loblolly pine plantation in Florida, USA on sandy loam soil [33]. The 81 to 146 $kg \cdot ha^{-1}$ annual N_{min} range of this study was somewhat higher than the 22 to 96 $kg \cdot ha^{-1}$ annual N_{min} range for a 14-year-old loblolly pine plantation in North Carolina, USA on a highly weathered, eroded clay loam receiving fertilizer and herbicide [2]. Across sites and years, annual N_{min} in this study was 6%, 8%, and 10% of total N measured in spring

2005 for the CONTROL, FERT, and FERTVS treatments, respectively. These proportions of annual N_{min} to total N are slightly higher than the 2%, 3%, and 6% for control, fertilizer, and fertilizer with herbicide treatments, respectively, conducted in a non-thinned 14-year-old loblolly pine plantation in North Carolina, USA [2]. However, proportions of annual N_{min} to total N of pre-thin CONTROL, FERT, and FERTVS treatments in this study were closer to those of the North Carolina loblolly pine plantation [2], respectively averaging 4%, 6%, and 8% across sites.

Annual N_{min} was increased by fertilization and to a greater extent by the combination of fertilization and vegetation control irrespective of timing relative to thinning and across a gradient of soil conditions. These results suggest that annual N supply can be increased by fertilization in the year of application in mid-rotation loblolly pine plantations prior to or after thinning. Increases in N_{min} after forest fertilization are well-demonstrated [2,9,10,12], but this study is the first to the authors' knowledge to show increases in N_{min} irrespective of timing relative to thinning. Similar to this study, combination treatments of fertilization and vegetation suppression increased annual N_{min} more than fertilization alone when conducted in an unthinned mid-rotation loblolly pine plantation with basal area comparable with that of the stands prior to thinning in this study [2].

The increased annual N_{min} in response to FERT and FERTVS treatments suggested that a priming effect occurred, but not all mechanisms conventionally defined for that effect occurred. The priming effect for N has been defined as extra release of soil-derived N in response to substances added to the soil [34]. The increased annual N_{min} of the FERT and FERTVS treatments are consistent with this definition. Soil microbes are often directly or indirectly responsible for priming effects, and increases in N_{min} after fertilization have been attributed to reduced microbial N immobilization when soil C:N ratios are below 30:1 [9,34]. All sites in this study had C:N ratios below 30:1, but the soil microbial parameters were relatively unaffected by treatments. The soil C_{mic} and dehydrogenase activity data was variable, so the procedures conducted may not have been sensitive enough to detect changes in soil microbial biomass and activity. There are also priming effect pathways that may function independent of soil microbes [34]. Labile C has been identified as an important parameter affecting the priming effect, particularly in response to forest management treatments such as vegetation control that alter organic matter inputs to soil. Such treatments can reduce labile C, which in turn reduces microbial N immobilization and increases N_{min} [35]. The greater annual N_{min} of the FERTVS treatment relative to the FERT and CONTROL treatments is consistent with increasing N_{min} in response to reducing vegetation C inputs. Labile C was lower for the FERTVS treatment relative to the CONTROL treatment at the Dodson and Lucky sites, but an opposite trend was observed at the Oakdale site. The contrasting trend at Oakdale may have been influenced by its under- and mid-story vegetation. Oakdale had sweetgum (*Liquidambar styraciflua* L.) greater than 8 cm in diameter as part of the mid-story, and it was resistant to the basal spray of triclopyr used in this study. Multiple applications of triclopyr were required to suppress sweetgum trees of that diameter. Oakdale also had abundant Chinese tallow tree that was resistant to triclopyr basal spray, requiring multiple applications. Chinese tallow tree sprouting also occurred once the mid-story was reduced, and the sprouts were relatively resistant to broadcast glyphosate and directed spray of triclopyr [7]. This herbicide resistance made suppression take several months, which may have affected labile C inputs.

As with fertilization and fertilization with vegetation control, thinning impacted annual N_{min} across all site conditions. Thinning nearly doubled annual N_{min} across all sites and treatments, although this comparison was made with non-thinned stands receiving higher than average annual precipitation and thinned stands that received lower than average annual precipitation. Studies of loblolly pine harvesting have demonstrated N_{min} increases after clearcutting, with N_{min} increases attributed to increased soil temperatures after harvest as well as mixing of organic matter with mineral soil during harvest [36,37]. Recently thinning was shown to increase N_{min} of *Cryptomeria japonica* forests, with N_{min} increases related to increases in soil temperature [38]. This study similarly observed that soil temperature was significantly increased by 2–3 °C after thinning across all sites and treatments. Soil temperature was also positively correlated with N_{min}.

Although annual N_{min} trends were similar across years and sites, monthly patterns in N_{min} were more affected by site conditions. Monthly N_{min} was relatively variable, which made differences among treatments, if present, difficult to detect. When fertilization was done prior to thinning, only at the Dodson site was a difference in average monthly N_{min} observed. At Dodson, a difference in monthly N_{min} was found only with the FERTVS treatment. Dodson had the highest understory biomass among all sites prior to thinning [7], which may have tempered the effect of the FERT treatment on monthly N_{min}. Average monthly N_{min} was increased by fertilization at both the Dodson and Oakdale sites. At each site average monthly N_{min} of the FERT and FERTVS treatments were similar; vegetation control may not have been as necessary in the year of thinning to foster increases in monthly N_{min}.

Monthly exchangeable N (NH_4-N, NO_3-N, total exchangeable N) trends were affected by timing of treatments and site conditions. The dominant form of exchangeable N was NH_4-N, so total exchangeable N trends were nearly identical to NH_4-N trends. Loblolly pine plantation soils are typically characterized by dominance of NH_4-N as exchangeable N [39]. Fertilization conducted prior to thinning had little effect on NH_4-N and total exchangeable N, with increases occurring only in the two months after fertilization at the two sites with lower soil N, Dodson and Lucky. When fertilization was done after thinning monthly NH_4-N and total exchangeable N concentrations increased after fertilization at all sites, but responses were different at each site. At the Lucky site, NH_4-N and total exchangeable N increases occurred only in the two months after fertilization. At Dodson, fertilization after thinning consistently increased monthly NH_4-N and total exchangeable N, but only when combined with vegetation suppression. Both fertilization treatments increased monthly NH_4-N at the Oakdale site, with greater increases occurring when combined with vegetation suppression. This gradient in NH_4-N and total exchangeable N increases across the sites was likely related to the gradient in total soil N and C:N ratios of the sites. Total soil N affects N transformations in soil, with transformations occurring more readily in soils with higher total soil N [40]. Nitrogen transformations in soil are also affected by C:N ratios, with transformations occurring more readily in soils with lower C:N ratios [40].

Monthly NO_3-N trends were affected by timing of fertilization and site conditions. Changes in monthly NO_3-N were relatively low when fertilization was conducted prior to thinning. Only at the Oakdale site were monthly NO_3-N concentrations elevated by fertilizer treatments, and the duration of response was the four-month period after application. As mentioned above, the higher soil N and lower

C:N ratio of the Oakdale site likely made it more prone to changes in N. Soil NO_3-N concentrations similarly increased in response to fertilization and fertilization and vegetation control in an unthinned mid-rotation loblolly pine plantation in North Carolina, USA [39]. Average monthly NO_3-N was changed by fertilization at all sites when conducted post-thinning. Both fertilization treatments increased NO_3-N at the Lucky site, and the FERTVS treatment increased monthly NO_3-N more than the FERT treatment at the Dodson and Oakdale sites.

The increases in NO_3-N and the ratio of NO_3:total exchangeable N have implications for how the timing of fertilization matched site N demand. Increases in soil NO_3-N concentrations after fertilization can be indicative of low plant uptake of applied N [41]. Increases in the proportion of NO_3-N in the total exchangeable N pool in soil are also an index of N saturation [12]. The increases in NO_3-N and NO_3:exchangeable N ratios at the Oakdale site when fertilized before and after thinning suggest that soil N was sufficient for the site to meet stand N demand, particularly after thinning. Foliage nutrient testing determined that foliage N concentrations at the Oakdale site were above critical values [7]. Soil N of the Oakdale site was likely enhanced by the presence of wax myrtle, a N-fixing shrub, in its understory [7]. Wax myrtle added 1.9 kg $N \cdot ha^{-1} \cdot year^{-1}$ in a slash pine (*Pinus elliotti* Engelm. var. *elliotti*) plantation in Florida, USA [42]. The increases in soil NO_3-N and the NO_3:exchangeable N ratio at the Lucky site and the increases in soil NO_3-N at the Dodson in response to FERTVS treatment conducted post-thinning suggest that this treatment exceeded stand N demand. These results show that avoiding fertilization pre- or post-thinning at an N-sufficient site and applying herbicide in the year of thinning and fertilization at the range of site conditions in this study could minimize NO_3-N concentration increases, which can help in reducing N leaching loss potential and improving fertilizer use efficiency [33].

5. Conclusions

For the range of site conditions in this study, fertilizing at mid-rotation stimulated increases in annual soil N supply irrespective of timing relative to thinning and the effect was more pronounced when vegetation control was conducted in concert with fertilization. This finding suggests management flexibility in the timing of fertilization at mid-rotation to increase soil N supply. However, site conditions were important considerations for minimizing the potential for NO_3-N leaching and matching stand N demand. The site with the highest total soil N and the lowest C:N ratio was more prone to NO_3-N increases after fertilization whether done prior to or after thinning. At all sites, combining vegetation suppression with fertilization promoted increases in NO_3-N when done after thinning. As such, combining vegetation suppression with fertilization in the year of thinning may have increased soil N supply to levels that exceeded stand N demand.

Acknowledgments

The authors gratefully acknowledge funding from the USDA Agenda 2020 program and the USDA Forest Service, Southern Research Station, the cooperation of Weyerhaeuser NR, International Paper, and Boise, and all research technicians and students that assisted in field and laboratory to make this project possible.

Author Contributions

Michael Blazier led the writing, data analyses, and co-conceived the development of the project; Andrew Scott co-conceived the development of the project, collaborated on the analyses and interpretation of the data and writing; Ryan Coleman conducted much of the field and laboratory procedures and collaborated on the writing and data analyses.

References

1. Allen, H.L.; Dougherty, P.M.; Campbell, R.G. Manipulation of water and nutrients: Practice and opportunity in southern U.S. pine forests. *Forest Ecol. Manag.* **1990**, *30*, 437–453.
2. Gurlevik, N.; Kelting, D.L.; Allen, H.L. Nitrogen mineralization following vegetation control and fertilization in a 14-year-old loblolly pine plantation. *Soil Sci. Soc. Am. J.* **2004**, *68*, 272–281.
3. Dickens, E.D.; Moorhead, D.J.; McElvany, B. Pine plantation fertilization. *Better Crops* **2003**, *87*, 12–15.
4. Jokela, E.J. Nutrient management of southern pines. In *Slash Pine: Still Growing and Growing*; General Technical Report SRS-6; Dickens, E.D., Barnett, J.P., Hubbard, W.G., Jokela, E.J., Eds.; USDA Forest Service, Southern Research Station: Asheville, NC, USA, 2004; pp. 27–35.
5. Albaugh, T.J.; Allen, H.L.; Fox, T.R. Historical patterns of forest fertilization in the southeastern United States from 1969 to 2004. *South. J. Appl. For.* **2007**, *31*, 129–137.
6. Zerpa, J.L.; Fox, T.R. Controls of volatile ammonia losses from loblolly pine plantations fertilized with urea in the Southeast USA. *Soil Sci. Soc. Am. J.* **2011**, *75*, 257–266.
7. Blazier, M.A.; Scott, D.A. Nitrogen distribution within the soil-plant-microbial system in response to pre-thinning fertilization treatments in Louisiana. In Proceedings of the 13th Biennial Southern Silvicultural Research Conference, Memphis, TN, USA, 28 February–4 March 2005; Conner, K.F., Ed.; General Technical Report SRS-92; USDA Forest Service, Southern Research Station: Asheville, NC, USA, 2006; pp. 129–134.
8. Jokela, E.J.; Dougherty, P.M.; Martin, T.A. Production dynamics of intensively managed loblolly pine stands in the southern United States: A synthesis of seven long-term experiments. *For. Ecol. Manag.* **2004**, *192*, 117–130.
9. Scott, D.A.; Burger, J.A.; Kaczamarek, D.J.; Kane, M.B. Nitrogen supply and demand in short-rotation sweetgum plantations. *For. Ecol. Manag.* **2004**, *189*, 331–343.
10. Maimone, R.A.; Morris, L.A.; Fox, T.R. Soil nitrogen mineralization potential in a fertilized loblolly pine plantation. *Soil Sci. Soc. Am. J.* **1991**, *55*, 522–527.
11. Li, Q.; Allen, H.L.; Wilson, C.A. Nitrogen mineralization dynamics following the establishment of a loblolly pine plantation. *Can. J. For. Res.* **2003**, *33*, 364–374.
12. Fenn, M.E.; Poth, M.A.; Terry, J.D.; Blubaugh, T.J. Nitrogen mineralization and nitrification in a mixed-conifer forest in southern California: Controlling factors, fluxes, and nitrogen fertilization response at a high and low nitrogen deposition site. *Can. J. For. Res.* **2005**, *35*, 1464–1486.
13. Allen, H.L.; Schlesinger, W.H. Nutrient limitations to soil microbial biomass and activity in loblolly pine forests. *Soil Biol. Biochem.* **2004**, *36*, 581–589.
14. Soderstrom, B.; Baath, E.; Lundgren, B. Decrease in soil microbial activity and biomass owing to nitrogen amendments. *Can. J. Microbiol.* **1983**, *29*, 1500–1506.

15. Thirukkumaran, C.M.; Parkinson, D. Microbial respiration, biomass, metabolic quotient and litter decomposition in a lodgepole pine forest floor amended with nitrogen and phosphorus fertilizers. *For. Ecol. Manag.* **2002**, *159*, 187–201.
16. Blazier, M.A.; Hennessey, T.C.; Deng, S. Effects of fertilization and vegetation control on microbial biomass carbon and dehydrogenase activity in a juvenile loblolly pine plantation. *For. Sci.* **2005**, *51*, 449–459.
17. Bechtold, J.S.; Naiman, R.J. Soil texture and nitrogen mineralization potential across a riparian toposequence in a semi-arid savanna. *Soil Biol. Biochem.* **2006**, *38*, 1325–1333.
18. Kissel, D.E.; Cabrera, M.L.; Vaio, N.; Craig, J.R.; Rema, J.A.; Morris, L.A. Rainfall timing and ammonia loss from urea in a loblolly pine plantation. *Soil Sci. Soc. Am. J.* **2004**, *68*, 1744–1750.
19. Eno, C.F. Nitrate production in the field by incubating the soil in polyethylene bags. *Soil Sci. Soc. Am. Proc.* **1960**, *24*, 277–279.
20. Bremner, J.M. Inorganic forms of nitrogen. In *Methods of Soil Analysis, Part 2*; Black, C.A., Ed.; American Society of Agronomy: Madison, WI, USA, 1965; pp. 1179–1237.
21. Carlson, R.M. Automated separation and conductimetric determination of ammonia and dissolved carbon dioxide. *Anal. Chem.* **1978**, *50*, 1528–1531.
22. Hart, S.C.; Stark, J.M.; Davidson, E.A.; Firestone, M.K. Nitrogen mineralization, immobilization, and nitrification. In *Methods of Soil Analysis Part 2, Microbiological and Biochemical Properties*; Weaver, R.W., Angele, S., Bottomly, P., Eds.; Soil Science Society of America: Madison, WI, USA, 1994; pp. 985–1018.
23. Jenkinson, D.S.; Powlson, D.S. The effects of biocidal treatments on metabolism in soil-I. Fumigation with chloroform. *Soil Biol. Biochem.* **1976**, *8*, 167–177.
24. Jenkinson, D.S.; Powlson, D.S. The effects of biocidal treatments on metabolism in soil-V. A method for measuring soil biomass. *Soil Biol. Biochem.* **1976**, *8*, 209–213.
25. Sparling, G.P.; West, A.W.; Whale, K.N. Interference from plant roots in the estimation of soil microbial ATP, C, N, and P. *Soil Biol. Biochem.* **1985**, *17*, 275–278.
26. Jenkinson, D.S. Determination of microbial biomass carbon and nitrogen in soil. In *Advances in Nitrogen Cycling in Agricultural Ecosystems*; Wilson, J.R., Ed.; CAB International: Wallingford, UK, 1988; pp. 368–386.
27. Lenhard, G. The dehydrogenase activity in soil as a measure of the activity of soil microorganisms. *Z. Pflanzenernäh Düng Bodenkd* **1956**, *73*, 1–11.
28. Alef, K. Dehydrogenase activity. In *Methods in Applied Soil Microbiology and Biochemistry*; Alef, K., Nannipieri, P., Eds.; Academic Press: San Diego, CA, USA, 1995; pp. 228–231.
29. Tabatabai, M.A. Soil enzymes. In *Methods of Soil Analysis Part 2: Microbiological and Biochemical Properties*; Bigham, J.M., Ed.; Soil Science Society of America Inc.: Madison, WI, USA, 1994; pp. 775–833.
30. Camiña, F.; Trasar-Cepeda, C.; Gil-Sotres, F.; Leirós, C. Measurement of dehydrogenase activity in acid soils rich in organic matter. *Soil Biol. Biochem.* **1998**, *30*, 1005–1011.
31. Zou, X.M.; Ruan, H.H.; Fu, Y.; Yang, X.D.; Sha, L.Q. Estimating soil labile organic carbon and potential turnover rates using a sequential fumigation-incubation procedure. *Soil Biol. Biochem.* **2005**, *37*, 1923–1928.

32. Southern Climate Impacts Planning Program (SCIPP). Historical Climate Trends Tool. Available online: http://www.southernclimate.org/data.php (accessed on 30 December 2014).
33. Lee, K.; Jose, S. Nitrogen mineralization in short-rotation tree plantations along a soil nitrogen gradient. *Can. J. For. Res.* **2006**, *36*, 1236–1242.
34. Kuzyakov, Y.; Friedel, J.K.; Stahr, K. Review of mechanisms and quantification of priming effects. *Soil Biol. Biochem.* **2000**, *32*, 1485–1498.
35. Allen, H.L. *Thirteenth Annual Report*; North Carolina State University Forest Nutrition Cooperative: Raleigh, NC, USA, 2001; p. 8.
36. Vitousek, P.M.; Matson, P.A. Disturbance, nitrogen availability, and nitrogen losses in an intensively managed loblolly pine plantation. *Ecology* **1985**, *66*, 1360–1376.
37. Piatek, K.B.; Allen, H.L. Nitrogen mineralization in a pine plantation fifteen years after harvesting and site preparation. *Soil Sci. Soc. Am. J.* **1999**, *63*, 990–998.
38. Zhuang, S.; Wang, J.; Sun, X.; Wang, M. Effect of forest thinning on soil net nitrogen mineralization and nitrification in a *Crytomeria japonica* plantation in Taiwan. *J. For. Res.* **2014**, *25*, 571–578.
39. Gurlevik, N. Stand and Soil Responses of a Loblolly Pine Plantation to Midrotation Fertilization and Vegetation Control. Ph.D. Thesis, North Carolina State University, Raleigh, NC, USA, January 2002.
40. Knoepp, J.D.; Vose, J.M. Regulation of nitrogen mineralization and nitrification in Southern Appalachian ecosystems: Separating the relative importance of biotic *vs.* abiotic controls. *Pedobiologia* **2007**, *51*, 89–97.
41. Adeli, A.; Rowe, D.E.; Read, J.J. Effects of soil type on bermudagrass response to broiler litter application. *Agron. J.* **2006**, *98*, 148–155.
42. Permar, T.A.; Fisher, R.F. Nitrogen fixation and accretion by wax myrtle (*Myrica cerifera*) in slash pine (*Pinus elliotti*) plantations. *For. Ecol. Manag.* **1983**, *5*, 39–46.

13

Growth and Nutrient Status of Foliage as Affected by Tree Species and Fertilization in a Fire-Disturbed Urban Forest

Choonsig Kim [1,*], Jaeyeob Jeong [2], Jae-Hyun Park [1] and Ho-Seop Ma [3]

[1] Department of Forest Resources, Gyeongnam National University of Science and Technology, Jinju 660-758, Korea; E-Mail: pjh@gntech.ac.kr

[2] Centre for Environmental Risk Assessment and Remediation, University of South Australia, Adelaide, SA 5095, Australia; E-Mail: jeojy003@mymail.unisa.edu.au

[3] Department of Forest Environmental Resources, Gyeongsang National University, Jinju 660-701, Korea; E-Mail: mhs@gnu.ac.kr

* Author to whom correspondence should be addressed; E-Mail: ckim@gntech.ac.kr;

Academic Editors: Heinz Rennenberg and Mark A. Adams

Abstract: The aim of the present study was to evaluate the growth and macronutrient (C, N, P, K) status in the foliage of four tree species (LT: *Liriodendron tulipifera* L.; PY: *Prunus yedoensis* Matsumura; QA: *Quercus acutissima* Carruth; PT: *Pinus thunbergii* Parl.) in response to fertilization with different nutrient ratios in a fire-disturbed urban forest located in BongDaesan (Mt.), Korea. Two fertilizers ($N_3P_8K_1$ = 113:300:37 kg·ha^{-1}·year^{-1}; $N_6P_4K_1$ = 226:150:37 ha^{-1}·year^{-1}) in four planting sites were applied in April 2013 and March 2014. The growth and nutrient responses of the foliage were monitored six times for two years. Foliar growth and nutrient concentrations were not significantly different ($p > 0.05$) in response to different doses of N or P fertilizer, but the foliage showed increased N and P concentrations and content after fertilization compared with the control ($N_0P_0K_0$). Foliar C and K concentrations were little affected by fertilization. Foliar nutrient concentrations and contents were significantly higher in PY and LT than in PT. The results suggest that the foliar N and P concentration could be used as a parameter to assess the nutrient environments of tree species restored in a fire-disturbed urban forest.

Keywords: carbon; forest fire; nitrogen; phosphorus; potassium; restoration; urban soil

---

1. Introduction

Foliage analysis has received considerable research attention because the nutrient concentrations of foliage have been accepted as adequate indicators of growth and soil fertility at sites in forest stands [1–3]. Generally, the nutrient responses of foliage are commonly used as a tool to assess the nutrient requirements and deficiencies [4].

Urban forests play an important role in enhancing ecosystem service with many demands, such as recreational service, esthetics and biodiversity [5]. However, many urban forest ecosystems face challenging natural and anthropogenic influences, such as forest fire and air pollutions [6]. For example, soils disturbed through forest fire exhibit a myriad of nutritional problems, such as nitrogen (N) and phosphorus (P) deficiency through increased nutrient leaching, surface runoff, and soil erosion [7–8]. The loss of plant nutrients and destabilization of soils in fire-disturbed urban forests might inhibit the rooting and growth of newly planted tree species [9]. In addition, forest management practices, such as nutrient additions, are required to supply sufficient nutrients to optimize the growth of newly planted tree species in urban forests burned by forest fire.

Fertilization stimulates tree growth through positive effects on leaf area and foliar photosynthetic rate [10,11] with increased nutrient concentrations in the living components of trees [12]. However, the status of foliar nutrients was dependent on the type [13] and dose [14] of the fertilizer, tree species [12,15] and many environmental resource factors, such as soil properties and water supply [1,16]. Foliar growth and nutrient status might be important factors in determining responses to fertilization because foliar nutrient concentrations are sensitive to soil nutrient conditions, although the concentrations were much more dependent on species than on sites [1,13,14]. In addition, the application of a suitable fertilizer ratio after foliage analysis, considering the soil environmental conditions and tree growth characteristics, is one of the most effective ways to reduce cost and fertilizer waste.

Although several studies have investigated the changes in soil properties following fires in forest ecosystems in Korea [12,14], limited information is available on urban forest landscapes. In addition, the quantification of foliar nutrient status following fertilization is critical for the actual and potential vegetation restoration of a fire-disturbed urban forest. The aim of the present study was to examine the growth and nutrient responses of foliage based on the compound ratio of fertilizer from four tree species planted in a fire-disturbed urban forest. We hypothesized that the growth and nutrient concentrations of foliage may correspond to the fertilization with different nutrient ratios by tree species-related differences.

2. Experimental Section

2.1. Site

The study site was located in BongDaesan (Mt.) of the Ulsan Metropolitan city, located in southeastern Korea (Figure 1). This mountain was a frequent forest fire area (37 times from 2003 to 2011), primarily resulting from arson. The dominant forest soils are a slightly dry brown forest soil (mostly Entisols or Inceptisols, United States Soil Classification System) derived from granite parent rocks. The annual average precipitation and temperature in this area are 1277 mm·year^{-1} and 14.1 °C, respectively.

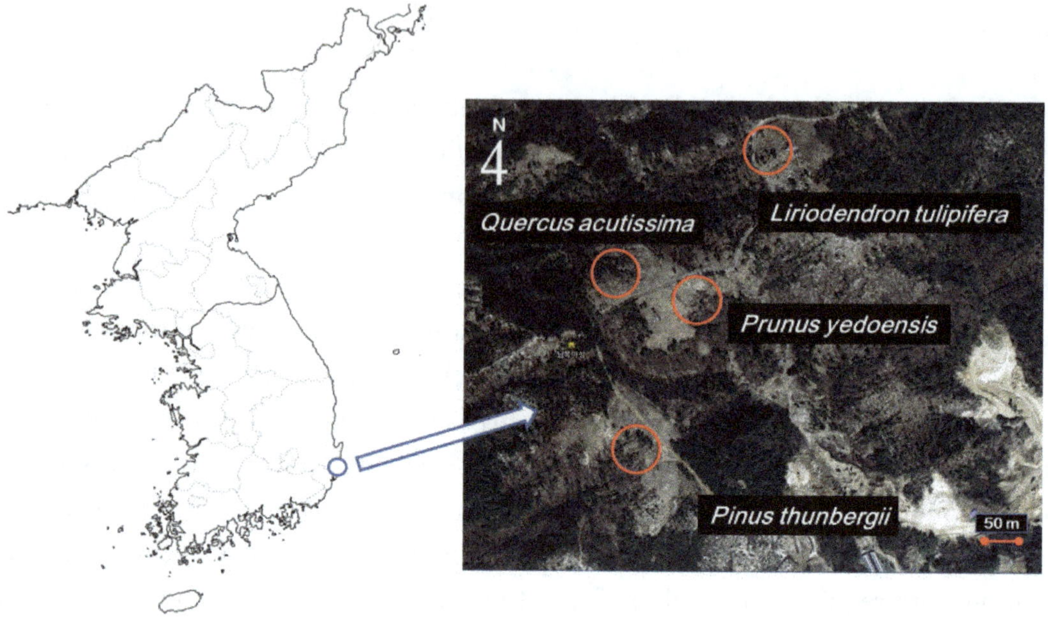

Figure 1. Location of tree planting sites in a fire-disturbed urban forest.

2.2. Methods

This study comprised a completely randomized design with 2 blocks involving total 24 plots (two fertilizer ($N_3P_8K_1$ and $N_6P_4K_1$) and one control ($N_0P_0K_0$) treatments × four different tree planting species (*Liriodendron tulipifera* L. (LT); *Prunus yedoensis* Matsumura (PY); *Quercus acutissima* Carruth (QA); and *Pinus thunbergii* Parl. (PT)) × two blocks) in a fire-disturbed urban forest. The fire resulted from arson in the spring of 2008. In the spring of 2009, the two-year-old (1-1) seedlings of four tree species were planted under a similar site environmental condition on gentle slopes (5–15°) within a close situated distance among tree planting sites from burned forests (Figure 1). Experimental plots consisting of three deciduous hardwood (LT, PY, QA) and one evergreen coniferous plantations (PT) were located adjacent to each other (Figure 1), as sampling bias due to differing site components can be reduced by using a sampling scheme with identical designs for adjacent sites [17]. The stand characteristics and soil properties before the fertilization are shown in Tables 1 and 2, respectively. The soil properties of the study site were similar to a severely burned forest soil in Korea [14].

Table 1. Stand characteristics before fertilization in the study site.

Tree Species	Location	Elevation (m)	Stand Density (trees·ha^{-1})	DBH † (mm)	Height (m)
Liriodendron tulipifera	35°32′34.92″ N 129°26′22.92″ E	154	1666 (160)	28.6 (2.8)	2.87 (0.14)
Prunus yedoensis	35°32′26.48″ N 129°26′16.22″ E	121	1267 (160)	49.4 (2.3)	4.10 (0.13)
Quercus acutissima	35°32′26.77″ N 125°26′15.15″ E	118	1533 (66)	26.4 (1.1)	2.22 (0.10)
Pinus thunbergii	35°32′16.36″ N 125°26′17.56″ E	98	1600 (206)	27.3 (3.7)	2.31 (0.11)

Standard errors in parenthesis. † DBH: diameter at breast height (1.2 m).

Table 2. Soil property before fertilization in the study site.

Tree Species	B.D. †	Sand	Silt	Clay	C	N	P	K^+	Ca^{2+}	Mg^{2+}
	$g \cdot cm^{-3}$	%					$mg \cdot kg^{-1}$	$cmol_c \cdot kg^{-1}$		
Liriodendron tulipifera	0.90 (0.01)	49 (4.7)	40 (4.0)	10 (0.7)	3.1 (0.50)	0.15 (0.02)	6.0 (2.0)	0.16 (0.02)	1.60 (0.28)	0.50 (0.09)
Prunus yedoensis	0.89 (0.01)	47 (1.3)	43 (0.7)	10 (1.2)	3.6 (0.70)	0.17 (0.02)	1.2 (0.5)	0.29 (0.11)	1.54 (0.39)	0.56 (0.13)
Quercus acutissima	0.88 (0.02)	43 (1.8)	47 (2.9)	10 (1.2)	1.9 (0.04)	0.08 (0.01)	9.9 (0.7)	0.09 (0.01)	0.43 (0.06)	0.19 (0.02)
Pinus thunbergii	0.91 (0.01)	63 (8.7)	28 (8.1)	9 (0.7)	1.9 (0.01)	0.07 (0.01)	6.5 (2.6)	0.08 (0.02)	0.38 (0.11)	0.16 (0.03)

Standard errors in parenthesis. † B.D.: bulk density.

Treatment plots (plot size = 5 × 5 m) of each tree species were randomly assigned with a 1 m buffer zone between each plot on the same facing slopes and aspects to minimize spatial variation in soil properties. The combination of fertilizer ratios ($N_3P_8K_1$ (113 kg N $ha^{-1} \cdot year^{-1}$, 300 kg P $ha^{-1} \cdot year^{-1}$, 37 kg K $ha^{-1} \cdot year^{-1}$) and $N_6P_4K_1$ (226 kg N $ha^{-1} \cdot year^{-1}$, 150 kg P $ha^{-1} \cdot year^{-1}$, 37 kg K $ha^{-1} \cdot year^{-1}$)) was based on the guideline of fertilization after forest fire in Korean forests [9]. Urea, fused superphosphate, and potassium chloride fertilizers were used as sources of N, P, and potassium (K), respectively. Fertilizers were applied for two years (in April 2013 and March 2014) by hand across each plot.

Fresh foliar (current-year-old) samples were collected at six times (27 June, 23 August, and 17 October, 2013 and 26 June, 22 August, and 18 October, 2014) for two years with pruners from the mid-crown of three trees per each treatment plot (Figure 2). The foliar samples, in plastic zipper bags, were transported to the laboratory and foliage was separated from twigs or small branches. For each treatment, three repetitions of 10 leaves were counted and weighted. Leaf area was measured by fresh foliar samples by using scanned leaf meter (CI-202 area meter CID, Inc., Camas, WA, USA). The specific leaf area of the foliage was determined as the leaf area (cm^2) and dry weight (g) of the ratio [16]. The foliar samples were oven-dried at 65 °C for 48 h, and the dried samples were ground in a Wiley mill and passed through a 40-mesh stainless steel sieve. Foliar carbon (C) and N concentrations from the ground materials were determined using an elemental analyzer (Thermo Scientific, Flash 2000, Italy). Foliar P and K concentrations were determined through dry ashing 0.5 g of dry foliage at 470 °C for 4 h, digesting the ash with 3 mL of concentrated 5 M HCl, diluting the digest with 0.25 mL of concentrated HNO_3 and 3 mL concentrated 5 M HCl, and measuring the concentrations via ICP (Perkin Elmer Optima 5300DV).

Figure 2. Morphological characteristics of foliage by fertilizer compound ratios of tree species collected on August 2014: (**a**) *Liriodendron tulipifera* L.; (**b**) *Prunus yedoensis* Matsumura; (**c**) *Quercus acutissima* Carruth; and (**d**) *Pinus thunbergii* Parl.

2.3. Statistical Analysis

All data were analyzed using the PROC MIXED procedure of SAS [18] to determine the significance of main fixed effects (tree species (S), fertilizer treatment (F), sampling month (M)) and their interactions (S × F, S × M, F × M, S × F × M), whereas the sampling years of foliage were considered a random effect. The following model was used to describe the data analysis (Equation (1)).

$$Y_{ijk} = u + S_i + F_j + M_k + (SF)_{ij} + (SM)_{ik} + (FM)_{jk} + (SFM)_{ijk} + e_{ijk} \quad (1)$$

where u is the overall mean effect, S is the different tree species (i = 1, 2, 3, 4), F is the fertilizer treatment (j = 1, 2, 3), and M is the sampling month (k = 1, 2, 3). When significant differences were observed, the treatment means were compared using Tukey's test at $p < 0.05$.

3. Results and Discussion

3.1. Results

3.1.1. Growth Response

Foliar growth responses (leaf area, dry weight, and specific leaf area) were significantly ($p < 0.05$) affected by tree species and fertilization (Table 3). The foliar leaf area and dry weight showed a significant two-factor interaction between tree species and fertilization (Figure 3). The leaf area of LT, PY and PT increased after the fertilization, but the leaf area of QA was not affected by fertilization

(Figure 3). The foliar dry weight of LT and QA was significantly higher in the $N_6P_4K_1$ treatment than in the $N_0P_0K_0$ treatment (Figure 3).

Table 3. *P*-value from results of three-way ANOVA on tree species, fertilization and sampling month on growth and nutrient responses of foliage from tree species planted in a fire-disturbed urban forest. Bold values denote significance at $p < 0.05$.

Component	Leaf Area	Specific Leaf Area	Dry Weight	Nutrient Concentration				C/N Ratio	Nutrient Content			
				C	N	P	K		C	N	P	K
Species (S)	**<0.001**	**<0.001**	**<0.001**	**<0.001**	**<0.001**	**<0.001**	**<0.001**	**<0.001**	**<0.001**	**<0.001**	**<0.001**	**<0.001**
Fertilizer (F)	**<0.001**	**0.049**	**0.002**	0.444	**<0.001**	**<0.001**	0.666	**<0.001**	**0.003**	**<0.001**	**<0.001**	0.158
Month (M)	**0.003**	0.309	**<0.001**	**<0.001**	**<0.001**	**<0.001**	**<0.001**	**<0.001**	**<0.001**	0.066	**<0.001**	**<0.001**
S × F	**0.002**	0.545	**0.018**	0.231	**0.032**	0.264	0.986	**0.037**	**0.030**	**0.007**	**0.003**	0.707
S × M	0.234	**0.012**	0.269	**<0.001**	**0.016**	0.770	0.126	0.424	0.187	0.278	**0.002**	**<0.001**
F × M	0.861	0.981	0.903	0.419	0.614	0.275	0.699	0.834	0.917	0.945	0.807	0.994
S × F × M	0.999	0.685	0.625	0.830	0.978	0.999	0.867	0.940	0.607	0.950	0.659	0.963

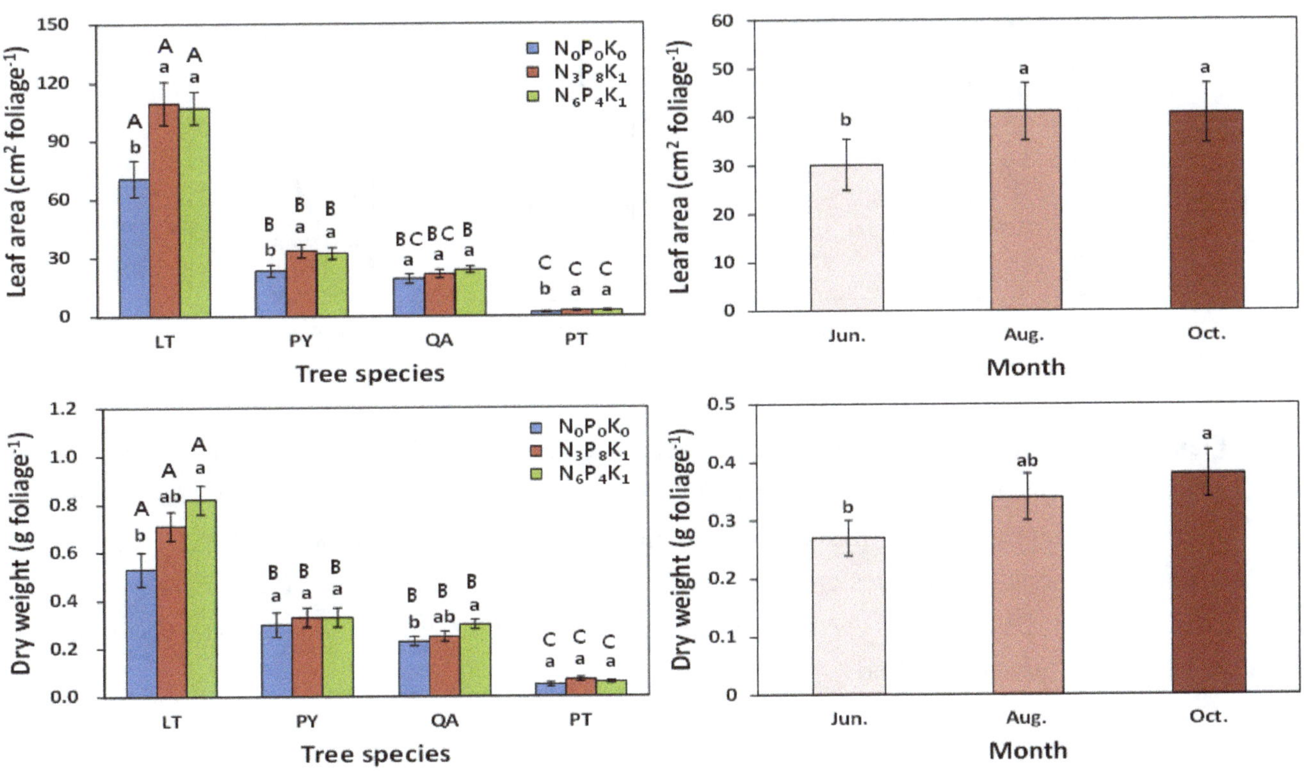

Figure 3. Leaf area and dry weight of foliage by tree species and sampling month in a fire-disturbed urban forest (mean ± standard error). Treatment means with the same lower case letter among fertilizer or sampling month treatments and treatment means with the same upper case letter among tree species are not significantly different at $p < 0.05$. LT: *Liriodendron tulipifera* L.; PY: *Prunus yedoensis* Matsumura; QA: *Quercus acutissima* Carruth; and PT: *Pinus thunbergii* Parl.

A significant main effect of the foliar leaf area and dry weight on the sampling month, which was lower in June than in August and October (Figure 3), was observed. The specific leaf area was marginally affected by fertilization ($p = 0.049$), with a significant two-factor interaction between tree species and sampling month (Figure 4).

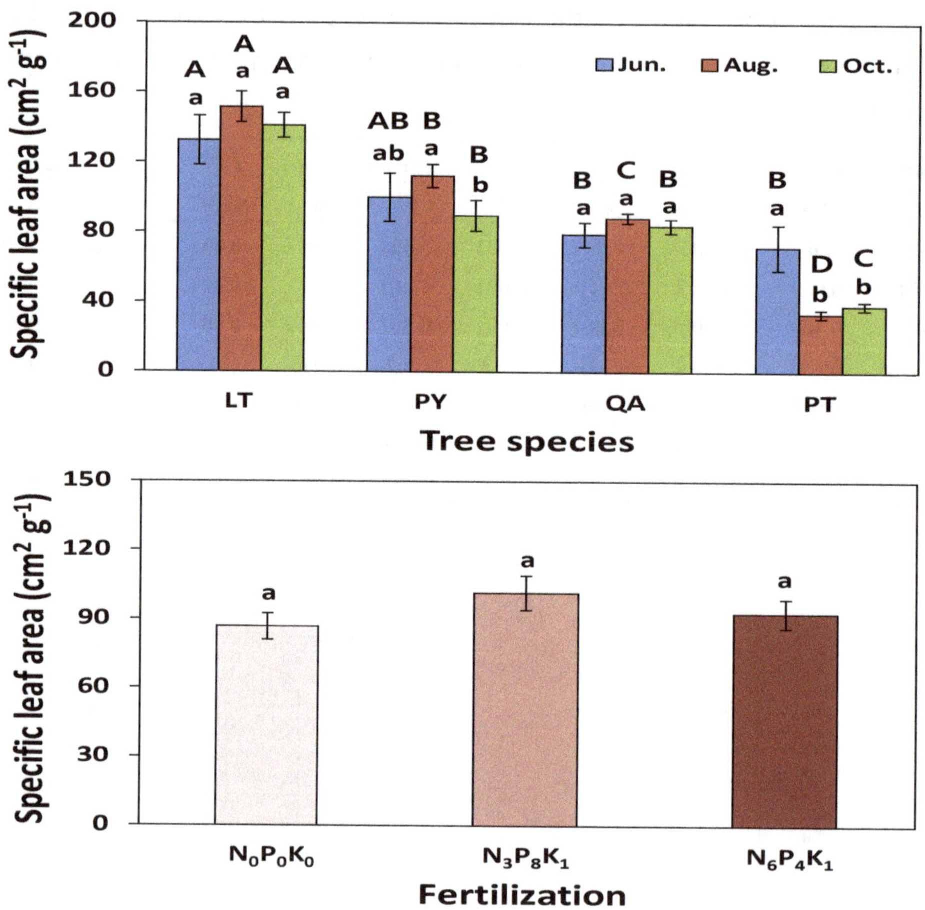

Figure 4. Specific leaf area of foliage by tree species and fertilization in a fire-disturbed urban forest (mean ± standard error). Treatment means with the same lower case letter among sampling month or fertilizer treatments and treatment means with the same upper case letter among tree species are not significantly different at $p < 0.05$. LT: *Liriodendron tulipifera* L.; PY: *Prunus yedoensis* Matsumura; QA: *Quercus acutissima* Carruth; and PT: *Pinus thunbergii* Parl.

3.1.2. Nutrient Responses

There was no significant fertilization effect of the foliar C concentration, while the C concentration showed a two-factor interaction between tree species and sampling month (Table 3, Figure 5). The foliar C content among tree species was highest in LT, followed by PY, QA and PT, while the content of LT and QA was significantly higher in the $N_6P_4K_1$ treatments compared with the $N_0P_0K_0$ treatments. The foliar N concentration significantly differed with the sampling month (Figure 5). The foliar N concentration of four tree species was lower in October or in the $N_0P_0K_0$ treatment than in June or in the fertilizer treatments (Figure 5), but the concentration did not significantly differ between the different N ratio of the fertilizer, such as the $N_3P_8K_1$ and $N_6P_4K_1$ treatments. The foliar N content was similar to foliage C content of the four tree species (Figure 5).

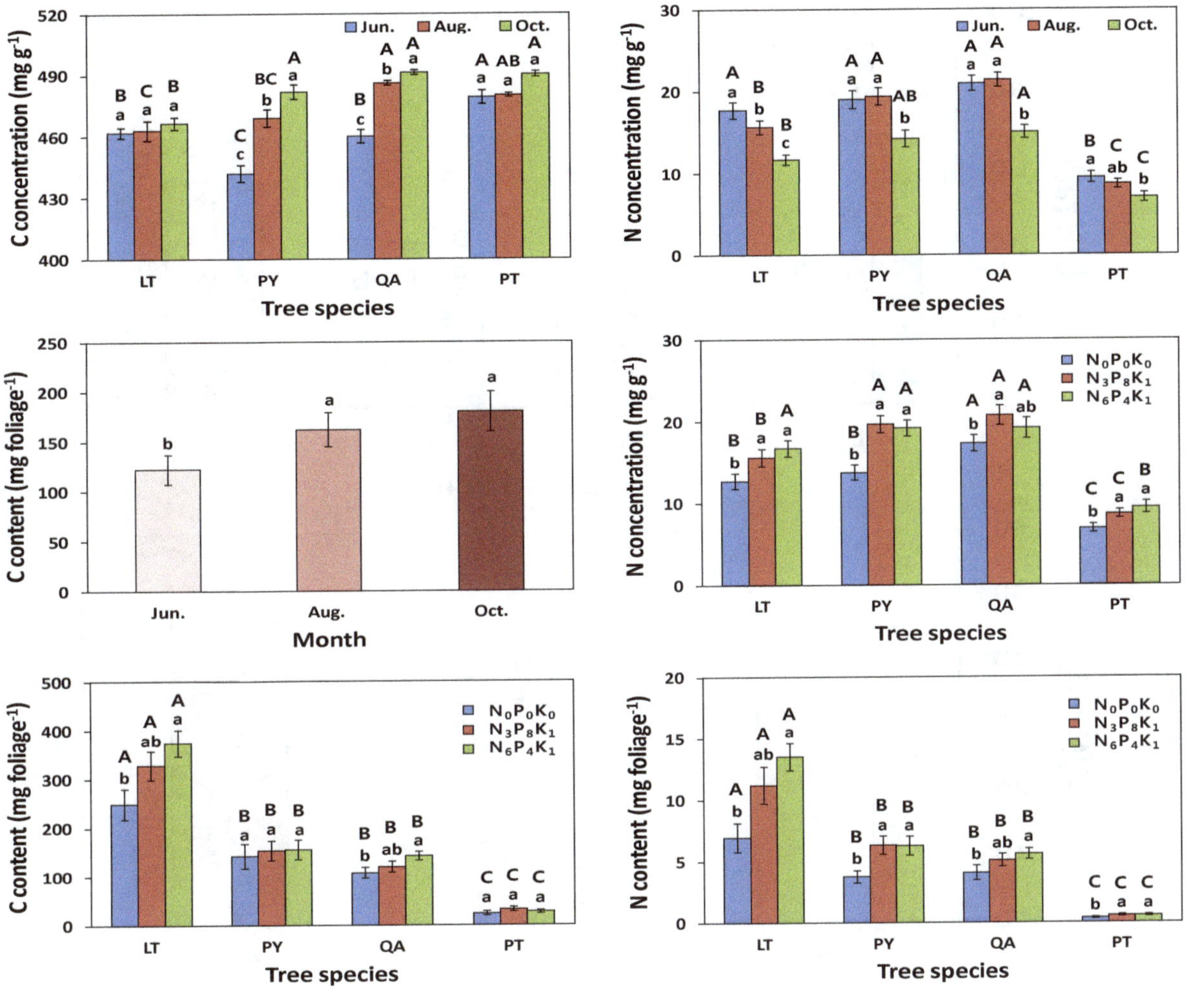

Figure 5. Carbon and nitrogen concentration and content of foliage by tree species and sampling month in a fire-disturbed urban forest (mean ± standard error). Treatment means with the same lower case letter among sampling month or fertilizer treatments and treatment means with the same upper case letter among tree species are not significantly different at $p < 0.05$. LT: *Liriodendron tulipifera* L.; PY: *Prunus yedoensis* Matsumura; QA: *Quercus acutissima* Carruth; and PT: *Pinus thunbergii* Parl.

A significant two-factor interaction between the tree species and fertilizer treatments on the C/N ratio of foliage was observed (Figure 6). The C/N ratio in all tree species was significantly lower in the fertilizer than in the $N_0P_0K_0$ treatments, regardless of the compound ratio of fertilizer, while the C/N ratio was significantly higher in conifers (PT) than in broadleaf tree species (LT, PY, and QA). The C/N ratio was also highest in October, followed by August and June (Figure 6).

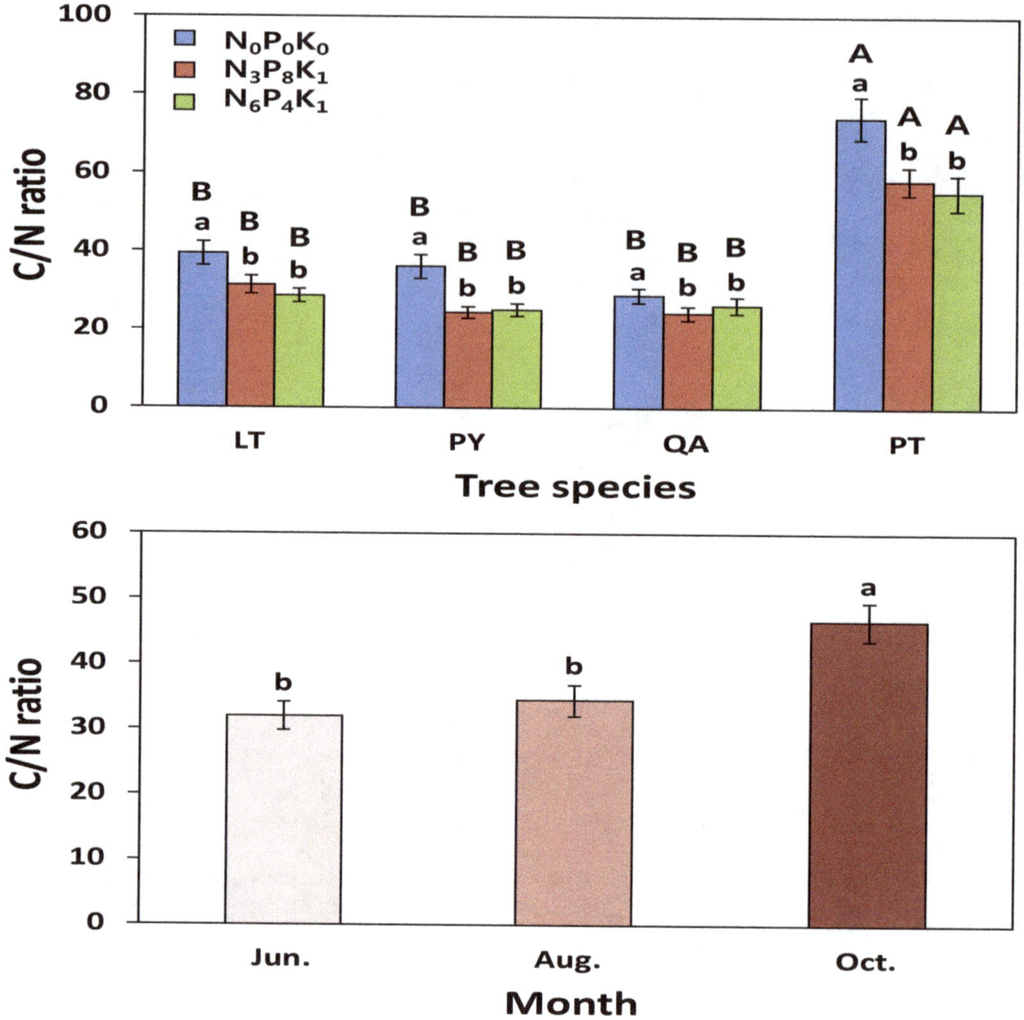

Figure 6. Carbon and nitrogen ratio of foliage by tree species and sampling month in a fire-disturbed urban forest (mean ± standard error). Treatment means with the same lower case letter among fertilizer or sampling month treatments and treatment means with the same upper case letter among tree species are not significantly different at $p < 0.05$. LT: *Liriodendron tulipifera* L.; PY: *Prunus yedoensis* Matsumura; QA: *Quercus acutissima* Carruth; and PT: *Pinus thunbergii* Parl.

The foliar P concentration was significantly affected by tree species, fertilization and sampling month with no two- or three-way interaction (Table 3, Figure 7). The foliar P concentration was significantly higher in PY than in other tree species. The foliage of PY showed a significantly high P concentration compared with the other three tree species (Figure 7). The foliar P concentration was highest in August, followed by October and June (Figure 7). There was a significant two-factor interaction between the tree species and fertilizer treatments or between the sampling month and tree species on foliar P content (Figure 7). In contrast to the foliar P concentrations, the foliar K concentration and content were little affected by fertilization (Table 3), while the concentration and content were higher in October than in June, except for PT. The foliar K concentration was significantly higher in PY than in the other three tree species (Figure 8).

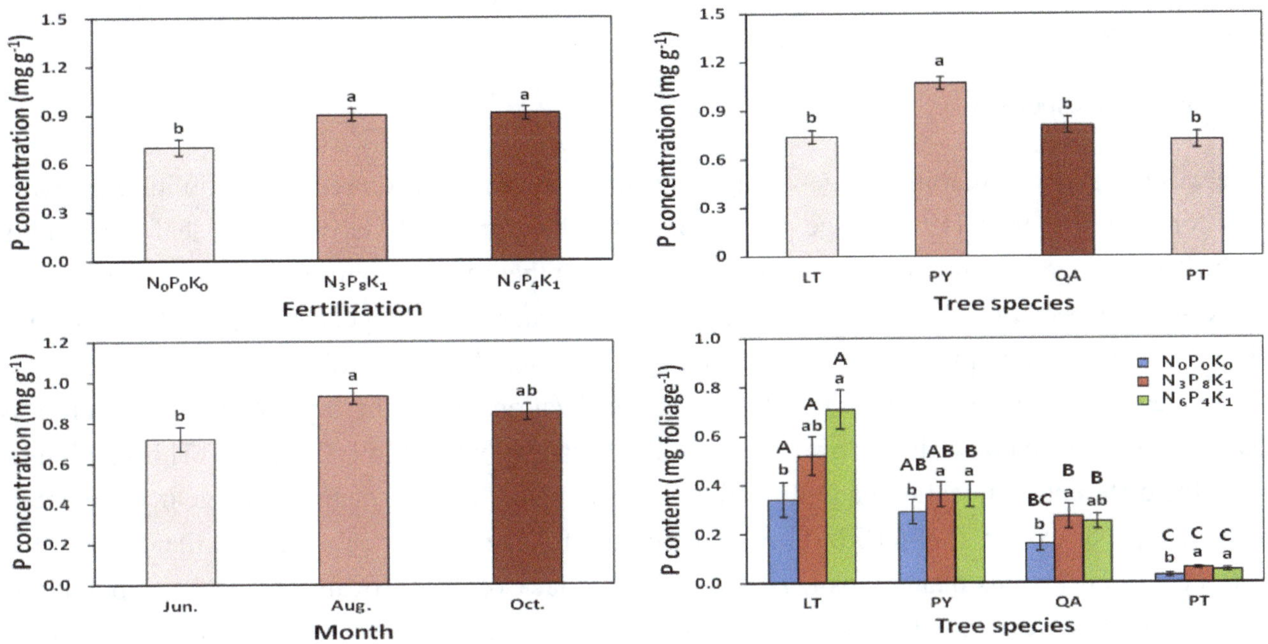

Figure 7. Phosphorus concentration and content of foliage by fertilization, tree species, and sampling month in a fire-disturbed urban forest (mean ± standard error). Treatment means with the same lower case letter among fertilizer, tree species or sampling month treatments and treatment means with the same upper case letter among tree species are not significantly different at $p < 0.05$. LT: *Liriodendron tulipifera* L.; PY: *Prunus yedoensis* Matsumura; QA: *Quercus acutissima* Carruth; and PT: *Pinus thunbergii* Parl.

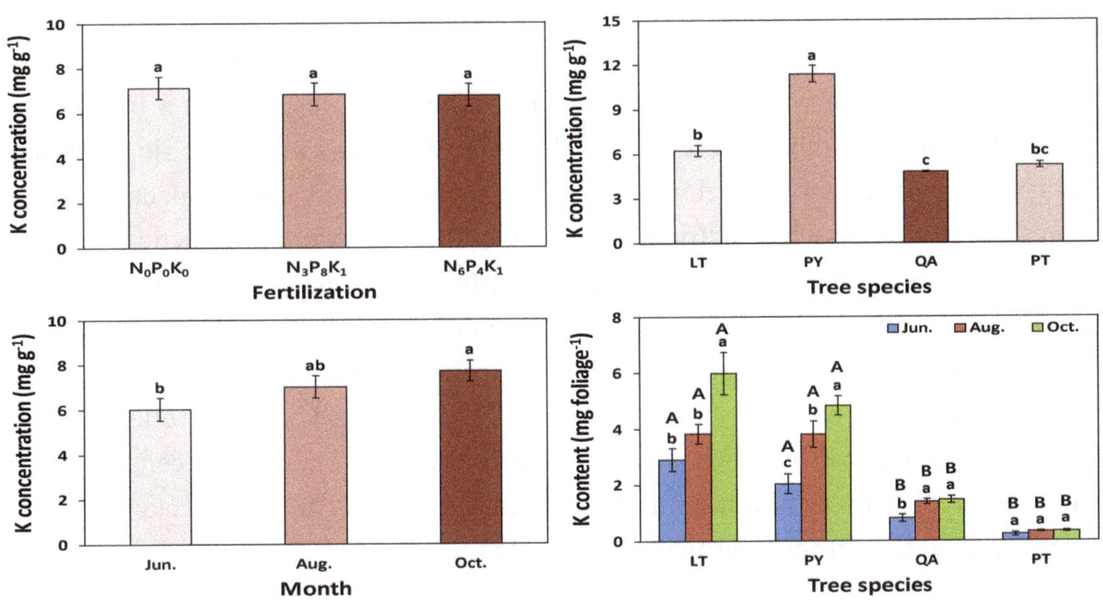

Figure 8. Potassium concentration and content of foliage by tree species, fertilization and sampling month in a fire-disturbed urban forest (mean ± standard error). Treatment means with the same lower case letter among fertilizer, tree species or sampling month treatments and treatment means with the same upper case letter among tree species are not significantly different at $p < 0.05$. LT: *Liriodendron tulipifera* L.; PY: *Prunus yedoensis* Matsumura; QA: *Quercus acutissima* Carruth; and PT: *Pinus thunbergii* Parl.

3.2. Discussion

3.2.1. Growth Responses

Because nutrient availability is generally considered the major resource factor limiting growth in many forest tree species [19], foliage growth after fertilization would have greater leaf area and dry weight than foliage grown without fertilization. Although there was a significant tree species effect on foliage growth because of the morphological difference among tree species (Figure 2), fertilization induced different growth responses among tree species planted in a fire-disturbed urban forest. Most tree species showed increased leaf area following fertilization, while the dry weight and specific leaf area were not affected by fertilization, except for the foliar dry weight in LT and QA. This result might reflect the different C allocation patterns during the growth and development of foliage in different tree species [20]. For example, N fertilization increased the foliage biomass, rather than the dry weight of the foliage in eucalyptus plantations [21]. However, the lack of a significant fertilizer compound ratio on the morphological growth responses of foliage might reflect other micronutrient imbalances resulting from increased N fertilization [11] or multi-factorial influences that occur during the leaf life-span, such as light environment, nutrients, temperature and water supply [22]. Seasonal patterns of foliar leaf area and dry weight could be attributed to foliage maturation or resource allocation between woody (roots, stem wood, and branches) and photosynthetic compounds [20]. The high leaf area and dry weight in August and October could reflect the maturation of foliage, whereas the low leaf area in June indicated that the foliage in the four tree species was not fully elongated during this season.

3.2.2. Nutrient Responses

The foliar C concentration was a poor indicator of the fertilizer response in a fire-disturbed urban forest ($p = 0.444$). Similarly, other studies reported that the inter- and intra-specific variations of the C concentration in tree species were determined through genetic and environmental factors [23,24], rather than resource factors, such as nutrient availability [25]. For example, genetic differences among tree species might result in different C concentration of foliage, which was significantly higher in PT than in LT (Figure 5). In addition, the foliage C concentration was generally increased in the heavy litter fall season (October) compared with the growing season (June–August) because of low mineral concentrations of foliage by nutrient resorption (Figure 5). The foliar C content reflected differences in the foliar dry weight induced after fertilization rather than the foliar C concentration among tree species and reflected differences in foliage maturation associated with the translocation of carbohydrates and other cellular materials to active growing tissues [15].

As expected, the foliar N concentration and content among tree species were lower in conifers (PT) than in broadleaf tree species (LT, PY, and QA). The foliar N concentration and content of the four tree species increased following fertilization, while insignificant increases in foliage N concentration occurred with different doses of N fertilizer ($N_3P_8K_1$ and $N_6P_4K_1$). A high foliar N concentration and content with fertilization likely reflects the increased uptake of available N in soil depths [15,19], as tree species with high N availability in mineral soil produce high foliage N concentration [26]. The low foliar N concentration and content observed under $N_0P_0K_0$ treatment suggested the N deficiency and limited tree growth in a fire-disturbed urban forest. However, a similar foliar N concentration of the fertilizer compound ratio might result from luxury N consumption from a high dose of N fertilizer ($N_6P_4K_1$) compared with $N_3P_8K_1$ fertilizer, although the foliar N concentration after fertilization was regulated through factors, such as the combined effect of the available soil N status, tree growth, and climate factors. The foliar N concentration and content were affected by sampling months because of the resorption of foliar N in October or the dilution effect due to the increased relative accumulation of carbohydrates with increasing dry weight [4]. Similarly, the high C/N ratio in October reflected foliar N resorption before heavy litter fall [27]. Also, C/N ratio was higher under nutrient deficient conditions ($N_0P_0K_0$) compared with trees grown under improved nutrient conditions ($N_6P_4K_1$ and $N_3P_8K_1$).

The foliar P concentration and content of the four tree species were increased after fertilization, suggesting that the four tree species were P deficient in the fire-disturbed urban forest. Similarly, the foliage of eucalypt forests showed an increased P concentration following P fertilizer application [28], reflecting enhanced uptake and mineralization in the rhizosphere. However, the foliar P concentration was not responsible for the increased dose of the P in the fertilizer ($N_6P_4K_1$ and $N_3P_8K_1$), indicating that much of the P uptake at a high dose might be associated with luxury consumption. The difference of foliage P content among tree species is similar to those observed in dry weight of foliage.

The K fertilizer had minor effects on the foliar K concentration and content of the tree species except of PY, suggesting that this nutrient might not play an important role in limiting growth with N or P applications. Similar foliar K concentrations following fertilization could be affected through a diluting effect via increased foliar dry weight with N or P supply levels compared with the $N_0P_0K_0$ treatment. Additionally, the K in foliage could be increasingly leached after rain from the tree canopy in the different fertilizer treatments because higher fluxes of K from throughfall were observed at more fertile sites compared with sites with a poorer nutrient status [27]. However, the significant high K concentration and content of foliage in PY could be influenced by high soil K availability (0.29 cmolc·kg^{-1}) compared with other tree planting sites of 0.08–0.16 cmolc·kg^{-1} (Table 2). This result supports that the K concentration in sweetgum (*Liquidambar styraciflua* L.) foliage was associated with the inherent soil chemical properties rather than fertilization [29].

4. Conclusions

The growth characteristics, such as leaf area, dry weight and specific leaf area in foliage, were not affected by fertilizer compound ratios, whereas fertilization produced high foliar N or P concentrations of four tree species planted in a fire-disturbed urban forest. In contrast to N and P, foliar C and K concentrations and contents in tree species were little affected by fertilization. The foliar N, P, and K concentrations substantially decreased during senescence (October) compared with the growing season (June, August), except for the increased foliar C concentration. There was no clear effect on the foliar nutrient status from different compound ratios (e.g., $N_3P_8K_1$ and $N_6P_4K_1$) of fertilizer on different tree species because of the luxury consumption at high N and P doses. These results suggest that a new compound ratio of fertilizer is needed to optimize the vegetation growth of trees planted in fire-disturbed urban forests because of tree species differences on foliage N and P concentration and content following fertilization.

Acknowledgments

This study was carried out with the support of "Forest Science & Technology Projects (Project No. S211212L030320)" provided by Korea Forest Service.

Author Contributions

Choonsig Kim led the writing; Jaeyeob Jeong performed the experiments; and Jae-Hyun Park and Ho-Seob Ma analyzed the data.

References

1. Weetman, G.F.; Wells, C.G. Plant analyses as an aid in fertilizing forests. In *Soil Testing and Plant Analysis*, 3rd ed.; Westerman, R.L., Ed.; Soil Science Society of America: Madison, WI, USA, 1990; pp. 659–690.
2. Barron-Gafford, G.A.; Will, R.E.; Burkes, C.; Shiver, B.; Teskey, R.O. Nutrient concentrations and content, their relation to stem growth, of intensively managed *Pinus taeda* and *Pinus elliottii* stands of different planting densities. *For. Sci.* **2003**, *49*, 291–300.
3. Tausz, M.; Trummer, W.; Wonisch, A.; Goessler, W.; Grill, D.; Jiménez, M.S.; Morales, D. A survey of foliar mineral nutrient concentrations of *Pinus canariensis* at field plots in Tenerife. *For. Ecol. Manag.* **2004**, *189*, 49–55.
4. Silfverberg, K.; Moilanen, M. Long-term nutrient status of PK fertilized Scots pine stands on drained peatlands in North-Central Finland. *Suo* **2008**, *59*, 71–88.
5. Pakaki, D.E.; Carreir, M.M.; Cherrier, J.; Grulke, N.E.; Jennings, V.; Princetl, S.; Pouyat, R.V.; Whitlow, T.H.; Zipperer, W.C. Coupling biogeochemical cycles in urban environments: Ecosystem services, green solutions, and misconceptions. *Front. Ecol. Environ.* **2011**, *9*, 27–36.
6. Escobedo, F.J.; Nowak, D.J. Spatial heterogeneity and air pollution removal by an urban forest. *Landsc. Urban Plan.* **2009**, *90*, 102–110.
7. Kim, C.; Koo, K.S.; Byun, J.K.; Jeong, J.H. Post-fire effects on soil properties in red pine (*Pinus densiflora*) stands. *Forest Sci. Tech.* **2005**, *1*, 1–7.
8. Xue, L. Li, Q.; Chen, H. Effects of a wildfire on selected physical, chemical and biochemical soil properties in a *Pinus massoniana* forest in south China. *Forests* **2014**, *5*, 2947–2966.
9. Kim, Y.S.; Byun, J.K.; Kim, C.; Park, B.B.; Kim, Y.K.; Bae, S.W. Growth response of *Pinus densiflora* seedlings to different fertilizer compound ratios in a recently burned area in the eastern coast of Korea. *Landsc. Ecol. Eng.* **2014**, *10*, 241–247.
10. Garrison, M.T.; Moore, J.A.; Shaw, T.M.; Mika, P.G. Foliar nutrient and tree growth response of mixed-conifer stands to three fertilization treatments in northeast Oregon and north central Washington. *For. Ecol. Manag.* **2000**, *132*, 183–198.

11. Amponsah, I.G.; Comeau, P.G.; Brockley, R.P.; Lieffers, V.J. Effects of repeated fertilization on needle longevity, foliar nutrition, effective leaf area index, and growth characteristics of lodgepole pine in interior British Columbia, Canada. *Can. J. For. Res.* **2005**, *35*, 440–451.
12. Miller, J.H.; Allen, H.L.; Zutter, B.R.; Zedaker, S.M.; Newbold, R.A. Soil and pine foliage nutrient responses 15 years after competing-vegetation control and their correlation with growth for 13 loblolly pine plantations in the southern United States. *Can. J. For. Res.* **2006**, *36*, 2412–2425.
13. Kim, C.; Ju, N.K.; Lee, H.Y.; Lee, K.S. Effects of growth, carbon and nitrogen response of foliage in a red pine stand. *Korean J. Soil Sci. Fertil.* **2013**, *46*, 9–15.
14. Kim, C.; Byun, J.K.; Park, J.H.; Ma, H.S. Litter fall and nutrient status of green leaves and leaf litter at various compound ratios of fertilizer in sawtooth oak stands, Korea. *Ann. For. Res.* **2013**, *56*, 339–350.
15. Mugasha, A.G.; Pluth, D.J.; Macdonald, S.E. Effects of fertilization on seasonal patterns of foliar mass and nutrients of tamarack and black spruce on undrained and drained minerotrophic peatland sites. *For. Ecol. Manag.* **1999**, *116*, 13–31.
16. Bauer, G.; Schulze, E-D.; Mund, M. Nutrient contents and concentrations in relation to growth of *Picea abies* and *Fagus sylvatica* along a European transect. *Tree Physiol.* **1997**, *17*, 777–786.
17. Oksanen, L. Logic of experiments in ecology: Is pseudoreplication a pseudoissue? *Oikos* **2002**, *94*, 27–38.
18. SAS Institute Inc. *SAS/STAT Statistical Software*; Version 9.1; SAS Institute Inc.: Cary, NC, USA, 2003.
19. Binkley, D.; Fisher, R. Nutrient management. In *Ecology and Management of Forest Soils*, 4th ed.; Wiley-Blackwell: Hoboken, NJ, USA, 2013; pp. 254–275.
20. Gough, C.M.; Seiler, J.R.; Maier, C.A. Short-term effects of fertilization on loblolly pine (*Pinus taeda* L.) physiology. *Plant Cell Environ.* **2004**, *27*, 876–886.
21. Laclau, J.P.; Almeida, J.S.R.; Gonçalves, J.L.M.; Saint-André, L.; Ventura, M.; Ranger, R.; Moreira, R.M.; Nouvellon, Y. Influence of nitrogen and potassium fertilization on leaf lifespan and allocation of above-ground growth in *Eucalyptus* plantations. *Tree Physio.* **2009**, *29*, 114–124.
22. Wilson, P.J.; Thompson, K.; Hodgson, J.G. Specific leaf area and leaf dry matter content as alternative predictors of plant strategies. *New Phytol.* **1999**, *143*, 155–162.
23. Bert, D.; Danjon, D. Carbon concentration variations in the roots, stem and crown of mature *Pinus pinaster* (Ait.). *For. Ecol. Manag.* **2006**, *222*, 279–295.
24. Zhang, Q.; Wang, C.; Wang, X.; Quan, X. Carbon concentration variability of 10 Chinese temperate tree species. *For. Ecol. Manag.* **2009**, *258*, 722–727.
25. Jeong, J.; Park, J.-H.; Kim, J.-I.; Lim, J.-T.; Lee, S.-R.; Kim, C. Effects of container volumes and fertilization on red (*Pinus densiflora*) and black pine (*Pinus thunbergii*) seedlings growth. *Forest Sci. Tech.* **2010**, *6*, 80–86.
26. Sariyildiz, T.; Anderson, J.M. Variation in the chemical composition of green leaves and leaf litters from three deciduous tree species growing on different soil types. *For. Ecol. Manag.* **2005**, *210*, 303–319.
27. Hagon-Thorn, A.; Varnagiryte, I.; Nihlgard, B.; Armolaitis, K. Autumn nutrient resorption and losses in four deciduous forest tree species. *For. Ecol. Manag.* **2006**, *228*, 33–39.

28. O'connell, A.M.; Grove, T.S. Influence of nitrogen and phosphorus fertilizers on amount and nutrient content of litterfall in a regrowth eucalypt forest. *New For.* **1993**, *7*, 33–47.
29. Scott, A.; Burger, J.A.; Kaczmarek, D.J.; Kane, M.B. Growth and nutrition response of young sweetgum plantations to repeated nitrogen fertilizer on two site types. *Biomass Bioenerg.* **2004**, *27*, 313–325.

Permissions

All chapters in this book were first published in MDPI; hereby published with permission under the Creative Commons Attribution License or equivalent. Every chapter published in this book has been scrutinized by our experts. Their significance has been extensively debated. The topics covered herein carry significant findings which will fuel the growth of the discipline. They may even be implemented as practical applications or may be referred to as a beginning point for another development.

The contributors of this book come from diverse backgrounds, making this book a truly international effort. This book will bring forth new frontiers with its revolutionizing research information and detailed analysis of the nascent developments around the world.

We would like to thank all the contributing authors for lending their expertise to make the book truly unique. They have played a crucial role in the development of this book. Without their invaluable contributions this book wouldn't have been possible. They have made vital efforts to compile up to date information on the varied aspects of this subject to make this book a valuable addition to the collection of many professionals and students.

This book was conceptualized with the vision of imparting up-to-date information and advanced data in this field. To ensure the same, a matchless editorial board was set up. Every individual on the board went through rigorous rounds of assessment to prove their worth. After which they invested a large part of their time researching and compiling the most relevant data for our readers.

The editorial board has been involved in producing this book since its inception. They have spent rigorous hours researching and exploring the diverse topics which have resulted in the successful publishing of this book. They have passed on their knowledge of decades through this book. To expedite this challenging task, the publisher supported the team at every step. A small team of assistant editors was also appointed to further simplify the editing procedure and attain best results for the readers.

Apart from the editorial board, the designing team has also invested a significant amount of their time in understanding the subject and creating the most relevant covers. They scrutinized every image to scout for the most suitable representation of the subject and create an appropriate cover for the book.

The publishing team has been an ardent support to the editorial, designing and production team. Their endless efforts to recruit the best for this project, has resulted in the accomplishment of this book. They are a veteran in the field of academics and their pool of knowledge is as vast as their experience in printing. Their expertise and guidance has proved useful at every step. Their uncompromising quality standards have made this book an exceptional effort. Their encouragement from time to time has been an inspiration for everyone.

The publisher and the editorial board hope that this book will prove to be a valuable piece of knowledge for researchers, students, practitioners and scholars across the globe.

List of Contributors

Ing-Marie Gren
Department of Economics, Swedish University of Agricultural Sciences, 75007 Uppsala, Sweden

Catherine E. Lovelock
School of Biological Science, The University of Queensland, St Lucia, Queensland 5072, Australia

Lisa J. Duckett and Ilka C. Feller
Smithsonian Environmental Research Center, 647 Contees Wharf Rd, Edgewater, MD 21037, USA

Lorae T. Simpson
Smithsonian Environmental Research Center, 647 Contees Wharf Rd, Edgewater, MD 21037, USA
Smithsonian Marine Station at Fort Pierce, 701 Seaway Dr, Fort Pierce, FL 34949, USA

Ana Paula Pulito and Claudio Roberto Silva
Fibria S.A., Aracruz CEP: 29197-900, Brazil

José Leonardo de Moraes Gonçalves, José Henrique Tertulino Rocha and Ayeska Hübner
Department of Forest Science, Luiz de Queiroz College of Agriculture (ESALQ), University of São Paulo (USP), Piracicaba CEP:13418-900, Brazil

Philip J. Smethurst
CSIRO, Private Bag 12, Hobart TAS 7001, Australia

José Carlos Arthur Junior
Forest Science and Research Institute (IPEF), Piracicaba, CEP: 13400-970, Brazil

Clayton Alcarde Alvares
Department of Forestry and Environmental Resources, North Carolina State University, Raleigh, NC 27695, USA

Luiz Fabiano de Moraes and José Luiz Gava
Suzano, Itapetininga CEP: 18207-780, Brazil

Aline Cristina Miranda
Department of Forest Science, "Julio de Mesquita Filho" Universidade Estadual Paulista, Botucatu CEP: 18603-970, Brazil

Marcos Yassuo Kamogawa
Mathematics, Statistics and Chemistry Department, Luiz de Queiroz College of Agriculture (ESALQ), University of São Paulo (USP), Piracicaba CEP:13418-900, Brazil

Raul Chaves
Duratex, Agudos CEP: 17120-000, Brazil

Fredrik From and Annika Nordin
Department of Forest Genetics and Plant Physiology, Umeå Plant Science Centre, Swedish University of Agricultural Sciences, Skogsmarksgränd 1, 90183 Umeå, Sweden

Joachim Strengbom
Department of Ecology, Swedish University of Agricultural Sciences, 75007 Uppsala, Sweden

Ania Kobylinski
Natural Resources and Environmental Studies (NRES) Graduate Program, University of Northern British Columbia (UNBC), 3333 University Way, Prince George, BC V2N 4Z9, Canada

Arthur L. Fredeen
NRES Institute; Ecosystem Science and Management Program, UNBC, 3333 University Way, Prince George, BC V2N 4Z9, Canada

Ah Reum Han and Pil Sun Park
Department of Forest Sciences, Seoul National University, Seoul 151-921, Korea

Yunmi Park
Division of Special-purpose Trees, Korea Forest Research Institute, Suwon 441-847, Korea

Byung Bae Park
Department of Environment and Forest Resources, Chungnam National University, Daejeon 305-764, Korea

Jaeeun Sohng
Department of Forest Sciences, Seoul National University, Seoul 151-921, Korea
School of Forestry & Environmental Studies, Yale University, New Haven, CT 06511, USA

Mi-Ae Jeong
Department of Forest Sciences, Seoul National University, Seoul 151-921, Korea
Department of Landscape Architecture, Seoul National University, Seoul 151-921, Korea

Richard Hrivnák and Ivan Jarolímek
Institute of Botany, Slovak Academy of Sciences, Dúbravská cesta 9, SK-845 23 Bratislava, Slovakia

Michal Slezák
Institute of Botany, Slovak Academy of Sciences, Dúbravská cesta 9, SK-845 23 Bratislava, Slovakia
Department of Biology and Ecology, Faculty of Education, Catholic University, Hrabovská cesta 1, SK-034 01 Ružomberok, Slovakia
Department of Phytology, Faculty of Forestry, Technical University in Zvolen, T. G. Masaryka 24, SK-960 53 Zvolen, Slovakia

Benjamín Jarčuška
Institute of Forest Ecology, Štúrova 2, SK-960 53 Zvolen, Slovakia

Judita Kochjarová
Institute of Botany, Slovak Academy of Sciences, Dúbravská cesta 9, SK-845 23 Bratislava, Slovakia
Botanical Garden, Comenius University, SK-038 15 Blatnica, Slovakia

Yang Cao and Yunming Chen
State Key Laboratory of Soil Erosion and Dryland Farming on the Loess Plateau, Northwest A & F University, Yangling 712100, China
Institute of Soil and Water Conservation, Chinese Academy of Sciences and Ministry of Water Resources, Yangling 712100, China

Zareen Khan, Shyam L. Kandel, Daniela N. Ramos, Gregory J. Ettl, Soo-Hyung Kim and Sharon L. Doty
School of Environmental and Forest Sciences, College of the Environment, University of Washington, Seattle, WA 98195, USA

Tae Kyung Yoon
Department of Environmental Science and Engineering, Ewha Womans University, Seoul 03760, Korea

Nam Jin Noh
River Basin Research Center, Gifu University, Gifu 501-1193, Japan

Haegeun Chung
Department of Environmental Engineering, Konkuk University, Seoul 05029, Korea

A-Ram Yang
Forest Practice Research Center, Korea Forest Research Institute, Pocheon 11186, Korea

Yowhan Son
Department of Environmental Science and Ecological Engineering, Graduate School, Korea University, Seoul 02841, Korea
Department of Biological and Environmental Science, Qatar University, Doha, Qatar

Xianming Dou
School of Environment Science and Spatial Informatics, China University of Mining and Technology, Xuzhou 221116, China
School of Resources and Earth Sciences, China University of Mining and Technology, Xuzhou 221116, China

Baozhang Chen
School of Environment Science and Spatial Informatics, China University of Mining and Technology, Xuzhou 221116, China
State Key Laboratory of Resources and Environmental Information System, Institute of Geographic Sciences and Nature Resources Research, University of Chinese Academy of Sciences, Beijing 100101, China

T. Andrew Black and Rachhpal S. Jassal
Faculty of Land and Food Systems, University of British Columbia, Vancouver, BC V6T 1Z4, Canada

Mingliang Che
State Key Laboratory of Resources and Environmental Information System, Institute of Geographic Sciences and Nature Resources Research, University of Chinese Academy of Sciences, Beijing 100101, China
University of Chinese Academy of Sciences, Beijing 100049, China

Michael A. Blazier and Ryan Coleman
Hill Farm Research Station, Louisiana State University Agricultural Center, Homer, LA 71040, USA

D. Andrew Scott
Southern Research Station, USDA Forest Service, Normal, AL 35762, USA

Choonsig Kim and Jae-Hyun Park
Department of Forest Resources, Gyeongnam National University of Science and Technology, Jinju 660-758, Korea

Jaeyeob Jeong
Centre for Environmental Risk Assessment and Remediation, University of South Australia, Adelaide, SA 5095, Australia

Ho-Seop Ma
Department of Forest Environmental Resources, Gyeongsang National University, Jinju 660-701, Korea

Index

www.ingramcontent.com/pod-product-compliance
Lightning Source LLC
LaVergne TN
LVHW060519210926
848485LV00005B/1107